ON THE TEACHING OF LINEAR ALGEBRA

Mathematics Education Library

VOLUME 23

The titles published in this series are listed at the end of this volume.

ON THE TEACHING
OF LINEAR ALGEBRA

Edited by

JEAN-LUC DORIER
Laboratoire Leibniz,
Grenoble, France

KLUWER ACADEMIC PUBLISHERS
DORDRECHT / BOSTON / LONDON

Library of Congress Cataloging-in-Publication Data

ISBN 0-7923-6539-9

Published by Kluwer Academic Publishers,
P.O. Box 17, 3300 AA Dordrecht, The Netherlands.

Sold and distributed in North, Central and South America
by Kluwer Academic Publishers,
101 Philip Drive, Norwell, MA 02061, U.S.A.

In all other countries, sold and distributed
by Kluwer Academic Publishers,
P.O. Box 322, 3300 AH Dordrecht, The Netherlands.

Printed on acid-free paper

Printed in the Netherlands.

TABLE OF CONTENTS

FOREWORD TO THE ENGLISH EDITION

A first version of this book was published, in French, under the title, *L'Enseignement de l'Algèbre Linéaire en Question*, by La Pensée Sauvage Édition (Grenoble) in 1997. It was the result of a collaboration between French, Brazilian, Moroccan and North American research teams for which the issue of linear algebra teaching was a central research interest.

The present English edition is not just a mere translation of the original. Most chapters have been revised according to new results, the report of a new research work has been integrated into the last chapter and Sierpinska's contribution has been totally renewed. Moreover, the introduction and the conclusion have been renewed and adapted to a more international audience.

Most contributors being French native speakers, the language issue has been quite complex.

We wish to thank here the 'équipe DIDIREM' (Paris 7 University), The IREM of Paris 7, the 'équipe d'Analyse' (Paris 6 University - CNRS), the IUFM of Versailles and Concordia University for their financial support.
We wish also to express our deep gratitude to Astrid Defence, Linda Northrup, Anna Sierpinska and Caroline West, for their precious help in translating and rereading our work.

Jean-Luc Dorier

PREFACE[1]

ANDRÉ REVUZ[2]

During the sixties, at a conference in Zürich, I made the acquaintance of a charming old man who was none other than Plancherel - of Plancherel's theorem - and who, during a very interesting conversation, insisted on the fact that of all the teaching he had done that of linear algebra seemed to be by far the most difficult for students to understand. Thirty years later the situation does not seem to have changed very much and we can assure Plancherel that he is in good company.

There is, however, in the eyes of mathematicians, a particular paradox: of all mathematical theories, and I don't think that this is an illusion, the theory of linear algebra appears to be one of the simplest and the difficulties attendant on its instruction are out of proportion with its intrinsic difficulties.

So, what is going on and what can one do about it? One radical solution to the problem is to say : anyone who can't quickly master linear algebra is incapable of doing mathematics and therefore doesn't interest us. This is certainly what many mathematicians think and what some of them don't hesitate to admit. Unfortunately, this reply is not acceptable. A first reason might be that applications of linear algebra are at the same time very varied and very significant and that it is important that outside the domain of mathematics many people know how to use it to good avail.

But there is a second more fundamental reason: mathematics is not the sole property of mathematicians who would in its name form a sect or a secret society. Fulfilling their role in society is a moral exigency and, for more selfish reasons, the best way for them to defend their discipline from attacks from those who, rightly or wrongly, have had the impression of having been kept out of it.

This amounts to saying that we need a teaching of mathematics that would be more and more efficacious, while at the same time we have to be honest enough to admit that so far it has only been so to a minor degree. (Mathematics is not the only subject in this situation but this is not to be regarded as an excuse). And it must be admitted that this is not an easy task for which all the good intentions and ideas, apparently reasonable, yet which have not been submitted to a control of reality, are of little help.

A common preconception among mathematicians is that in order to teach mathematics well, all that is necessary is to know the subject well. The teaching of linear algebra provides a striking counter example. The theory is well developed, those who teach it know it personally very well ... yet the students do not understand.

The contrary preconception, held by educationalists who believe that with an appropriate pedagogy anything can be taught, is just as disastrous. There again linear algebra provides a counter example : it is a resistant mathematical matter whose difficulty will not be overcome by any amount of educational know how, no matter how elaborate it is, if it is not based on a sound understanding of the mathematical content; that is, the ideas, the concepts, and the methods at work in the theory.

In order to make progress in the teaching, one must first of all consider it as an object of study. That is precisely the mission of 'la didactique' of mathematics whose motto could well be : « How do students acquire a mathematical frame of mind? »

The didactic study exposed in the present work is examplary on more than one count:

- the mathematical content is present throughout.

- it is the result of team work, or several teams at work, who collaborate without losing their autonomy. This gives rise to a multiplicity of viewpoints, an enlargement of experiential fields, and criticisms of results.

- it is a carefully controlled experimental work whose results are presented with rigorous honesty.

- finally, the conclusions distinguish between results that can be reasonably considered as established and those which call for a more profound study and further experimentation. Besides which, the authors are wise enough to believe that there is more than one way to teach linear algebra and they refuse to provide recipes for teaching that some readers might be expecting from them.

Some of my fellow mathematicians have admitted to me that they would believe in the didactic theory (la didactique) were it capable of providing teachers with precise, easy to follow advice. But this is precisely what it must not do: the application of recipes without understanding their origin leads to the worst misconceptions (hundreds of examples abound). The best recommendations, the best 'instructions' in the world, are, at best, totally ineffective.

If it is true that to teach well one needs to possess a mathematical culture that is alive, it is also true that one needs to possess a vibrant didactic culture, which is neither the application of recipes, nor a routine pompously qualified as experience, but one that arises from the work of the didacticians, with whom all teachers should collaborate.

As regards the mathematical content, the first chapter highlights a very interesting historical and epistemological study of linear algebra. This study clearly demonstrates the relevance of Jacques Hadamard's remark that I never tire of citing : « Natural ideas are those which come last. » Of course 'natural' here does not mean spontaneous, but adapted to the nature of the object under study. It is an apparent paradox attached to the development of mathematics that the greatest fecund ideas are intrinsically simple, but it took a long time and a great deal of effort to bring them to light.

But this poses a formidable problem for the teacher :

Is it necessary to have the student relive the historical development until the present day? In most cases this would be a fastidious job, difficult to realize, since one does not know all the details of the evolution, and it would also mean allowing the student to follow incorrect reasoning and to get lost in dead end trails.

Should we, on the other hand, bluntly present the theory of linear algebra as a finished product in its intrinsically 'simple form'? Experience would suggest otherwise : the simplicity is not that obvious to the student.

So where does the solution lie?

To a large extent, it lies, no doubt, in what is presented in this work under the title of 'meta lever', a method which it is certainly interesting to develop and further refine. There exists in mathematics courses a strange prudery which forbids one to ask questions such as, « Why are we doing this? », « At what is the objective aimed? », whereas it is usually easy to reply to such questions, to keep them in mind, and to show that one can challenge these questions and modify the objectives to be more productive or more useful. If we don't do this we give a false impression of a gratuitous or arbitrary interpretation of a discipline whose rules are far from being unmotivated or unfounded.

One must also consider the time aspect. Simple ideas take a long time to be conceived. Should we not therefore allow the students time to familiarize themselves with new notions? And must we not also recognize that this length of time is generally longer than that of the official length of time accorded to this teaching and that we should be counting in years? When the rudiments of linear algebra were taught at the level of the lycée (college level), the task of first year university teachers was certainly easier : for sure the student's knowledge was not very deep, however it was not negligible and it allowed them to reach a deeper understanding more quickly.

The effects of the fact that the teaching of proofs has disappeared from secondary school teaching - unbelievable, but alas true - will quickly be felt at the first year university level and subsequently in higher cycles.

I would like to see an indepth didactic study done on the time factor in the learning process. Without it ever being actually said, all teaching is organized as if pupils were immediately supposed to understand and assimiliate what they were being taught and to be able to immediately pass the periodic tests. It is not surprising, under these conditions, that ideas and culture have given way to techniques and algorithms.

It is clear that the learning time is not uniform. There are periods when nothing seems to sink in, when the teacher can have the impression that the students are marking time or even going backwards, and other periods in which everything becomes easy. It is the first of these periods which are crucial and which should form the object of an indepth didactic study because it is during these periods of time that the seeds of comprehension are sewn. To, then, (as is often the case) make students carry out series of repetitive exercises is fastidious, not in any way a positive reinforcement for those who have already understood, and a total waste of time for those who have not yet understood. It is during these difficult moments that what the authors of this book have called the 'meta lever' can be the most useful; not that it will solve all problems as if by magic, but that it can act as a lever to unblock the thoughts if it is placed at the correct spot with the correct amount of leverage. With regards to this notion, the authors tread prudently, which is their way in all fields, but I, personally, am persuaded that this is a most fruitful path.

That which is clear is that the utilization of the 'meta lever' runs contrary to an implicit ideology which rules in teaching and which perverts it. I quote two passages from the book:

- « There is an apparent paradox: in order to improve their skills, we develop a more stringent strategy towards the students, with the risk, at least provisionally, of lowering their results in exams. »

- « The replacement of the logic of success by the logic of learning appears to be very costly; in other words, are we not likely to make learning more difficult by being more exacting? »

In reality, this is an old debate: as opposed to an ideal teaching process whose aim is the mastery of what one has learned, we have a teaching system obsessed with the results of class tests, of term exams, of contests, and which easily falls victim to last minute cramming. If the spirit which animates it is founded on the 'logic of success', it must be stated that it is only a short term success; one in which, for the majority of students, deeper understanding, assured only by a 'logic of learning', is sacrificed. It is certainly much easier for the teacher; it is wholly in accordance with the spirit of competition which corrupts our society, and by necessity the teaching that accompanies it; but finally it is a catastrophe as regards its true results. There are only a rare number of students who are able to rise to a level of 'logic of learning' by themselves. All real improvements in the teaching of mathematics must necessarily lie in the deepening and diffusion of the 'logic of learning'. If it is costly, it is because it presupposes a deep change in mentality in the teacher : he/she is no longer the prophet of truths that have to be transmitted, but a free spirit calling forth other free spirits to develop and assume their responsibilities. This presupposes the observance of seeds of ideas in the minds of the students (with a greater frequency and relevance than that with which they are usually accredited) and to give them the means in which to ripen.

Certainly, there are no strict guidelines for managing a class in this manner. This presupposes a sufficiently deep knowledge of the mathematics that we wish the students to discover and a similarly deep knowledge of the different roads to access mathematics. In other words, it presupposes that 'la didactique' of mathematics has done its job and has let its job be known. Didacticians should therefore pursue their efforts. They should not stop because of the fears expressed in the second citation above. And I say to the authors of this book : « You are on the right track. Go ahead! ».

[1] Translated by Astrid Defence and Anna Sierprinska.
[2] André Revuz is a French mathematician who played an important role during the reform of Modern Mathematics especially regarding the teaching of linear algebra (see end of the first part of this book). He also played a leading role in the development of research in math education in France.

INTRODUCTION

The introduction of the theory of vector spaces in the teaching of mathematics at secondary level education has been, in many countries, one of the main issues of the reform of modern mathematics in the 1960s. It not only changed radically the teaching of geometry, which had been for centuries the core of mathematics education, but it also represented a new approach in the teaching of algebra. During the reform, the study of linear equations, which had until then constituted the main part of linear algebra, became merely a tool. Although still a central part of linear algebra, it was overshadowed by the formal theory which was to become the model for all linear problems, even infinite-dimensional, in a unified and generalized approach. Moreover, linear models became increasingly central paradigms within mathematics as well as in other sciences.

The teaching of vector space theory was therefore, in many senses, emblematic of the reform of modern mathematics. One of the challenges of this reform was to make mathematics accessible more directly to more students. In this sense, the theory of vector spaces appeared as a model of simplicity. It seemed, for instance, easy and powerful to interpret the solution set of a system of numerical or differential linear equations as the inverse image of a point by an affine transformation. However, it did not take very long, after the reform was implemented, to realize that what was regarded as so easy was in fact a source of serious cognitive difficulties. More or less rapidly, depending on the country, the reform had progressively been abandoned in the early 1980s, and the theory of vector spaces disappeared from secondary education. It is now taught only in first year of university and mostly in science-orientated curricula. The content of the course may be very formal, limited to $\mathbb{R}^n$ and matrix calculus, or only introduced within a geometrical setting. However, teachers in charge of a linear algebra course are very often frustrated and disarmed when faced with the inability of their students to cope with ideas that they consider to be so simple. Usually, they incriminate the lack of practice in basic logic and set theory or the impossibility for the students to use geometrical intuition. On the other hand, students are overwhelmed by the number of new definitions and the lack of connection with previous knowledge.

The purpose of this book is not to give a miraculous solution to overcome these difficulties. It will present a substantial overview of research works, which consist in diagnoses of students' difficulties, epistemological analyses and experimental teaching, offering local remediation. Nevertheless, these works' main results are new questions, problems and difficulties. Yet, this should not be interpreted as a failure. Improving the teaching and learning of mathematics cannot consist in one remediation valid for all. Cognitive processes and mathematics are far too complex for such an idealistic simplistic view. It is a deeper knowledge of the nature of the concepts, and the cognitive difficulties they enclose, that helps teachers make their teaching richer and more expert; not in a rigid and dogmatic way, but with flexibility. Therefore, the purpose of this book is to inform mathematics teachers in charge of linear algebra courses, as well as researchers in mathematics education, about the main results of research works on the teaching and learning of linear algebra. In France, like in North America, research in mathematics education at university level focused, in a first stage, mostly on teaching and learning of Calculus. Research about the teaching and learning of linear algebra started at the beginning of the 1980s, and gradually became a major issue in the 1990s. In this book, we will try to expose, in a maximum of detail, research works which have played an initiating role in the field. We will also give an overview of more recent developments. Of course, it is impossible to be exhaustive, thus we apologize for research works that we have not mentioned.

The first part of this book is not directly devoted to the teaching of linear algebra. It presents, on the basis of a historical survey, an epistemological analysis on the nature of linear algebra. Through the study of original works, Dorier analyzes the evolution of linear algebra from the first theoretical results on systems of linear equations (around 1750), until the final elaboration of the axiomatic theory of vector spaces and the first attempts at teaching these concepts. In particular, the author points out and analyzes different phases in the process of unification and generalization that led to the modern theory. He shows the difficulties, sometimes important, that each new phase had to overcome before being accepted, especially regarding the axiomatic approach.

This historical part is in interaction with the didactical research, not only because the historical material nourishes the didactical analysis but also because the didactical concern gives some specific orientation in the epistemological analysis of the historical context.

The second part of the book, devoted to educational issues, is divided into 8 chapters.

The first four chapters present a synthesis of a research program led by Dorier, Robert, Robinet and Rogalski, who have been working, as a team, on the issue of linear algebra, since 1987.

The first chapter is devoted to the first investigations made by the group, their evolution and the first conclusions and perspectives of research to which they led to. It is mostly a diagnosis, made in ordinary conditions, of the teaching of linear algebra in French science universities. It gives an overview of the main errors and difficulties of the students and, in relation with the historical analysis, it allows for a better understanding of how the unifying and generalizing nature of linear algebra is a source of the learning and teaching difficulties. In this sense, the main issue raised in this chapter concerns what the authors have named *the obstacle of formalism*.

In the second chapter, Robert introduces the concept of *level of conceptualization*. This theoretical approach allows a new type of analysis of the difficulties encountered by the students in the learning of linear algebra.

In the third chapter, Rogalski gives a global description of the *teaching project* with which he has experimented for several years. This project, corresponding to the experimental aspect of the research, is based on the first diagnoses, but it also tests new hypotheses.

The fourth chapter is devoted to the presentation of the *meta lever*: a new teaching tool, elaborated in order to try to make students overcome the obstacle of formalism, which is central in the experimentation. Three types of experimental use of this teaching tool are presented as illustrations. The issue concerning the difficulties encountered in the evaluation of these experimentations is addressed in the conclusion of this chapter, which also presents a synthesis of the results of the whole research project, the difficulties that remain unsolved, and the perspectives.

The teaching of linear algebra in North America has different characteristics than in France. It is usually less formal, and the model of $\mathbb{R}^n$ and matrix calculus are more central. Nevertheless, this teaching presents some difficulties, which may be partly different from those in France, but are globally quite similar. Three chapters of this book are devoted to North American research works about the teaching and learning of linear algebra.

In the fifth chapter, Harel presents the recommendations made by the *Linear Algebra Curriculum Study Group*. He gives his personal interpretation of these recommendations through three teaching/learning principles: the concreteness principle, the necessity principle and the generalizibility principle. Although this approach is presented in the context of a North American perspectives, it bears several similarities with the issue raised in the preceding work concerning the unifying and generalizing nature of the theory of vector space.

In the sixth chapter, Hillel distinguishes several modes of description (or language) in use in linear algebra: the abstract mode, the algebraic mode and the geometric mode. He analyzes, not only how these different modes of description function and can be used, but also how they interact, and especially how it is possible (or not) to move from one mode to the other. Moving between the algebraic and the formal modes of representation has been noticed as one specific difficulty. Hillel analyzes this problem in activities concerning the matrix representations of an operator in different bases.

In the sixth chapter, Sierpinska analyzes some aspects of students' reasoning in linear algebra. She bases her work on several teaching experiments at Concordia University, in Montreal, between 1993 and 1999. In some sessions the students used the dynamic geometry software, Cabri Geometry II. Some tutoring sessions were also analyzed. From her analysis of these sequences, and of the history of linear algebra, she is led to distinguish between practical and theoretical thinking. Students think in practical rather than theoretical ways and Sierpinska points out several cases in which this is a source of difficulties in the learning of linear algebra. In the second stage of her work, Sierpinska distinguishes between three modes of reasoning in linear algebra, corresponding to its three interacting languages : the visual geometric language, the arithmetic language and the structural language. The author illustrates with examples students' reluctance to enter into the structural mode of thinking and, in particular, their inability to move flexibly between the three modes.

In the eight chapter (the final one), five different works are presented. Four are doctoral dissertations (one being still in progress) and are all more or less connected with the research program presented in the first four chapters.

The first two are quite closely connected and are also related to Hillel's and Sierpinska's contributions. In the first, Pavlopoulou, using Duval's approach, distinguishes three registers of semiotic representations in linear algebra : graphic, tabular and formal. Her hypothesis is that the ability of operating in each register and, even more, of translating from one into the other, is essential to the understanding of concepts in linear algebra. This work is closely connected with Hillel's approach.

The second work, by Alves-Dias, addresses the general issue of cognitive flexibility through the question of change of register but also mathematical setting and viewpoint. Alves-Dias's hypothesis is that some complex cognitive processes cannot be reduced to problems of conversions. She analyzes in detail the question of change of representations for vector subspaces between parametric and Cartesian equations. In her experimentation, she shows that, even if changes of register and setting are important in this mathematical task, the cognitive activity cannot be reduced to them.

The third work presents the results of several surveys led over two years by a research team in Rennes. These surveys present a diagnosis of students' difficulties and some possible remediations. This work is complementary to the first chapter.

In the fourth work, Behaj is interested in the way knowledge is structured in the learner's mind in order to be memorized. He also wants to see in which way the structure of a course proposed by a teacher can influence the structuring in the students' mind. He made several interviews with teachers and pairs of students at least in second year of university. The mathematical subject he chose was linear algebra. Therefore, his work gives a valuable material about teachers' practice and the evolution of students' knowledge in linear algebra after two, three or four years at university.

Finally, in the fifth work, Chartier analyzes the role and the place of geometry in the teaching of linear algebra. As a theoretical framework, she uses Fischbein's work on mathematical intuition (i.e., distinguishing between analogical and paradigmatic models) to interpret the position of geometry towards linear algebra in the historical context, in textbooks, and in teachers' and students' practice.

The structure chosen for the book results from our concern to preserve the identity of each research work. Nevertheless, this choice may give the reader the feeling of a collection of isolated works without connection. Most of the authors, however, have collaborated over several years, but it is not always easy to reflect, in the details of each presentation, the interactions they had. In the conclusion, we try to give a synthetic overview of all the works presented, focusing on the common issues.

Jean-Luc Dorier

PART I

EPISTEMOLOGICAL ANALYSIS
OF THE GENESIS
OF THE THEORY OF VECTOR SPACES

JEAN-LUC DORIER

PART 1
EPISTEMOLOGICAL ANALYSIS
OF THE GENESIS OF THE THEORY
OF VECTOR SPACES

1. INTRODUCTION

In its modern version, linear algebra is essentially based on the theory of vector spaces. In France, this theory is usually taught in its axiomatic version according to Bourbaki's presentation, in such a hypothetico-deductive manner that the logical presentation of concepts might make teachers believe that its content is very simple. This is reinforced by the standardization of its vocabulary and tools, due to an extensive use of vector space theory in many fields, within or outside mathematics. This recently acquired universality - the axiomatic approach became widely used only after 1930 - should not hide a long development stretching over several centuries with difficult phases of unification.

Our aim is to try to draw out the most important stages of this evolution in order to counter-balance the somewhat over-estimated feeling of perfection attached to linear algebra[1]. Beyond this attempt to reconstruct the historical development, in order to challenge usual organizations and presentations of the concepts of linear algebra in teaching, this work aims at providing material in order to initiate didactical analyses. Moreover, the question of interaction between research in history and in mathematics education will be discussed at several stages in this work as well as in the second part of this book.

As an introduction, we would like to put forward, through a somewhat simplistic illustration, one of the epistemological aspects of linear algebra, which turned out to be essential to the didactical side of our research work (see the second part of this book).

Let us introduce the following three equations:

(1) $x f(x) - f''(x) = (x^2+1)$;

(2) $u_{n+2} - 3 u_{n+1} = u_n +1$;

3

DORIER J.-L. (ed.), *The Teaching of Linear Algebra in Question*, 3–81.
©2000 *Kluwer Academic Publishers. Printed in the Netherlands.*

$$(3) \begin{cases} 3x + 4y - t = 0 \\ 2y + 5z + t = 2 \end{cases}.$$

Of course, each of these equations has to be solved with specific tools, with respect to very distinct fields of mathematics. Yet, in all three cases, it is possible to state that the set of solutions is *an affine two-dimensional linear manifold*. Making such a statement is not just a way of showing erudition, it is both practical and essential since it suggests a precise solving method: *search for a particular solution, and for two independent solutions of the homogeneous corresponding equation*. Then the general solution is the sum of the particular solution and any linear combination of the two solutions of the homogenous equation.

This example points out one of the key aspects of linear algebra - that is, a mathematical theory that covers in a unified way (therefore in a formal and abstract setting), all linear problems. In fact, the possibility of modeling so many situations in nearly all parts of mathematics and other sciences is one of the main advantages of linear algebra. Yet, one could argue that this is quite normal, since linearity is always the easiest case, and one starts with linear problems before attacking more difficult questions. Nevertheless, it should be remembered that it was only during the 19th century that examples such as (1) and (2) were solved as equations, in the sense that they were interpreted as relations between objects via operations with known properties. In many ways, linear algebra had no object before the turn of the 20th century, or at least only existed in an embryonic state. Of course, such a statement applies to many other mathematical fields. But there is a difference. For instance, the history of the theory of integration can be traced to a time before the concept of function was fully constituted, but the evolution of the theory of integration and of the concept of function happened in a dialectical process: they came to maturity side by side, each benefiting from the progress of the other. It is much less clear that such a dialectical progress happened in the case of linear algebra. Vector space theory does not really constitute new material, rather a new way of 'looking at' old problems and of organizing mathematical knowledge. With the axiomatic theory, new problems in infinite dimension (initially mostly in functional analysis at the beginning) were solved, but the most valuable benefit came from the improvement in solving old problems. Also, a better cohesion of mathematics was reached, and differences could be overcome by putting forward similar structures between very distinct fields. This historical fact has a strong impact on teaching, especially at the first year university level, since precisely because of the level of mathematics with which the students enter university it is impossible to introduce them to infinite dimensional problems whose solution depends on the application of the axiomatic theory. Therefore, the unifying and generalizing aspects of vector space theory becomes an unavoidable aspect of its teaching and learning; should one say of its acceptance ? An historical account of linear algebra must therefore take this dimension into account. It must not only trace the origins of concepts but also try to understand how each of them evolved, until it attained its highest degree of generality as it is known in the modern theory. This implies estimating the importance of all the different stages of evolution with regard to the maturity of the concepts involved, as well as to the potential for generalization they bear at each stage.

Regarding methodology, the historical research presented here, from an epistemological perspective, covers a period from 1750 until very recent times. Our first goal was to trace the origins and evolution of the following concepts: vector, vector space, linear dependence and independence, (set of) generator(s), basis, rank, dimension, and linear application. In most cases, we have used original texts, however, in order to have a better basis on which to evaluate the evolution of linear algebra in relation to the rest of mathematics, we have also used monographs on related or transversal subjects. Moreover, we have sometimes made reference to papers (including some of our own) published in historical journals or books, where some specific aspects are developed in more detail than was possible here.

The first part of this book is divided into two main sections:

The first section deals with the analytical and geometrical origins of linear algebra until such time as a first unified corpus of tools and methods was built around the theory of determinants. In the first subsection, we show how the role of the study of systems of linear equations, served as a framework, upon which an embryonic theory of linearity was built. This point was essential for us, not only because it deals with the origins of the theory, but also because it shows how the most elementary concepts of linear algebra took decades of reflection on questions that today seem so obvious to us (if we can conceive of them at all). It was also important to insist on this period, when the theory of determinants was the framework for all questions dealing with linearity, because it is likely to be unusual for a modern reader.

The second subsection is devoted to the difficult question of the epistemological relations between certain parts of geometry and linear algebra. It seems important to us in this matter to question some stereotypes which tend to over-evaluate, or at least misjudge, the natural importance of the geometric model in the constitution of vector space theory.

As a counterpart, the third subsection deals with Grassmann's theory of extension. This theory starts from a reflection on geometry but explores higher dimensions and can be considered in many ways as the first formal theory of linearity in finite dimension. The richness of this work, which only had a slight influence on further historical development, led us to give it a particular part in our analysis.

In the fourth subsection we try to draw out the frame of the first stage of unification of finite dimensional problems of linearity. This is characterized by a corpus of objects and methods from varied origins, centered essentially around the theory of determinants and geometrical analogy. Moreover, we will see what separates this phase, which took place in the second half of the 19th century, from a formal approach in terms of algebraic structure.

The second section deals with the emergence and constitution of formal and axiomatic approaches. In the first subsection, we comment on the first axiomatic approaches, presenting their origins, their methods and what motivated their authors, as well as the reasons why they failed to be used.

In a second subsection, we analyze the evolution in the treatment of linear problems in infinite dimension. This is closely related to the history of what preceded the emergence of functional analysis. Moreover, it is interesting to see how infinite determinants were still in use when the first axiomatic approaches had been discovered a few years previously. We will see the reasons why these two poles in the development of linear algebra did not interact until 1920.

The last subsection is devoted to the final phase, from 1930, in the elaboration of the modern axiomatic theory of vector spaces. The respective influence of geometrical, algebraic and analytical origins in this process is discussed. Finally we will give a brief account of the first teaching attempts of the theory of vector spaces at various levels, in different countries. This will lead us to the second part of this book and the question of the possible interaction between historical research and mathematics education.

2. ANALYTICAL AND GEOMETRICAL ORIGINS

2.1. Systems of Numerical Linear Equations: First Local Theory of Linearity

Techniques for solving systems of linear equations can be found in very ancient civilizations all around the world. Yet until the middle of the 18th century, they were *ad hoc* tools, since mathematicians (in a broad sense) essentially needed to solve systems of linear equations. Some type of classification may have been suggested, but equations were not a subject for theoretical investigation independent of the solving question. This type of approach remained dominant for quite a while, but progressively, qualitative and descriptive studies appeared and could be separated from the solving of specific equation or system of equations.

2.1.1. Euler and the Dependence of Equations
Concerning linear equations, one of the first significant examples can be found in a text by Leonhard Euler, entitled, *Sur une Contradiction Apparente dans la Doctrine des Lignes Courbes,* from 1750 (Euler 1750). Euler discusses Cramer's paradox[2]. The study of this problem led him to question the fact that any system of n linear equations with n unknowns has a unique solution. This fact was at that time implicitly supposed to be true by all. Euler starts with the case of two equations, and gives the following example:

$$3x - 2y = 5 \text{ and } 4y = 6x - 10 \text{ ;}$$

Here is what he says about it:

> On verra qu'il n'est pas possible d'en déterminer les deux inconnues x et y, puisqu'en éliminant l'une x, l'autre s'en va d'elle-même et on obtient une équation identique, dont on est en état de déterminer rien. La raison de cet accident saute d'abord d'aux yeux, puisque la seconde équation se change en $6x - 4y = 10$, qui n'étant que la première $3x - 2y = 5$ doublée, n'en diffère point.[3] (ibid., 226)

Such a statement was not a great revelation for the mathematicians of Euler's time: for them the fact that two equations could be identical was not worthy of notice. Until then, no theory of linear systems had been needed - only practical rules for solving them were important. The novelty of Euler's text is that it does not give new methods for solving systems of linear equations: it offers instead, a descriptive and qualitative approach to their study.

Let us now look in more detail at what Euler says. The importance of his statement is that, although he finally claims that the second equation is obviously the double of the first, it was not this fact that indicated that the system was undetermined; rather it came about by a process of elimination. This proves that solving the system is still the main priority.

For $n=3$, Euler gives two examples, one with two similar equations and one in which one equation is the double of the sum of the other two. He does not attempt to solve either one of the systems but concludes:

> Ainsi quand on dit que pour déterminer trois inconnues, il suffit d'avoir trois équations, il y faut ajouter cette restriction, que ces trois équations diffèrent tellement entr'elles, qu'aucune ne soit déjà comprise dans les autres.[4] (ibid., 226)

For $n=4$, Euler says that in some cases even two unknowns may remain undetermined and he gives the following example with four equations:

$$5x + 7y - 4z + 3v - 24 = 0 \,,$$
$$2x - 3y + 5z - 6v - 20 = 0 \,,$$
$$x + 13y - 14z + 15v + 16 = 0 \,,$$
$$3x + 10y - 9z + 9v - 4 = 0 \,,$$

> elles ne vaudroient que deux. car ayant tiré de la troisième la valeur de
> $$x = -13y + 14z - 15v - 16$$
> et l'ayant substituée dans la seconde pour avoir :
> $$y = \frac{33z - 3v - 52}{29} \quad et \quad x = \frac{-23z + 33v + 212}{29} \,,$$
> ces deux valeurs de x et de y étant substituées dans la première et la quatrième équation conduiront à des équations identiques[5], de sorte que les quantités z et v resteront indéterminés.[6] (ibid., 227)

Therefore, here again, the proof is based on solving techniques of elimination and substitution. The criterion of dependence is the fact that unknowns remain undetermined. Moreover, Euler does not mention any linear relation between the equations, although two are quite obvious, for instance: (1) - (2) = (4) et (1) - 2x(2) = (3). As a conclusion he gives a general statement for any n:

> Quand on soutient que pour déterminer n quantités inconnues il suffit d'avoir n équations qui expriment leur rapport mutuel, il y faut ajouter cette restriction que toutes les équations soient différentes entr'elles, ou qu'il n'y en ait aucune qui soit renfermée dans les autres.[7] (ibid., 228)

For a modern reader such expressions as 'an equation is comprised in others' or 'an equation is contained in others' immediately means that the equations are linearly dependent. However, a more careful reading of what Euler does and means shows that in his mind these expressions refer to an 'accident' at the end of the solving process leading to the fact that some unknown quantities remain undetermined. Of course, several times (but not always), he shows that such an accident comes from linear relations between the equations, but that is not the criterion which is pertinent for him: his approach is consistent with that of others of his time, i.e. solving the system of equations is the first goal. The difference between the linear dependence property and Euler's 'being comprised (contained)' property may seem minor, yet, we will see that it had a very important incidence in the further development of linear algebra. Indeed, Euler's viewpoint ties him to the setting of equations, while linear dependence is a more general concept, valid for a large class of objects, for which only the possibility of making linear combinations is required. Therefore, it is important for our goal to make the distinction: we will call *inclusive dependence,* this property of an equation contained in others. Until the second half of the 19th century, inclusive dependence rather than linear dependence was the dominant conception in problems dealing with linear equations.

Euler's explicitation of the exceptional case of inclusive dependence in square systems of linear equations was an important change of viewpoint in the approach to linear equations. His text contains even more interesting new ideas on the subject. This can already be seen in the example given above with four equations in which he expresses their the complementarity between the number of linear relations between the equations and the number of unknowns left undetermined (in other words the 'size' of the set of solutions): « Il peut même arriver que deux équations soient déjà comprises dans les deux autres et alors il n'y aura que deux équations qui restent dans le calcul et par conséquent deux inconnues resteront indéterminées. » (« It is even possible that two equations are already comprised in the two others, then only two equations will remain in the calculations and consequently two unknowns will remain undetermined »). This can be qualified as a first empirical insight of the relationship between the number of independent equations, the size of the set of solutions and the number of parameters necessary to represent it. These ideas would be formalized later through the concepts of rank and duality. At the end of his text, when dealing more precisely with Cramer's paradox for $n=4$, Euler gives an explanation including some similar insights:

> Quand deux lignes du quatrième ordre s'entrecoupent en 16 points, puisque 14 points, lorsqu'ils conduisent à des équations toutes différentes entr'elles, sont suffisants pour déterminer une ligne de cet ordre, ces 16 points seront toujours tels que trois ou plusieurs des équations qui en résultent sont déjà comprises dans les autres. De sorte que ces 16 points ne déterminent plus que s'il n'y en avoit que 13 ou 12 ou encore moins et partant pour déterminer la courbe entièrement on pourra encore à ces 16 points ajouter un ou deux points.[8] (ibid., 233)

As a conclusion to our analysis of Euler's contribution, we would like to underline the fact that his approach is embedded in the solving paradigm: he does not offer any theoretical approach of linear dependence. Yet his work is essential, not only because it solved Cramer's paradox, but also because it put forward simple intuitive facts with important implications.

2.1.2. Cramer and the Birth of the Theory of Determinants

In the same year, Gabriel Cramer published a treatise entitled *Introduction à l'Analyse des Courbes Algébriques.* Except for a letter by Leibniz, from 1693, but only published for the first time in 1850, this text was the first one in which a specific notation was given in order to write a system of linear equations with non-specified coefficients[9]. It also established a rule in order to express the solution of a square system as a function of its coefficients[10] - using what will later be known as the determinant.[11] From Cramer's work, determinants became widely used and constituted a thriving branch of mathematics. The theory of determinants became the inevitable framework for the study of systems of linear equations, consequently the Eulerian approach ceased to exist. After Cramer, very few of the numerous works on determinants involved any sort of qualitative study on systems of linear equations. Improvements in the explicitation of Cramer's rule of calculation, and increasingly difficult calculations of specific determinants were the two main streams of development. However, even though determinants were an efficient tool for the study of linear equations, they also introduced a certain complexity due to the technical approach inherent to their use. Consequently intuitive approaches, like Euler's (op. cit.), were no longer possible and a considerable change occurred in the study of systems of linear equations.

2.1.3. The Concept of Rank

In fact, for nearly a century, the questions related to undetermined and inconsistent systems of linear equations were neglected, whereas it is only through these questions that one can approach the notions of linear dependence and independence, and rank. From around 1840 to 1879, within the theory of determinants, the concept of rank took shape. In the context of linear equations, rank is an invariant which determines the size of the set of solutions (minimal number of generator / maximal number of independent solutions) and, by a process of duality, the number of relations of dependence (minimal number of equations describing the set of solutions / maximal number of independent equations). To create the concept of rank containing these different aspects, mathematicians had to overcome several obstacles and change their point of view on certain elementary notions. We analyzed this process in detail in (Dorier 1993); we shall just give the main outlines here.

One can list three main sources of obstacles and difficulties :
1. the recognition of the invariance which was, if not unseen, at least assumed without necessity of proof.
2. the possibility of the same definition of dependence between equations and n-tuples.
3. the anticipation of the concept of duality and the consideration of all the systems of equations which have the same set of solutions.

Of course, these three sources of difficulties are not independent, and the progress overcoming any one of them influenced that overcoming the other two. As was pointed out in Euler's work, he concept of inclusive dependence remained but was also rapidly connected to the evanescence of the main determinant of a square system of linear equations. Moreover, the notion of minor allowed a determination of the 'size' of the set of solutions in relation to the maximal number of independent equations. In fact, the maximal order r of non-evanescent minors in a system of p linear equations in n unknowns gives the number $n-r$ of arbitrary unknowns to be chosen to describe the set of solutions of a consistent system and represents the maximal number of independent equations in the system. Such ideas had become well known by the middle of the 19th century.[12] The classical method consisted in isolating the part of the equations corresponding to a nonzero minor of maximal order and then in using Cramer's rule, with the other unknowns as parameters appearing in the second members. This manipulation required theoretical justifications, which prior discoveries on determinants made explicit. Moreover, this first phase opened new ways to a more systematic investigation into systems of equations. However, it was more the search for a practical method than concern for theoretical achievement that led to this first step. Henry J. S. Smith's approach pointed out a change of point of view which marked a fundamental step: In a paper of 1861 (Smith 1861), he shows that the maximal order of a nonzero minor is also related to the maximal number of independent solutions. However, this did not help, directly, to describe the set of solutions better. In this sense, it is more a theoretical than a practical result and showed the slight but crucial change of approach marked by Smith: He was not only interested in giving ways to solve systems of equations, but also studied them from a theoretical basis.

Between 1840 and 1879 the concept of rank was, therefore, implicitly central to the description of systems of linear equations. With the use of determinants, an analogous treatment of the question of dependence on equations and n-tuples was possible. Yet the technicality of demonstrations involving determinants seems to have made it difficult to provide a clear and concise overview of all the relations of invariance and duality involved. For instance, the type of work developed in this period was very different from Euler's (op. cit.). The tools were more sophisticated, but their use required such technique that intuition was averted.

2.1.4. Georg Ferdinand Frobenius

Frobenius was one of the first (if not the first) to have introduced a crucial change in the approach of linear problems. In a text from 1875 entitled, *Über das Pfaffsche Probem*, he defines, in modern terms, the notions of linear dependence and independence simultaneously for equations and n-tuples, making the link with the conception of inclusive dependence quite clear:

> Mehrere particuläre Lösungen
>
> $A_1^{(\chi)}, \ldots, A_n^{(\chi)}$, $(\chi = 1, \ldots, k)$
>
> sollen daher *unabhängig* oder *verschiedenen* heissen, wenn $c_1 A_\alpha^{(\chi)} + \ldots + c_k A_\alpha^{(\chi)}$ nicht für $\alpha = 1, \ldots, n$, verschwinden kann, ohne dass $c_1, \ldots, c_k$ sämmtlich gleich Null sind, mit anderen Worten, wenn die k linearen Formen $A_1^{(\chi)} u_1 + \ldots + A_n^{(\chi)} u_n$ unabhängig sind.[13] (Frobenius 1875, 223).

By considering equations and n-tuples as the same kind of objects with regard to linearity, he makes a crucial step toward the modern concept of vector. Then he defines the notion of basis[14] of solutions and introduces the notion of *associated ('zugeordnet oder adjungiert') system* to a given system: that is, a system of linear equations whose coefficients are the components of the elements of any basis of solutions of the initial system. In modern terms, with regard to the space of solutions, an associated system represents a basis (of linear forms) of the orthogonal space, therefore this notion is related to the concept of duality. With this new tool it is easy, then, to prove that the maximal order of nonzero minors (for any system of linear equations) is an invariant, not only regarding the system, but also regarding the set of solutions (all systems having the same set of solutions have the same maximal order of nonzero minor).

He also establishes that this number is an invariant regarding the 'size' of the set of solutions, as well as regarding the maximal number of independent equations. For the first time, Frobenius' approach was totally different from the solving process coming from Cramer; there was no more arbitrary separation between principal and secondary unknowns and equations and, above all, the concept of linear dependence had replaced inclusive dependence. In this text the concept of rank was finally defined in all its generality and characteristics. Nevertheless, it was only in 1879 that Frobenius himself named the concept as the maximal order of nonzero minor:

Wenn in einer Determinante alle Unterdeterminanten $(m+1)$ten Grades verschwinden, die $m$ten Grades aber nicht sämmtlich Null sind, so nenne ich *m den Rang* der Determinante.[15] (Frobenius 1879, 1)

Mathematicians of the time rapidly understood the importance of the concept of rank. Not only did it avoid the use of long circumlocutions, but it also allowed much easier and clearer solutions to many problems dealing with determinants. Frobenius showed the efficiency of his new concept in many of his writings dealing with quadratic forms or differential equations, of which the two papers quoted here are good samples.

2.1.5. Further Development

Nevertheless, as is visible in Frobenius' definition quoted above, the notion of rank was still intrinsically bound to the context of the theory of determinants. This was not only true for its definition, but also for the type of properties in which it was used (in the statement as well as in the proof of the theorems). In 1886, Alfredo Capelli and Giovanni Garbieri, in their *Corso di Analisi Algebrica*, showed that a system of rank k is equivalent to a triangular system with exactly k nonzero diagonal terms. In their proof they applied a method of elimination using minors of the system as coefficients but, even if their approach used technical results of the theory of determinants, the new result they introduced pointed out an invariant linked to the rank which had nothing to do with determinants. This approach is very important in effective methods for solving systems of linear equations - for instance, in numerical analysis or linear programming. It will be particularly developed when computers are capable of solving systems of a large number of equations and unknowns.

Thus, the study of systems of linear equations led to the explicitation of two key concepts of linearity : rank and duality. In this process, the unification of linear properties common to both equations and their solutions played a crucial role, a decisive step toward unification, and, therefore, toward the modern concept of vector as an element of a linear structure. In fact, Frobenius laid the basis for a first theoretical approach to linearity, even if the context of determinants now appeared to be a little too technical to make intuitive ideas effective. From a didactical viewpoint, the historical analysis shows the consistency of starting with the study of systems of linear equations when teaching linear algebra. But it also gives reasons for finding tools other than determinants, which would avoid the technical difficulties they engender. It makes one wonder how the concepts of rank and duality would have taken shape if the determinants had not been introduced so early, through a more intuitive approach like Euler's. We will see in the second part of this book that alternative didactical approaches can be offered in which the Gaussian elimination is used as a method for solving systems of linear equations.

2.2. Geometry and Linear Algebra: a Two-Century-Long Process of Complex Reciprocal Exchanges

The question of the relationship between linear algebra and geometry cannot be discussed without referring to the different backgrounds of culturally related teaching traditions that may vary from one country to another. In France, for instance, during the period of Modern Mathematics (the New Maths) (roughly 1968-1985), geometry in secondary teaching was introduced mostly through the filter of vector space theory, whose main results in finite dimension were taught in high school. Euclid's axioms were replaced by those of affine spaces with a strong emphasis on the study of linear and affine transformations. However, several mathematicians never agreed with this approach, and it became one of the most widely discussed issues when opposing (or defending) the teaching of Modern Mathematics. Therefore, since the counter-reform, the syllabus concerning geometry has been profoundly modified. Although the fundamentals of Cartesian and vectorial geometry are still taught, the teaching of geometry is now essentially centered on a synthetic approach. For instance, transformations are more often introduced through the study of their action on figures than through their properties as transformations on a set of points with structural features. In the late eighties, this change started to affect the teaching of linear algebra at university. Many mathematics departments decided to create a sophomore course on Cartesian and vector geometry as a prerequisite for the more formal course on vector space theory. Nevertheless, this attitude shows that geometry is seen by most mathematicians as a natural context for introducing the main concepts of linearity. This idea was explicit in Dieudonné's famous book *Algèbre linéaire et géométrie élémentaire* (Dieudonné 1964) and we will show, at the end of this paper, the crucial role it played in the reform of Modern Mathematics in secondary teaching in France. Therefore, there is a strong cultural emphasis (at least in France) that suggests a natural relationship between geometry and linear algebra. Nevertheless, from a historical viewpoint, the question is far from being so simple.

Since geometry, unlike algebra, is intrinsically linked to visual perception, it is potentially a source for intuitive thinking. Moreover, it has a long tradition of classical problems and forms of questioning that go back to the birth of Mathematics in ancient Greece. The introduction of algebra in geometry represented, through the use of the analytical method, a radically different approach to these problems in that they could now be solved with more systematic methods. For instance, the classification of curves could be carried out through the classification of Cartesian equations. Moreover, Cartesian geometry had the potential to break the limitation of study to three dimensions, even though this generalization did not come into effect for nearly another two centuries after Descartes. It was also from a certain complementarity between algebra and geometry that the idea of creating a new type of algebra, operating on geometric entities and not on numbers representing them, emerged. This was to be an essential factor not only in the development of geometry but also in the development of vector calculus and linear algebra in the second half of the 19th century.

One of the myths (supported by traditional teaching) about the natural link between geometry and linear algebra comes from the extensive use of common vocabulary in the two fields. For instance, the fact that the linear structure is called a 'vector space' automatically certifies the geometrical origins of the theory. Another reason for assuming a natural link comes from the use of geometrical representation: e.g. the sum of two vectors can be represented as the diagonal of a parallelogram (as in the parallelogram of velocities and forces, a very ancient type of representation used in physics to symbolize the combined action of two velocities or forces applied at a same point). However, there is a big gap between the traditional use of this representation and the modern interpretation of the algebraic sum of vectors (see Crowe 1967, 2).

2.2.1. Analytic Geometry

Historically, the link between geometry and linear problems is a result of the discovery of analytic geometry, made independently by René Descartes (1637) and Pierre de Fermat (1643)[16];. Their goal was to apply the power of algebra to geometry:

> Tous les problèmes de géométrie peuvent facilement se réduire à des termes tels qu'il n'est besoin par la suite que de connaître la longueur de quelques lignes droites pour les construire.
> Et comme toute l'arithmétique n'est composée que de quatre ou cinq opérations qui sont, l'addition, la soustraction, la multiplication, la division et l'extraction de racines, qu'on peut prendre pour une espèce de division, de même n'a-t-on autre chose à faire en géométrie, en ce qui concerne les lignes qu'on cherche, pour les préparer à être connues, que d'en ajouter d'autres ou d'en ôter. Ou bien en ayant une, que je nommerai l'unité pour la rapporter d'autant mieux aux nombres, et qui peut ordinairement être prise à discrétion, puis en ayant encore deux autres, en trouver une quatrième qui soit à l'une de ces deux comme l'autre est à l'unité, ce qui revient à multiplier. Ou bien en trouver une quatrième qui soit à l'une des deux comme l'unité est à l'autre, ce qui revient à diviser. Ou enfin trouver une, deux ou plusieurs moyennes proportionnelles entre l'unité et quelque autre ligne, ce qui revient à extraire la racine carrée ou cubique, etc...
> Et je ne craindrai pas d'introduire ces termes d'arithmétique en la géométrie afin de me rendre plus intelligible.[17] (Descartes 1638, 297-298)

The power of such a new method was quite obvious for the authors, thus Fermat ended his treatise by stating : « Nous avons donc embrassé dans un exposé bref et lucide tout ce que les anciens ont laissé inexpliqué sur les lieux plans et solides. » (« We have therefore embraced in a short and clear report all which the Ancients had left unexplained about plane and solid loci »). The analytical method became rapidly successful among mathematicians of the time, who acknowledged its power of simplification and unification by using it to solve a large number of new and old geometry problems. The algebraization of geometry resulting from the new method had very important consequences on its organization. One of the most fundamental, from the point of view of our research, is that linearity in geometry became much more essential. In this context indeed, it is quite natural to consider linear equations as the most basic type of equation, which makes straight lines the most basic curves in geometry which was not that clear until then, as straight lines and circles constituted traditionally a same category called plane loci[18]. Moreover, in analytic geometry, the classification of curves being based on their equations, the search for invariants led naturally to an analysis of the effects of the change of coordinates; that is, what was then referred to as linear substitutions or, today, more commonly known as linear transformations on the curves.

2.2.2. Leibniz's Criticism

In spite of this wide and immediate success of the analytical method in geometry, some mathematicians expressed criticism toward the new method. Indeed, they could not accept that to solve a geometrical problem something as totally external to geometry as algebra operating on numbers was used. The philosophical aspect of this position, shared to different degrees by many mathematicians, was also sustained by a practical one: i.e., when solving a geometrical problem with the analytical method the choice of a system of coordinates is generally completely free, but of course it has a crucial impact on the feasibility of the calculations. Moreover, the analytical method operates on numbers and equations therefore hides the geometrical reality and prevents any intuitive control. If one can prove with it, the analytical method does not give the reasons for a result to be the way it is. Gottfried Wilhelm Leibniz was one of the most critical toward Descartes' approach. Here is what he wrote to Christian Huyghens, in a letter dated 8th September 1679 :

> Je ne suis pas encor content de l'Algèbre, en ce qu'elle ne donne ny les plus courtes voyes, ny les plus belles constructions de Geometrie. C'est pourquoy lorsqu'il s'agit de cela, je croy qu'il nous faut encor une autre analyse proprement géométrique ou linéaire[19], qui nous exprime directement situm, comme l'Algèbre exprime magnitudinem. Et je croy d'en avoir le moyen et qu'on pourroit représenter des figures et mesme des machines et mouvements en caractères, comme l'Algèbre represente les nombres ou grandeurs.[20] (Leibniz 1850, 1:382)

Leibniz tried to invent a geometrical calculus on the basis of his criticisms towards analytic geometry:

> J'ay trouvé quelques éléments d'une nouvelle caracteristique, tout à fait différente de l'Algèbre, et qui aura des grands avantages pour representer à l'esprit exactement et au naturel, quoyque sans figure, tout ce qui dépend de l'imagination. L'algèbre n'est autre chose que la caractéristique des nombres ou des grandeurs. Mais elle n'exprime pas directement la situation, les angles, et le mouvement, d'où vient, qu'il est souvent difficile de reduire dans un calcul ce qui est dans la figure, et qu'il est encor plus difficile de trouver des demonstrations et des constructions géometriques assez commodes lors meme que le calcul d'Algebre est tout fait. Mais cette nouvelle caracteristique suivant des figures de vue, ne peut manquer de donner en meme temps la solution et la construction et la demonstration géometrique, le tout d'une maniere naturelle et par une analyse. C'est à dire par des voyes déterminées.[21] (ibid., 1:384)

The geometrical analysis presented by Leibniz was based on a relation of congruence between n-tuples of points: two couples of points are congruent if their two points are at an equal distance from each other, two triplets are congruent if the two triangles they determine can be brought into coincidence, etc. He could, therefore, define a sphere as the set of points X such that AX is congruent to AB, a plane as the set of points X such that AX is congruent to BX, a circle as the set of points X such that ABX is congruent to ABC, a straight line as the set of points X such that AX is congruent to BX and to CX, etc. Leibniz applied his analysis to a few elementary geometrical problems, but he never really developed his approach and may have been aware of the limitations of his characteristic. The fact that the relation of congruence did not take into account the different directions in space nor the orientation, prevented this type of calculus from being operational beyond a certain point.[22] Leibniz's letter was published for the first time in 1833, at a time when the search for a geometrical calculus had not really progressed significantly.[23]

2.2.3. *Geometrical Representation of Imaginary Quantities*
A first step in this direction came from the possibility of representing geometrically the imaginary quantities (i.e. complex numbers). The principles of this representation were discovered independently by several persons whose goals were

either the search for a geometrical calculus or finding a legitimate status to the impossible square roots of negative numbers. The status of the latter, known as imaginary quantities, was very controversial until the first half of the 19th century. In a work from 1673, John Wallis had already tried to give a geometrical illustration of the square roots of negative numbers but his model, using surfaces under the water, was not really satisfactory, especially regarding the multiplication of such numbers (Wallis 1673). Within a few years, most certainly independently, five men, mostly unknown to the mathematical world, developed (with varying degrees of success) the principles of geometrical representations of imaginary quantities: Caspar Wessel in 1799, Adrien Quentin Buée in 1805, Jean Robert Argand in 1806, C.V. Mourey in 1828 and John Warren also in 1828.[24] Yet, it was only with Carl Friedrich Gauss' work, published in 1831, that these principles became widely known and used by mathematicians. thereby gaining a legitimate status for imaginary quantities within geometry. They provided an alternative model for applying algebra to geometry which was more intrinsic than the coordinates usually used with respect to the Cartesian plane. Some of the authors (especially Argand and Wessel) tried to extend their discovery to three dimensions and, therefore, to create operations on triplets of real numbers, but they all come to a dead end when it came to defining a multiplication of triplets.

2.2.4. *Möbius's Barycentric Calculus*
Around the same period, two mathematicians, August Ferdinand Möbius and Giusto Bellavitis created models of geometrical calculus. In 1827, Möbius created his *Barycentrische Calculus*, a kind of algebra operating on points of the geometric space. In the beginning of his treatise, Möbius insists on the necessity of distinguishing some characteristics relative to the orientation and direction of geometrical entities; he introduces a notion of directed line-segment and also notions of oriented triangle and pyramid. However, he defines the sum of two directed line-segments only when they are collinear. He generalizes this concept of sum for non-collinear oriented line-segments in 1843 in his *Elemente der Mechanik des Himmels*, being the first to present a linear structure of geometrical entities in space (Möbius 1843, 1-2).

The definition given by Möbius for a sum of weighted points could only be based on the sum of collinear directed line-segments and therefore is different from what a modern reader would expect. It is based on the following theorem:

Ist eine beliebige Anzahl = n von Puncten $A,B,C,$ ..., N mit resp. Coefficienten $a,b,c,...n$ gegeben, deren Summe nicht = 0 ist, so kann immer ein Punct S, und nur einer, -der Schwerpunct,- von der Beschaffenheit gefunden werden, dass, wenn man durch die gegeben Puncte und der Punct S nach einer beliebigen Richtung Parallelen zieht, und diese mit einer willkürlich gelegten Ebene schneidet, welches resp. in $A',B',C',$ ..., S' geschehe, dass dann immer
$$a.AA' + b.BB' + ... + n.NN' = (a + b + c + ... + n). \, SS',$$
und folglich, wenn die Ebene durch S selbst geht,
$$a.AA' + b.BB' + ... + n.NN' = 0$$
ist.[25] (Möbius 1827, 9-10)

Then he introduces the notation : $a.A + b.B + c.C + ... + n.N = (a + b + c + ... + n).S$. In the case when the sum of the coefficients equals zero, Möbius says briefly that the point is sent to infinity, but does not analyze this case in detail.[26]

Möbius did not intend to build an algebraic theory; he aimed at providing mathematicians and physicists with a tool in order to improve the solving of geometry or mechanics problems. This is why he was not interested in investigating specific cases which seem to have only a theoretical relevance. From this perspective, his work was quite a success, but it also induced a theoretical breakthrough. In particular, it inspired Karl von Staudt, in his discovery of projective coordinates, which were essential in order to separate projective geometry from metric considerations and, therefore, to better understand the nature of this very important approach to geometry.

2.2.5. Bellavitis's Calulus of Equipollences

On another hand, Giusto Bellavitis, with his *Calculus of Equipollences*, of which the first results were published around 1833, was the first to define something very close to the concept of geometrical vector as a class of equipollent couples of points - what he calls an *equipollence*. He defines the principles of addition of equipollences and multiplication by a number, and then the concept of multiplication of two equipollences. He admits that he was strongly inspired by Buée's memoir about the geometrical representation of imaginary quantities (see above). Here is a statement that gives a good idea of his approach:

> *Teorema fondamentale.* Nelle equipollenze si transportano i termini, si sostituiscono, si sommano, si sottrano, si moltiplicano, si dividono ec., in una parola si eseguiscono tutte le operazioni algebraiche, che sarebbero legittime se si trattasse di equazioni, e le equipollenze che ne risultano sono sempre esatte. Le equipollenze non lineari non possono riferirsi, come si disse al§5, se non se alle figure poste in un solo piano. [27]
> (Bellavitis 1835, 247)

Bellavitis showed the pertinence and the efficiency of his method through various applications to geometry and physics. Nevertheless, from a theoretical viewpoint, it can be said that his method was nothing but a transcription of Buée's ideas, although his replacing of imaginary quantities by real geometric entities established quite a different perspective regarding the development of a geometrical calculus. In particular, except for multiplication, Bellavitis' equipollences were three-dimensional. Moreover, as can be seen in the above quotation, Bellavitis was concerned with the algebraic structure of equipollences; he explicitly presented them as a new type of algebra having most of the properties of usual algebra.

2.3. Hamilton's Quaternions

We have said above that generalizing the geometric representation of imaginary quantities to three dimensions was halted because of the difficulty of defining a suitable multiplication of triplets. In fact, until the middle of the 19th century, the only model for algebra had been given by real numbers, therefore, a new type of algebra had to adhere to the (albeit implicit) commutative properties of addition and multiplication - what we now call the commutative field structure. In this sense, mathematicians tried naturally and implicitly to create a geometric calculus for space with an addition and a multiplication over $\mathbb{R}^3$, with the properties of a commutative field. Sir William Rowan Hamilton had been interested in the question of the geometric representation of imaginary quantities and the possible generalization to three dimensions for several years, when he discovered the *Quaternions* around 1843. He was the first one to make explicit the properties that addition and multiplication of triplets should satisfy. He wrote several notes about his research and his various trials. They show that he often changed his approach, from geometric to algebraic. Finally, he made a breakthrough towards his discovery of the quaternions by examining the geometric properties of the multiplication of imaginary quantities - represented by vectors. He claimed that this type of multiplication was based on the product of the lengths of each vector and the angle included by them. However, by trying to extend these ideas to three dimensions, he discovered that the included angle itself was not sufficient but that it was also necessary to consider the plane on which this angle was drawn. In other words, what was the rotation that indicated in which direction to turn (Hamilton 1866, 1:106-110). This was essential, indeed, if in three-dimensional space, the length is a scalar, like in the plane. On the contrary, a rotation in three dimensions is characterized not only by its angle (a scalar) but also by its axis, which depends on two parameters (two projective coordinates). This analysis made Hamilton realize that a three-dimensional geometric calculus should be based on quadruplets rather than triplets. Moreover, the non-permutability of rotations in their composition in three dimensions led him to abandon the commutativity of the multiplication of quadruplets. In 1844, Hamilton published the first elements of his theory of quaternions (Hamilton 1844) and spend almost all the rest of his working life promoting their use in all branches of mathematics and physics. His fame and personality, as well as his influence on scientific life in Great-Britain at the time, made his theory very popular. In particular, in electromagnetism, a conflict took place between quaternionists and those who preferred a vectorial approach based on Grassmann's geometrical calculus.[28] The role played by the different theories in the development of vector calculus is analyzed in detail in Michael J. Crowe's *A History of Vector Analysis* (Crowe 1967). As far as the theory of vector spaces is concerned, the discovery of quaternions had another consequence which was even more important: because of the non-commutativity of the multiplication, they provided the first type of algebra different from the model of the real numbers, thereby breaking the principle of permanence.[29] Hamilton's discovery opened a new area of research on what was called the systems of hypercomplex numbers and became known later as associative algebra. This new branch played an important role in the construction of Modern Algebra.[30]

All the discoveries about geometrical calculus modified the geometric approach and laid the foundation for the use of vectorial or quaternion methods in physics. Yet, what influence did this evolution have on the development of the theory of vector-spaces? Unlike what many people think, it is not possible to answer this question so easily. Indeed the influence may be, on the whole, not that important and in any case, much less direct than what is usually believed. Moreover this question is quite important from a didactic point of view since it puts into question the whole notion of teaching a course in linear algebra based on the concepts of vector geometry. . It is clear that vector geometry gives the theory of vector spaces a field of intuition of formal concepts in a more familiar setting, in particular through graphic representation and vocabulary, allowing changes of settings. Nevertheless, the theory of vector spaces did not emerge from vector calculus as a sort of natural progression. Things were much more complicated and the first formal approach took decades before being used in contexts quite different from geometry (infinite dimensional function spaces). At the end of the first half of the 19th century, geometry changed quite radically. The traditional framework of geometry exploded, creating different types of geometry having privileged relations with different parts of mathematics or other sciences. In this topographical modification of the connections within mathematics, the approach to linear problems was also greatly modified, creating a new type of relation between algebra and geometry.

2.4. Hermann Grassmann's Ausdehnungslehre

Hermann Grassmann's *Ausdehnungslehre* (Theory of Extension) plays a very specific role in the history of linear algebra. It was based on geometrical intuition, but it also adopted from the outset a very formalistic viewpoint (we will discuss what we mean by this later). It offered a totally new, global perspective of geometry, unifying its affine, vectorial and projective aspects. Nevertheless this work, misunderstood in its time, had only indirect and limited effects on the development of geometry and the theory of vector spaces. All these reasons lead us to give a privileged place to this theory in our analysis.

2.4.1. The Context
In 1844, Hermann Günther Grassmann, a German self-taught mathematician, totally unknown to mathematicians of his time, published the first part[31] of a new theory (Die Ausdehnungslehre) entitled *Die Lineale Ausdehnungslehre*. He wanted this to be a new branch of mathematics which would not limit itself to the context of geometry but of which geometry would be only one specific application. Apart from a few exceptions, the reception to Grassmann's work ranged from indifference to violent criticism. He was mostly reproached for lack of clarity in his presentation, especially regarding the long philosophical introduction that prevented most mathematicians from reading any further. Grassmann had chosen to adopt a very general viewpoint, trying to justify any new idea on a philosophical basis, and only introducing practical applications to geometry or mechanics at the end of each chapter, after the theoretical approach. Moreover, the theory started with a long list of new definitions and concepts for which Grassmann, as an expert linguist[32], had carefully chosen German rather than Latin names, unfamiliar even to German mathematicians. As a consequence few people read his theory in detail and those who did rejected it. In 1862, Grassmann published a totally re-mastered version. He suppressed the philosophical introduction and changed the presentation, cutting out many philosophical considerations, so that he followed mathematical tradition more closely. Nevertheless, he did not manage to make his ideas become better known.

It took a long time for Grassman's genius to be recognized, and even then only partly so (see the introduction of Schubring 1995). It was not until after his death that mathematicians and physicists began to understand the depth of his theory (yet usually limiting their view to the applications to geometry or physics). Nevertheless, around 1920, Grassmann's Ausdehnungslehre inspired Elie Cartan when he laid the foundations for modern multilinear algebra and differential geometry in which what was called Grassmann manifolds or simply, Grassmannians, played a central role. In spite of this limited influence on the evolution of linear algebra, the analysis of Grassmann's contribution is essential for historians. Indeed, in the context of the second half of the 19th century, when new kinds of bridges were established between algebra and geometry, Grassmann presented a very original epistemological approach which anticipated, in many ways, results rediscovered only half a century later. Moreover, in spite of certain weak aspects, Grassmann's views remain unique, therefore, our understanding of the nature of linear algebra can still be enriched from a close analysis of his contribution.

As we mentioned above, Grassmann was a self-taught mathematician; he learned mostly from his father and the books of his own library. In 1840, he was 31 years old. In order to be able to teach mathematics (he was already a secondary school teacher for other subjects), he wrote a memoir entitled, *Theorie der Ebbe und Flut*,[33] which would only be published after his death. He laid out the principles of a geometric calculus on which he had started to work as early as 1832. In his memoir he shows that this calculus allowed him to greatly simplify some of the results put forward by Lagrange in his *Mécanique Analytique* as well as to come up with original deductions to some of the problems in Laplace's *Mécanique Céleste*. In the foreword to his *Ausdehnungslehre*, however, Grassmann acknowledges the influence of the work of his father, Justus Grassmann:

> Den ersten Anstoss gab mir die Betrachtung des Negativen in der Geometrie; ich gewöhnte mich, die Strecken AB und BA als entgegengesetzte Grössen aufzufassen; woraus denn hervorging, dass, wenn A,B,C Punkte einer geraden Linie sind, dann auch allemal AB+BC=AC sei, sowohl wenn AB und BC gleichbezeichnet sind, als auch wenn entgegenstezt bezeichnet, d.h. wenn C zwischen A und B liegt. In dem letzteren Falle, waren nun AB und BC nicht als blosse Längen aufgefasst, sondern an ihnen zugleich ihre Richtung festgehalten, vermöge deren sie eben einander entgegengesetzt waren. So drängte sich der Unterschied auf zwischen der Summe der Längen und zwischen der Summe solcher Strecken, in deen zugleich die Richtung mit festgehalten war. Hieraus ergab sich die Forderung den letzten Begriff der Summe nicht bloss für den Fall, dass die Strecken gleich - oder entgegengesetzt - gerichtet waren, sondern auch für jeden andern Fall festzustellen. Dies konnte auf's einfachste geschehen, indem das Gesetz, dass AB+BC=AC sei, auch dann noch festgehalten wurde, wenn A,B,C nicht in einer gerade Linie lagen. - Hiermit war denn der erste Schritt zu einer Analyse gethan, welche in der Folge zu dem neue Zweige der Mathematik führte, der hier vorliegt. Aber keinesweges ahnte ich, auf welch' ein fruchtbares und reiches Gebiet ich hier gelangt war; velmehr schien mir jenes Ergebniss wenig beachtungswerth, bis ich dasselbe mit einer verwandten Idee kombinirte. Indem ich nämlich den Begriff des Produktes in der Geometrie verfolgte, wie er von meinem Vater[*] aufgefasst wurde, so ergab sich mir, dass nicht nur das Rechteck, sondern auch das Parallelogram überhaupt als Produkt zweier an einander stossender Seiten desselben zu betrachten sei, wenn man nämlich wiederum nicht das Produkt der Längen, sondern der beiden Strecken mit Festhaltung ihrer Richtungen auffasste.[34] (Grassmann 1844, v-vi)

As shown in this quotation, the ideas underlying Grassmann's conception of a geometric calculus were similar to those of his contemporaries, but he distinguished himself in the treatment he applied to this rather simple idea, as he created a much more general theory than any of his predecessors. This particularity of Grassmann's work cannot be understood without taking into account his philosophical and epistemological approach, not only as explained in the foreword, but permeating all of his work. This is essential for an understanding of his mode of presentation and contributes to both the richness and the difficulty of his approach. It is impossible for us, in this presentation, to give a precise analysis of Grassmann's work.

Nevertheless we will try to draw out its main characteristics in order to make our explanation as clear as possible. For more detail one can refer to (Lewis 1977), (Flament 1992), (Châtelet 1992), (Fearnley-Sander 1979 et 1982), (Otte 1989), (Schubring 1996) and (Dorier 1996).

2.4.2. The Philosophical Background

Grassmann starts his introduction by defining what science and mathematics mean for him and what place within these fields he envisions for his new theory. First of all, he distinguishes real sciences from formal sciences, the latter being divided into the Dialectic (or logic) and the Theory of Forms (or pure mathematics). For him, geometry should not be part of mathematics but merely an application of a formal theory, which would be a new branch of mathematics. Grassmann's work was especially inspired by the philosophies of Friedrich Schleiermacher whose classes he had followed as a student at the University of Berlin. From Schleiermacher's *Dialekti* [35], Grassmann adopted the use of contrasts as an essential component of the thinking process (see Lewis 1997). A contrast makes an opposition between two poles but they are two sides of a single idea. For instance, two objects cannot be thought of as equal if they have not previously been opposed in such a way that they could appear as distinct. Therefore the notion of contrast is not absolute but, on the contrary, reflects the relativity of thinking that leads systematically to analyzing things from opposite and complementary angles; an essential dialectical component of the thinking process. In the Theory of Forms, two types of contrasts are essential for Grassmann: discrete / continuous and equal / different. Combining these two contrasts, Grassmann obtains four branches of the Theory of Forms : Arithmetic (discrete-equal), Combinatoric Analysis (discrete-different), Theory of Functions (continuous-equal) and finally the Theory of Extension (continuous-different).

In order to have a clearer view of what Grassmann meant, we need to anticipate the beginning of the theory. In fact, Grassmann introduces a concept of space as a system of n-th order, step by step. A given first element generates a system of the first order by a continuous action (in both directions) of the same fundamental evolution. In the same way, by the continuous action of a different fundamental evolution, the elements of the system of first order generate a system of second order, etc., without any limitation in the order. Thus there is a notion of dynamic space; the dimension grows step by step. Therefore the concept of 'evolution' ('Aenderung') is a key concept in the understanding of the Ausdehnungslehre. This can be enlightened by the following quotation:

> Die Stellung der Geometrie zur Formenlehre hängt von dem Verhältniss ab, in welchem die Anschauung des Raumes zum reinen Denken steht. Wenn gleich wir nun sagten, es trete jene Anschauung dem Denken als selbstständig gegebenes gegenüber, so ist damit doch nicht behauptet, dass die Anschauung des Raumes uns erst aus der Betrachtung der räumlichen Dinge würde; sondern sie ist eine Grundanschauung, die mit dem Geöffnetsein unseres Sinnes für die sinnliche Welt uns mitgegeben ist, und die uns eben so ursprünglich anhaftet, wie der Leib der Seele. Auf gleiche Weise verhält es sich mit der Zeit und mit der auf die Anschauungen der Zeit und des Raumes gegründeten Bewegung, weshalb man auch die reine Bewegungslehre (Phorometrie) mit gleichem rechte wie die Geometrie den mathematischen Wissenschaften beigezählt hat.[36] (ibid., xxi)

Therefore, the concept of evolution, as a combination of time and movement, is based on a fundamental perception which "adheres to us as closely as body to soul". The mode of generation of the objects of the theory of extension creates new objects through continuous changes. The new objects depend on the fundamental (particular) evolutions through which they have been created. Through the particular / general contrast, Grassmann spends the first few paragraphs of his theory to show how this specific mode of generation can be overcome in a general approach of the objects.

One important aspect of the Ausdehnungslehre's introduction is Grassmann's explanation of his choice of presentation. Here again we can see Schleiermacher's influence and a dialectical use of contrasts. Grassmann contrasts philosophy which progresses from the general to the particular, to mathematics which progresses from the most elementary concepts to the more complex, thereby producing new and more general concepts by the conjunction of particulars. Then he derives from this analysis a common property that makes them scientific:

> Nun legen wir einer Behandlungsweise Wissenschaftlichen bei, wenn der Leser durch sie einestheils mit Nothwendigkeit zur Anerkennung jeder einzelnen Wahrheit geführt wird, andrerseits in den stand gesetzt wird, auf jedem Punkte der Entwickelung die Richtung des weiteren Fortschreitens zu übersehen.[37] (ibid., xxix-xxx)

This dual objective, rigor and the possibility of seeing the whole picture at once, is constantly present in the presentation adopted by Grassmann in 1844, especially at the beginning of the theory. It is also present through another type of contrast, between the rigor of exposition and the explicitation of intuition and analogy with geometry. This contrast determines the concept in a non-ending dialectical process. This choice, that Grassmann himself calls unusual in mathematics, gives a specific rhythm to the work. Instead of concise definitions, several concepts emerge from different perspectives in a slow and seemingly repetitive process (it could even seem contradictory in some cases). If this has been a reason for bad reception and difficulties in understanding Grassmann's ideas, it is also precious for historians since it sheds light on the whole process of invention that was at work in the development of this theory.

Immediately after the introduction, before he presents his new theory itself, Grassmann gives a *Survey of the General Theory of Forms ('Übersicht der allgemeinen Formenlehr')*. This theory refers to all branches of mathematics and introduces the concept of *conjunction ('Verknüpfung')*, which corresponds to the modern concept of operation. Indeed a modern reader would be tempted to interpret this section as an axiomatic presentation of the modern structures of group, ring and field. But Grassmann's viewpoint is quite different from an axiomatic presentation. His purpose was not to give a list of properties which would characterize different structures but to formalize the principles that follow from the pairs of general contrasts of equality and difference on one hand and connection and separation on the other. For instance, the property of associativity (Grassmann did not use this word) formalizes the possibility of considering as equal, two forms obtained by the connection of three elements. Therefore, Grassmann's approach was different from an axiomatic approach in that it proceeded in a systematic (combinatoric) examination of the possibilities so that various forms obtained by connections of others could be considered as equal or different. In this sense, the role played by this introductory chapter regarding the rest of the work was essential since it established the architectonic rules that would govern the formal process in the creation of new objects in the theory of extensions. This formal aspect, upon which the structure of *n*-th order systems was built, was in constant contrast - following a dialectical process - with what could be called the 'real' aspect of the theory: i.e., with the initial way of generating new elements and systems (Lewis 1977).

2.4.3. The Mathematical Content

In order to illustrate these ideas, we shall now show how Grassmann drew out the equivalent of the concepts of basis and dimension in the presentation of his theory. In the terminology of the Ausdehnungslehre, the concept of vector corresponds to the notion of *displacement ('Strecke')*. In the original mode of generation of a system of *n*-th order, *n* original fundamental methods of evolutions are specified in a certain order. Grassmann says that they have to be independent in the sense that they do not belong to any system of order less than *n*.. Therefore linear independence and generation are connected, under the same concept of fundamental methods of evolution, at the very roots of the theory, and the order of a system appears as the pre-concept of 'natural dimension'. At first, only displacements belonging to each of these original types of evolution are considered.[38] By use of the 'general / particular' contrast, Grassmann has to free the system of this particular mode of generation. To reach this goal, he uses another contrast, 'real / formal', in order to introduce new types of displacements. The first types of new displacements are the sums of displacements belonging to the original types of evolution. Grassmann starts with the sum of two displacements belonging to the same type of evolution and then of two displacements belonging to two different types of evolution. He bases his

definition on a geometrical perception and the additive rule $(\overrightarrow{AB} + \overrightarrow{BC} = \overrightarrow{AC})$. Then, he examines the significance of such a type of connection in the formal setting of his theory, using the *General Theory of Forms* as an architectonic framework. At first, Grassmann shows that the addition of displacements belonging to the fundamental types of evolution is associative (he does not use this word). Then, he considers the possibility of changing the order of the two displacements in the sum, which corresponds to a change in the order of the fundamental types of evolution. This formal modification has no meaning regarding the real aspect of the theory (because the order of the fundamental types of evolution is fixed), therefore Grassmann explains that, on a purely formal basis, it should be admitted that the change of order does not affect the result of the sum, and insists on the fact that it is a purely arbitrary modification. Nevertheless, he shows that this arbitrary choice is consistent with the real aspect of the theory. The way of generating a system of *m*-th order leads to the representation of a displacement *p* by the sum of *m* displacements belonging to the *m* fundamental evolutions $a, b, c,....$: $p = a_1 + b_1 + c_1 + ...$ Then he establishes the formula: $p_1 + p_2 = (a_1 + a_2) + (b_1 + b_2) + (c_1 + c_2) + ...$, which determines the addition of any two displacements; this is made possible by the preceding results on the commutativity and associativity of the addition of displacements belonging to the fundamental methods of evolution. On the other hand, it also goes back to the original mode of generation, as it symbolizes the way the sum is generated by the initial methods of evolution. It also marks a step forward in the search for independence of the system from its original way of generation. Indeed, a displacement can now be determined as the sum of other displacements, not necessarily belonging to the initial fundamental methods of evolution. §20 presents a refinement of this result as it describes conditions for a replacement of the initial methods of evolution. The introduction of this paragraph is clear on this matter:

> Durch die im vorigen § geführte Entwickelung ist die selbständige Darstellung der Systeme höherer Stufen vorbereitet. Nämlich es waren diese bisher als abhängig von gewissen zu Grunde gelegten Aenderungsweisen dargestellt, durch welche sie eben erzeugt wurden. Diese Abhängigkeit können wir in so fern aufheben, als wir zeigen können, dass dasselbe System m-ter Stufe durch je m Aederungsweisen erzeugbar sei welche demselben angehören und welche von einendaner unhabhängig sind (in dem Sinne von § 16), d. h. von keinem System niederer Stufe (als der m-en) umfasst werden.[39] (ibid., 30)

In this statement, the order (*Stufe*) appears as a general characteristic of the system which is not only tied to a particular choice of generators but is a general indicator of the size of a system. Moreover, some of the essential aspects of the concepts of basis and dimension are drawn out. At this stage of the theory, the concept of independence is briefly introduced in §16, where it concerns only fundamental evolutions. The generalization to all methods of evolution suggested by Grassmann at the end of the preceding quotation, is somewhat ambiguous. Indeed, what must be understood by: « *m* methods of evolution that are included in no system of lower order »? Does it mean in any system, or only in systems generated by some of the evolutions whose independence is discussed ? The reference to §16 makes it clearer that only the evolutions whose independence is discussed are involved. The fact that independent evolutions belong to no system at all of lower order can be deduced from this definition; yet this logical consequence is closely related to the result discussed in §20. It is therefore essential to make the distinction clear at this stage.

To prove the result quoted above, Grassmann uses a step by step method, showing that one fundamental method of evolution can be replaced by another which is independent of the remaining; this is what is now known as the exchange method ('Austauschsatz'[40]). The proof is quite simple, Grassmann makes it in a formal setting but the relations with the real aspect of addition explained in the previous paragraphs make the intuitive meaning visible at each stage. The key point is that if p is independent of $b, c,...$ then in the decomposition of p according to the initial methods of evolution : $p = a + b + c +...$, a is not zero. Therefore, any displacement a_1 belonging to a corresponds to a displacement p_1 belonging to p :

$p_1 = a_1 + b_1 + c_1 + ...$, hence : $a_1 = p_1 - b_1 - c_1 - ....$ Therefore in the generating process, any displacement belonging to a can be substituted by a sum of displacements belonging to $p, b, c, ...$ The possibility of reorganizing the order of the evolutions and the partition in similar displacements (cf. §17 to 19) ensures the exchangeability of p and a. Thus, the exchange theorem appears as a natural conclusion to the preceding paragraphs. Moreover, the dialectics between the 'formal' and the 'real' aspects is quite tangible as the proof of this result is both rigorous in its deductive form of presentation and meaningful in regard to the intuitive origin of the theory.

The exchange theorem is thus an essential issue in the foundations of the theory of extension; it also potentially includes most of the meaning of the concepts of basis and dimension. Yet, Grassmann did not clearly draw out all the implications from this result. For instance, his conclusion of the proof of the result announced in the beginning of §20 is somewhat vague:

> Und da man dies Verfahren fortsetzen kann, so folgt, dass man dasselbe System durch je m unabhängige Aenderungsweisen desselben erzugen kann [...].[41] (ibid., 32)

But the iteration of the exchange method requires at each step that the conditions of replacement be fulfilled. This is not that straightforward; it means that if $n(<m)$ initial methods of evolution $a, b, ..., c$ have been replaced by n new ones $p, q, ... r$, one of the $(m-n)$ remaining new methods of evolution (s) is independent of $p, q, ...r$ and $(m-n-1)$ of the remaining initial methods of evolution. This can be deduced from the independence of $p, q r, ... s$, but, even if it is only a technical point, there is here a failing in Grassmann's reasoning. Moreover, it is easy to deduce from §20, that no set of less than m methods would generate a system of m-th order[42]. Yet, Grassmann does not explicitly discuss this point either in §20 or further on, although it is implicitly admitted.

In spite of these imperfections, Grassmann's use of what will be known later as the exchange method is fundamental. Indeed, it is one of the ways of proving the invariance of the number of elements in any basis of a vector space, which does not make use of coordinates and is therefore the most intrinsic. We will see that none of Grassmann's direct followers made reference to this part of his work and therefore had a less elaborate presentation of the concept of basis.

We have only presented here a very small part of the content of the *Ausdehnungslehre*. This work presented most of the results of the theory of vector spaces in its particular mode of presentation, and even more. For instance, it established a result equivalent to the following formula (that we give in modern terms):

$$\dim (E + F) = \dim E + \dim F - \dim (E \cap F).^{43}$$

The Ausdehnungslehre also presented a new approach to the barycentric calculus (independent of Möbius' discovery which only became known to Grassmann after he wrote his treatise). Grassmann's presentation gave to the barycentric calculus better theoretical foundations. Grassmann also gave many applications of his theory to geometry, systems of linear equations and various fields of physics; he developed these applications in papers that he published in various journals after 1844. Nevertheless, he failed to obtain the recognition he was due.

2.4.4. *The 1862 Version*

In 1861, he finally decided to rewrite his theory entirely. It was not only a superficially modified version that he published in 1862, but a totally different presentation. Most of the philosophical background was erased and Grassmann used notations which conformed better to the mathematics of his time. For instance, the concept of number is used right from the beginning as well as linear combinations. In order to give a better idea of the changes, we give below a long extract from the beginning of the 1862 version.

Die Ausdehnungslehre
Vollständig und in strenger Form
[...] (A foreword of 8 pages, not reproduced here, follows)

Erster Abschnitt.
Die einfachen Verknüpfungen extensiver Grössen.

Kapitel 1.
Addition, Substraktion, Vervielfachung und Theilung extensiver Grössen.

1. Erklärung. Ich sage, Eine Grösse a sei aus den Grössen b, c ... durch die *Zahlen* β, γ, ..., abgeleitet, wenn

$a = \beta b + \gamma c + \ldots$

ist, wo β, γ,... relle Zahlen sind, gleichviel ob rational oder irrational, ob gleich Null oder verschieden von Null. Auch sage ich, a sei in diesem Falle numerisch abgeleitet aus b, c, ...

2. Erklärung. Ferner sage ich, dass zwei oder mehrere Grössen a, b, c, ... in einer Zahlbeziehung zu einander stehen, oder dass der Verein der Grössen a, b, c, ... einer Zahlbeziehung unterliege, wenn irgend eine derselben sich aus den übrigen numerisch ableiten lässt, also wenn sich zum Beispiel

$a = \beta b + \gamma c + \ldots$

setzen lässt, wo β, γ,... reelle Zahlen sind. Besteht der Verein nur aus Einer Grösse a, so soll nur in dem Falle gesagt werden, der Verein unterliege einer Zahlbeziehung, wenn $a = 0$. [...]

3. Erklärung. Einheit nenne ich jede Grösse, welche dazu dienen soll, um aus ihr eine Reihe von Grössen numerisch abzuleiten, und zwar nenne ich die Einheit eine ursprüngliche, wenn sie nicht aus einer anderen Einheit abgeleitet ist. Die einheit der Zahlen, also die Eins, nenne ich die absolute Einheit, alle übrigen relative. Null soll nie als Einheit gelten.

4. Erklärung. Ein System von Einheiten nenne ich jeden Verein von Grössen, welche in keiner Zahlbeziehung zu einander stehen, und welche dazu dienen sollen, um aus ihnen durch beliebige Zahlen andere Grössen abzuleiten.
[...]

5. Erklärung. Extensive Grösse nenne ich jeden Ausdruck, welcher aus einem Systeme von Einheiten (welches sich jedoch nicht auf die absolute Einheit beschränkt) durch Zahlen abgeleitet ist, und zwar nenne ich diese zahlen die zu den Einheiten gehörigen Ableitungszahlen jener Grösse [...].

6. Erklärung. Zwei extensive Grössen, die aus demselben System von Einheiten abgeleitet sind, addiren, heisst, ihre zu denselben Einheiten gehörigen Ableitungszahlen addiren, das heisst,

$\Sigma\, \alpha e + \Sigma\, \beta e = \Sigma\, (\alpha + \beta)\, e$.

7. Erklärung. Eine extensive Grösse von einer andern, aus demselben Systeme von Einheiten abgeleiteten subtrahiren, heisst di Ableitungszahlen der ersteren von den zu denselben Eiheiten gehörigen Ableitungszahlen der letzteren subtrahiren, das heisst,

$\Sigma\, \alpha e - \Sigma\, \beta e = \Sigma\, (\alpha - \beta)\, e$.

Anm. In Bezug auf die Klammerbezeichenung halte ich die Bestimmung fest, dass ein ohne Klammern geschriebenes Polynom oder Produkt aus mehreren Faktoren gleichbedeutend ist dem mit Klammern geschriebenen Ausdruck, in welchem alle Klammern gleich zu Anfang eintreten, also

$a + b + c = (a + b) + c$, $abc = (ab)c$

und so weiter.

8. *Für extensive Grössen a, b, c, gelten die Fundamentalformeln*:
1) $a + b = b + a$,
2) $a + (b + c) = a + b + c$,
3) $a + b - b = a$,
4) $a - b + b = a$.
Beweis.

[...]

9. Für extensive Grössen gelten die sämmtlichen Gesetzte algebraischer Addition und Substraktion.

Beweis. Denn diese Gesetzte können, wie bekannt, aus den vier Fundamentalformeln in Nr. 8 abgeleitet werden.

10. Erklärung. Eine extensive Grösse mit einer Zahl multipliciren heisst ihre sämmtlichen Ableitungszahlen mit dieser Zahl multipliciren, das heisst,

$$\Sigma\ \alpha e\ .\ \beta = \beta\ .\ \Sigma\ ae = \Sigma\ (\alpha\beta)e.$$

11. Erklärung. Eine extensive Grösse durch eine Zahl, die nicht gleich Null ist, dividiren, heisst, ihre sämmtlichen Ableitungszahlen durch diese Zahl dividiren, das heisst,

$$\Sigma\ \alpha e : \beta = \Sigma\ (\frac{\alpha}{\beta})\ e\ .$$

12. Für die Multiplikation und Division extensiver Grössen (a, b) durch Zahlen (α, β) gelten die Fundamentalformeln:

1) $a\ \beta = \beta\ a,$
2) $a\ \beta\ \gamma = a\ (\beta\ \gamma),$
3) $(a + b)\ \gamma = a\ \gamma + b\ \gamma,$
4) $a\ (\beta + \gamma) = a\ \beta + a\ \gamma,$
5) $a\ .\ 1 = a,$
6) $a\beta = 0$

dann und nur dann, wenn einander a = 0, oder β = 0,

$$7) a : \beta = a\ \frac{1}{\beta},\ \text{si}\ \beta\ {}^{>}_{<} 0.\ ^{*}$$

Beweis.

[...]

13. Für die Multiplikation und Division extensiver Grössen durch Zahlen gelten die algebraischen Gesetzte der Multiplikation und Division. [44]

Beweis Denn aus den Fundamentalformeln (1 bis 6) des vorhergehenden Satzes folgen in bekannter weise die sämmtlichen algebraischen Gesetzte der Multiplication, und durch Formeln (7) desselben Satzes wird die Division, ebenso wie in der Algebra, auf die Multiplikation zurückgeführt. Also gelten auch die algebraischen Gesetzte der Division für die Division extensiver Grössen durch Zahlen.

[...] [45] (Grassmann 1862, 13-16)

In this presentation the two lists of fundamental forms (§8 and 12) are very similar to the list of axioms of a vector space in its modern definition. Nevertheless, Grassmann's presentation was not axiomatic since the properties produced here are not given but deduced from the nature of the operations. The fact that the philosophical aspect of the 1844 version and the presentation of the general theory of forms were erased in the 1862 version makes the status of such properties ambiguous. In particular the properties related to subtraction (3 and 4 in §8) cannot make much sense if one does not refer to the general theory of forms (only present in the 1844 version) in which Grassmann explains how the concept of neutral element and opposite element can be deduced from the two properties given here. From a historical point of view, an analysis of the 1862 version is therefore subject to a previous analysis of the 1844 version. In this perspective, it is clear that, even with the absence of the philosophical background, the 1862 version still does not reveal an axiomatic axiomatic approach, because the formal aspect of the theory prevails and the dialectical process with the real aspect (as presented in the 1844 version) is hidden; but yet underlies the theory. Moreover, it is important to say here (we will develop this point below), that it was from his reading of Grassmann[46], that Peano deduced in 1888 the first axiomatic definition of a vector space.

In the 1862 version Grassmann also used the exchange method to prove that any set of n independent displacements generated the system of nth order. The consequences of this are more clearly expressed in the 1862 version than in the 1844 one, and the concepts of basis and dimension are expressed in all their generality.

Even with this new version, Grassmann never gained substantial recognition and his work was to have a very limited influence on the development leading to the theory of vector spaces. It was only around 1920, due in particular to Elie Cartan's work on exterior and multilinear algebra that the most innovative part of the Ausdehnungslehre would be revealed to the world of mathematicians (Cartan 1922). Yet, the Ausdehnungslehre is still considered as a difficult and somewhat obscure contribution to mathematics.

2.5. First Phase of Unification of Finite-Dimensional Linear Problems

It is now time for us to present a summary of the process that, starting from the study of systems of linear equations on the one hand, and geometric calculus on the other, led to the modern theory of vector spaces.

2.5.1. Analytical Origins

Linearity was first a numerical concept attached to linear equations. In this context, we have explained that from around 1750, the notion of determinant, crystallized questions concerning linearity for at least one and a half century. We have also seen how the passage from questions regarding the solving of equations to more qualitative and descriptive types of questions led to the establishment of the concepts that were to form the basis for a more theoretical approach to the solution of a system of linear equations. Within this framework, linearity appears in an implicit dual relationship between equations and n-tuples of solutions. Moreover, the first concepts of linearity, in particular the notion of rank, are developed through the explicitation of this duality via the unification made possible by a common concept of linear dependence on the two types of object (equations and n-tuples). Therefore, it would be erroneous to believe that the linearity of $\mathbb{R}^n$ a primitive concept since it only came into existence, at least theoretically, through the study of systems of linear equations in which the concept of the linearity of equations came first.

2.5.2. Geometrical Context

On the other hand, but at the same time, the analytical method made possible the use of algebra in geometry. Yet, this method was not based on the identification of geometrical space with $\mathbb{R}^3$, a concept which would only appear much later as a modern theoretical viewpoint. Indeed, Descartes and Fermat did not base their work on a correspondence between points and triplets or even between algebraic structures. They imported methods from algebra into geometry and the analogies were essentially between the problems. Moreover, at the beginning of analytic geometry, the systems of axes were dependent on the objects at stake in the problem. The most powerful analogy made possible by the methods of coordinates is the one between 'lines' (i.e. algebraic curves) and equations. In particular, this analogy opened new perspectives on the problem of classification of lines. In this sense, a central question is to identify all the lines having the same equation via a change of the system of coordinates (i.e. a linear substitution). Therefore the importance of linearity in geometry is due to the importation of algebraic methods into geometry. However, until the second half of the 19th century, at least, the geometrical interpretation of algebraic results was limited to three dimensions, and, even if technically the analytical method made possible the generalization of geometry to more than three dimensions, this was not investigated or accepted by most mathematicians. Although some mathematician were led to investigate problems with quadruplets or involving even more coordinates, they rejected their findings since they could not be applied to the 'reality' of a geometric situation. For instance, Möbius, in his barycentric calculus, defines two plane figures as equal and similar when, to each point of one corresponds a point of the other so that the distance between any two points of one is equal to the distance between the corresponding points of the other. He then proves that two such figures can be brought into coincidence by a displacement in space (a half revolution in space is necessary when the two figures do not have the same orientation in the plane). In order to generalize this result for three-dimensional figures, it would be necessary to consider a half revolution in a four-dimensional space when the two figures do not have the same orientation in space. Here is what Möbius says about this possibility:

> Zur Coicidenz zweier sich gleichen und ähnlichen Systeme im Raume von drei Dimensionen: A, B, C, D, ..., und A', B', C', D', ..., bei denen aber die Puncte D, E, ..., und D', E', ..., auf ungleichnamigen Seiten der Ebenen ABC und A'B'C' liegen, würde also, der Analogie nacvh zu schliessen, erforderlich sein, dass man das eine System in einem Raume von vier Dimensionen eine halbe Umdrehung machen lassen könnte. Da aber ein solcher Raum nicht gedacht werden kann, so ist auch die Coincidenz in diesem Falle unmöglich.[47] (Möbius 1827, 183)

2.5.3. New Perspectives

In the first half of the 19th century, some important changes occurred in the approach to geometry: the discovery of non-Euclidean geometries, the new approach, independent of metric relations, of projective geometry, etc. The convergence of these factors unhinged the model of traditional Euclidean geometry as the only possible geometry and offered a new perspective on the relation between mathematics and 'reality'. Under these circumstances, resistance to an extension to more than three dimensions could no longer be sustained. In 1846, Arthur Cayley made a decisive step:

> We can in fact without having recourse to any metaphysical notion in regard to the possibility of a space of four dimensions, reason as follows (all this can also be translated into language purely analytical): In supposing four dimensions of space it is necessary to consider *lines* determined by two points, *half-planes* determined by three points, and *planes* determined by four points (two planes intersect in a half plane, etc.). Ordinary space can be considered as a plane, and it will cut a plane in an ordinary plane, a half-plane in an ordinary line and a line in an ordinary point. (Cayley 1846, 217-218)

Therefore the field of geometry (geometries?) was extended to new areas and adherence to the reality of the physical world was no longer an absolute necessity. The analytical aspect of geometry opening to n-dimensional space, reinforces the relevance of algebraic methods in geometry. On the other hand, n-dimensional algebra can be interpreted geometrically. In this context all linear questions can be interpreted both algebraically and geometrically. Thus, in the second half of the 19th century, different trends in the treatment of linear questions were beginning to be unified around objects and methods which were theoretically better designed and in which similarities were made explicit.

So far, we have only given a partial account of the origins of this process. In particular, we did not mention anything about the study of bilinear and quadratic forms, whose classification goes back to the 18th century and played an important role in the development of linear systems. Indeed, in 1798, Euler was interested in representing whole numbers as a sum of squares. In this arithmetical context, he created a classificationof quadratic forms in $\mathbb{N}$, a form being in the same group as another if it can be transformed into the other through a linear substitution of its variables. For this purpose he introduced a notation for linear substitution which is very similar to a matrix, and, moreover, he established that the product of two linear substitutions is another linear substitution and gave its representation (something analogous to the product of two matrices) (Euler 1798, 306-309). Augustin Louis Cauchy later acknowledged Euler's inspiring notation when he established the fundamental results on the product of determinants: $det(AB) = det(A)det(B)$ (Cauchy 1815). Therefore, the study of linear substitutions (traditionally attached to questions of classification) is connected to the theory of determinants and constitutes the origins of the concept of matrix. Nevertheless, it took quite a long time before the concept of matrix was clearly distinguished from the concept of determinant, with which it was somehow confusedly amalgamated. Indeed, in order to have a concept of matrix, it was necessary to have the idea of an algebra operating on objects as complicated as tables of numbers. Hamilton's discovery of Quaternions was of course a decisive step in this direction. It opened the way for many research works on new types of algebras. Ferdinand Eisenstein (1844-45) and Arthur Cayley (1858) were the two most important figures in the discovery and the development of the theory of matrices. Yet, although matrices became widely and rapidly used in England, the situation was very different on the continent, especially in Germany, in spite of Eisenstein's contribution, which remained much less well-known than Cayley's. Indeed, Frobenius published, in 1878, a memoir in which he established results very similar to Cayley's but in terms of bilinear and quadratic forms, without using any matrix notation. The two viewpoints (that of matrices and that of bilinear and quadratic forms) provided tools from both the theory of determinants and from the study of linear equations to solve similar types of problems. This situation lasted until the beginning of the 20th century.

1) The theoretical corpus centered more on similarities of methods than objects, that took shape in the second half of the 19th century is an essential step in the unification and theorization of linearity. It had two essential characteristics that we summarize below:

It inherited several different trends coming, not only from algebraic and geometrical fields, but also from physics: in particular, the movements of planets or vibrating chords which led to the study of linear differential equations.

In all these different fields, similar main types of question (only superficially different) can be found which lead to a generalization of methodss. For instance, one type concerns what is now known as spectral theory. The origins of this field are varied:

- classification of bilinear and quadratic forms (in arithmetic, algebra or geometry through the study of conics and quadrics)
- systems of linear differential equations (in particular as models for problems of physics)
- search for invariants in different contexts
etc.

2) The theory of determinant is the framework that covers the different objects and methods of the corpus. Other important notions such as linear substitutions, bilinear and quadratic forms, matrices, etc. are directly related to it. However, geometry, essentially through its language and its modes of representation, constitutes another pole of convergence. The reinforcement of the geometrical aspect of this new type of algebra can be witnessed from the general spread (at the end of the 19th century) of the use of vector calculus; especially in the fields of mechanics and electromagnetism.

It is difficult to give a detailed account of the whole of linear algebra since it embraces a wide range of questions. The expression 'linear algebra' was introduced only at the beginning of the 20th century and was in the first period always completed by the adjective 'associative'. In this sense, it refers to what was also called hypercomplex numbers. The expression 'analytic geometry' was used by Cayley and James Sylvester, who were, along with Frobenius, among the mathematicians who did the most in order to make a theory of this corpus.

At the end of the 19th century the first axiomatic approaches were developed, but they did not come into their own right until the 1920s (see next chapter). Indeed the corpus of work that had been laid down by the end of the 19th century was too stable to be upset by the introduction of any new approaches at that time. It was the result of two centuries of studies during which a rich and strong network of relationships had been established between traditional fields of mathematics and physics, with tools and methods, which had proved to be efficient and had common features, even there was a lack of unification. In the next section, we will see, among other things, that such a stability and efficiency made the emergence of a new axiomatic approach difficult, even for infinite dimensional problems.

3. TOWARDS A FORMAL AXIOMATIC THEORY

The first axiomatic presentations of linear algebra and the crucial works on finite-dimensional linear problems are more or less contemporary (end of the 1880s). Nevertheless, the two aspects remained largely independent until at least 1920 and really started being unified only from 1930. However, it is quite striking to see that these two aspects of the recent history of linear algebra have coexisted for more than 40 years and yet have had so little influence on each other. On one side, the theoretical corpus, constituted for finite dimensional problems in the second half of the 19th century, was generalized, at the turn of the century, to infinite-dimensional problems by preserving the tools and the objects treating an infinite number of variables; while on the other side, some mathematicians tried (in vain) to impose a new axiomatic approach to linear problems, without trying to solve new problems but aiming, above all, at giving better theoretical foundations to the treatment of old problems. With the generalized notions of infinite system of linear equations and infinite determinant, mathematicians found inspiration in the solving of old problems in order to solve infinite-dimensional problems using analogy, even if some calculations reached a high level of technicality and some convergent questions were tedious and difficult to justify rigorously. On the other hand, mathematicians dealing with the axiomatic approach did not succeed in convincing their colleagues, partly because they failed to show the relevance of their methods to the most interesting linear problems of their time. Before we analyze how the two viewpoints finally met and gave birth to modern linear algebra, we are going to see how the first works proposing an axiomatic approach were developed and try to understand why they had a limited influence.

3.1. First Axiomatic Presentations of Linear Algebra

3.1.1. Giuseppe Peano

Giuseppe Peano was one of the first mathematicians to have tried to draw attention to Grassmann's work. In 1888, he published *Calcolo Geometrico secundo l'Ausdhenungslehre di H. Grassmann e precedutto dalle Operozioni della Logica Deduttiva* (Peano 1888), in which he presented the basic aspects of Grassmann's theory, still limited to geometry. Despite the fact that this is a much clearer presentation than Grassmann's, the limitation to geometry hides the general nature of the original. Surprisingly, in the last chapter of his book, Peano gave an axiomatic definition of what he called a 'linear system', which is the modern concept of vector space:

> **72.** Esisteno dei sistemi di enti sui quali sono date le seguenti definizioni:
> **1.** È definita l'*eguaglianza* di due enti **a** e **b** del systema, cioè è definita una proposizione, identica con **a=b**, laquale esprime una condizione fra due enti del sistema, soddisfatta da certe coppie di enti, e non da altre, e la quale soddisfa alle equazioni logiche:
> $(a=b) = (b=a), (a=b) \cap (b=c) < (a=c)$[48]
> **2.** È definita la *somma* di due enti **a** e **b**, vale a dire è definito un ente, indicato con **a+b**, che appartiene pure al sistema dato, e che soddisfa alle condizioni:
> $(a=b) < (a+c=b+c)$, **a+b=b+a**, **a+(b+c)=(a+b)+c**.
> E il valor comune dei due membri dell'ultima eguaglianza si indicherà con **a+b+c**.
> **3.** Essendo **a** un ente del sistema, ed *m* un numero intero e positivo, colla scrittura *m***a** intenderemo la somma di *m* enti eguali ad **a**. È facile riconoscere, essendo **a**, **b**... enti del sistema, *m*, *n*, ... numeri interi e positivi, che
> $(a=b) < (ma=mb)$; $m(a+b)=ma+mb$; $(m+n)a=ma+na$; $m(na)=(mn)a$; **1a=a**.
> Noi supporremo che sia attribuito un significato alla scrittura *m***a**, qualqunque sia il numero reale *m*, in guisa che siano ancora soddisfatte le equazioni precedenti. L'ente *m***a** si dirà *prodotto* del numero (reale) *m* per l'ente **a**.
> **4.** Infine supporremo che esista un ente del sistema, che diremo *ente nullo*, eche indicheremo con 0, tale che, qualqunque sia l'ente **a**, il prodotto del numero 0 per l'ente **a** dia sempre l'ente 0, ossia
> $0a = 0$.
> Se alla srittura **a-b** si attribuisce il significato **a+(-1)b**, si deduce:
> **a-a=0** ; **a+0=a** .
> DEF. *I sistemi di enti per cui sono date le definizioni 1, 2, 3, 4, in guisa da soddisfare alle condizioni imposte, disconsi* sistemi lineari.[49] (Peano 1888, 141-142)

The organization of Peano's treatise is somewhat surprising; both the first chapter (presenting the basics of deductive logic, introducing lots of new terms) and the last one seem disconnected from the rest of the treatise. The book could have been written without this introductory chapter and, on the other hand does not really prepare the reader for such a generalization as the one given at the end. However, the first chapter gave essential tools for the axiomatic presentation made in the last chapter, but it went much beyond what was needed in this context. If one is not aware of the real content of Grassmann's work ,it is difficult to link Peano's last chapter to the *Ausdehnungslehre*. Indeed, Peano's axiomatic definition was as much the result of his own reflection and work on logic and formalism as it was the result of his reading of Grassmann. In this sense, it is clear that Peano was concerned by the theoretical question of foundations in mathematics more than the possibility of solving new problems. Nevertheless, the comparison of Peano's system of axioms with Grassmann's 1862 version of the *Ausdehnungslehre* is striking because of the great similarity between the axioms and the fundamental forms given by Grassmann. We explained above in which sense Grassmann was very far from an axiomatic approach. Therefore, in spite of this similarity, it is important to understand the important evolution drawn by Peano in his work. Indeed, referring to our analysis of Grassmann's work, one can say that Peano separated the formal aspect (the architectonic given by the general theory of forms in the *Ausdehnungslehre*) from its dependence on the real aspect (the original mode of generation in Grassmann's theory). Instead of the former aspect Peano presented the bases of deductive logic. The result is that the last chapter of the *Calcolo Geometrico* was a bare axiomatic presentation whose simplicity not only made it a more powerful model for generalization, but also deprived the reader from keys that may have been essential. Moreover, although Peano gave the first axiomatic presentation of vector spaces, his contribution remained limited to basic notions and properties, and therefore, to very few applications.

Furthermore, even if Peano's formalism was more modern than Grassmann's, his presentation of some essential concepts was mathematically less satisfactory than Grassmann's, in particular, in the case for the concepts of dimension and basis. We have analyzed, above, Grassmann's approach in both versions of his theory. Here is the definition that Peano gave of dimension:

> DEF. Numere delle dimensioni d'un sistema lineare è il massimo numero di enti fra loro independenti che si possono prendere nel sistema.[50] (ibid., 143)

This definition conforms to the modern concept. However, this viewpoint gives rise to a difficulty. Indeed, although it is clear, with this definition, that any set of n independent vectors forms a basis (or a system of generators) in a linear system with n dimensions, it is not self-evident that there could not be a basis with less than n vectors. Peano did not bring out the question explicitly; in fact, it is quite clear that he implicitly assumed this result. It does not mean that he could not solve the problem but it shows that, in his approach, this was not a question that arose naturally. This problem may seem essentially technical and of no great importance, yet, in our opinion, it reveals very different conceptions of the notions of dimension and basis to Grassmann's. Indeed, in Grassmann's theory, we have shown that dimension is a constituent of the 'real' aspect because of the original fundamental set of methods of evolution, on which lies the identity of the system of m-th order. Therefore the dimension, being the order, is the measure of the degree of extension. In Peano's approach, the axiomatic induces a 'static' reference to geometry, in which the generation problem is absent. The elements are given and the dimension is no longer a measure of extension but rather a restriction in the degrees of freedom. This conception is more consistent with the geometric model, limited to the three dimensions of space. The plane is limited to two dimensions (width and length); a straight line has only one (length). The mode of generalization from the geometric model in Grassmann's theory being dynamic, it puts forward the question of generation and extension and leads to a very different approach of the concepts of dimension and basis, in which the exchange method plays a central role, and allows a richer determination of these concepts. Grassmann's use of dialectic of contrasts was a way of elucidating ambiguous or implicit facts that an axiomatic approach left in the dark.

The same year, Peano used the language and some results of his axiomatic approach to linear systems in an article in which he had to solve a system of n linear differential equations of first order in n variables. His solution was much more modern than any of his contemporaries: he used a Euclidean norm, eigenvalues, linear substitutions, etc. It is proof of the unifying and generalizing power of his axiomatic definition of linear systems; it also gave an interesting idea of the use of the geometric model in the generalization. However, this eight page text was the only use that Peano made of his theory; although, in 1888, he was only at the beginning of a long and fruitful career as a mathematician. Peano's *Calcolo Geometrico* succeeded in some ways in making some aspects of Grassmann's work better known in Italy, and also in France, but his axiomatic approach had a very limited influence.

3.1.2. Salvatore Pincherle

Salvatore Pincherle's work is partly in line with Peano's. From about 1890, Pincherle published several papers in which he used the axiomatic approach for problems with differential and integral equations. In 1901, with one of his students, Ugo Amaldi, he published a book entitled *Le Operazioni Distributive e le loro applicazioni all'analisi*, in which the authors presented an axiomatic theory of functional operators. Although Pincherle referred to Grassmann and Peano, his approach went far beyond the framework of geometry and placed itself in quite a general context using, in particular, infinite-dimensional linear spaces. In this sense, his work was quite unusual for its time, since the use of axiomatic theory for infinite dimension was not much investigated until after 1920 - as we will see in the next section. However, we chose to talk about Pincherle in this section for two main reasons. Firstly, even if his main goal concerned infinite-dimensional problems, he presented a very complete description of the theory in finite dimensions, especially concerning the concepts of dimension and basis (unlike Peano, and some successors). Secondly, his work had very little influence on the development of what would become functional analysis. The reasons for this lack of influence are partly due to the fact that such a general and formal axiomatic approach as proposed by Pincherle did not meet the concerns of most mathematicians in the first years of the 20th century. Indeed, the conception of a function as a series of coefficients was still dominant at that time, therefore, the tools and methods of the theoretical corpus, elaborated for finite-dimensional problems, could be generalized via the notions of infinite system of linear equations and infinite determinant, etc. This type of approach was quite successful because it was based on a rich and stable set of objects, tools and problems, which had not yet reached its limit and gave correct answers to problems of the time. On the other hand, Pincherle's type of approach seemed like an unnecessarily formalist viewpoint, contrary to intuition, that would have needed a total change in the usual approach to linear questions. We will comment on this duality between efficiency and theory in the next section.

Pincherle was aware of the novelty of his approach:

> In primo luogo, osservando che ogni funzione analitica di una variabile è individuata dai valori attribuiti ad un numero generalmente infinito ma numerabile di parametri, si possono considerare quelle classi di funzioni che contenegono tutte le combinazioni lineari dei loro elementi, ad esempio la ttalita delle funzioni regolari nell'intorno di uno stesso punto, come spazî ad un numero generalmente infinito, ma numerabile di dimensioni. Le operazioni distributive applicabili alle funzioni di una simile classe si presentano allora come le omografie negli spazi lineari ad un numero finito di dimensioni ; e questo concetto, tanto più se sussidiato da una notazione semplice ed espressiva, permettera di intuire in modo sintetico, e colla guida di continuate analogie colla geometria, molteplici relazioni di composizione, di scomposizione, di classificazione in gruppi, di transformazioni di diffate operazioni.[51]
> (Pincherle 1901, iii)

Therefore, Pincherle was not only in favor of an axiomatic approach to the study of linear operators in finite-dimensional spaces, but he also tried to generalize this to functional operators. This viewpoint was quite ahead of his time. The simple fact that he considered sets of functions that contain all the linear combinations of their elements was a totally new approach. On the contrary, he remained attached to an analytical viewpoint, by considering that a function is represented by an infinite series of parameters.

In the first chapter, Pincherle presented a very complete overview of *the general finite-dimensional linear set*. The axiomatic definition that he proposed was quite similar to Peano's. He also defined the notions of dimension and basis in the same way; yet, unlike Peano, he immediately made the connection with the question of generators. Indeed, he showed that in the linear set made of all the linear combinations of n independent elements, $n+1$ elements are always dependent. To prove this, he used the fact that any system of n homogeneous linear equations with $n+1$ unknowns always has a non trivial solution. This result led to a characterization of the bases ('sistema fundamentale') as the systems of n independent elements. Thus, Pincherle was more explicit than Peano concerning the concepts of basis and dimension. His approach was also very different compared to Grassmann since it did not use the exchange method. Moreover, Pincherle gave a definition of the concept of rank which does not use matrices or determinants:

> Dati gli elementi $\alpha_1, \alpha_2, ..., \alpha_n$, se fra essi passano r relazione lineari indipendenti e non più ($r<n$), l'insieme degli elementi $\alpha = a_1 \alpha_1 + a_2 \alpha_2 + ... + a_n \alpha_n$ conterrà $n-r$ elementi e non più linearmente indipendenti, e sarà pertanto ad $n-r$ dimensioni.[52]
> (Pincherle 1901, 9)

Previously, he had noticed that all the linear relations between elements α_i constitute a linear set. Finally, he made the connection with the rank of a matrix.

Subsequently, he studied the question of change of basis and introduced the notions of hyperplane and subspace, and studied their representation by equations. He stated and proved the theorem that gives the dimension of the sum of two subspaces as the sum of the dimension of each, minus the dimension of their intersection (he used, without proof, the theorem of the possibility of completion of any set of independent elements into a basis).

The next three chapters are devoted to the study of operators (*operazioni*); in particular, of linear operators (*operazioni distributive*). Most of the results presented here were known in terms of matrices or bilinear forms (especially concerning spectral theory). Pincherle established all of them in his new axiomatic approach, therefore fulfilling what Peano had only sketched. Moreover, several results were established without using the axiom of dimension, and thus Pincherle was able to transfer them directly when studying infinite-dimensional linear spaces.

In chapter 5, he introduced the linear set of numerical sequences (S). He showed that it satisfies the axioms and that there always exists a set of independent elements bigger than any given set of independent elements, and therefore that S is infinite-dimensional. Then he remarked that, formally, S is identical to the set of all numerical series and concluded:

> Per le considerazioni di natura puramente formale, non sarà necessario distinguere la convergenza della divergenza della serie, che avrà semplicemente l'ufficio di unire in un tutto la successione dei numeri a_0, a_1, ..., a_n, ... Ma per le applicazioni alla teoria delle funzioni, sarà necessario di tenere conto della convergenza di una serie come la.[53] (Pincherle 1901, 72)

Then he denoted S^o, the set of all power series whose radius of convergence is not zero, and showed that it is a linear subspace of S. He also used the notations S^r and S^∞ to designate the linear subspaces of power series with radius of convergence, not more than r and infinite, respectively. He also introduced the linear set of Laurent series: ($\sum_{n=-\infty}^{n=+\infty} a_n x^n$). He then studied several operators in detail, in a very formal approach, (differential operator, normal operators, difference operators, etc.). In several places, he insisted on the importance of the use of geometric language, but he did not use it much in practice.

Pincherle's contribution went far beyond Peano's first attempt, offering a unified axiomatic approach of the study of linear operators in finite and infinite-dimensional linear spaces. However, even though he was the author of the article on functional equations and operators in the French version of the 'encyclopédie des mathématiques pures et appliquées' (Pincherle 1912), in which he gave a very detail historical account and referred to his own work, Pincherle's work itself did not have much influence.

3.1.3. Cesare Burali-Forti and Roberto Marcolongo

Burali-Forti and Marcolongo are also direct successors of Peano. Together, or separately, they published several papers about vectorial methods and their applications to mathematics and physics, and in which they refer to Grassmann and Peano. In a book published in 1909, they gave an axiomatic definition of linear systems which was similar to that of Peano, although in some points less complete. The essential between their work and that of Peano's was the fact that their definition was given at the beginning of the book, and was applied to the study of 'vectorial homographies'[54]. Therefore, even if, on the theoretical side, their work was very similar to Peano's, and did not make much progress, their contribution is interesting in the sense that, unlike Peano, they showed, in practice, the functionality of the axiomatic approach: they used its concepts and vocabulary in a concrete geometrical framework. However, the usefulness of such a formal and abstract theory in only one specific context where it is not indispensable can be questioned. Here is how the authors explained their choice:

> Le omografie vettoriali sono contenute nelle più generali trasformazioni lineari, ben note come calcolo di determinanti e matrici per mezzo di coordinate. Noi pero ci siamo proposto lo scopo precipuo di presentare le omografie come *enti assoluti* e non come tachigrafi delle coordinate. Messo cosi in piena luce il carattere geometrico-assoluto del quale le omografie sono suscettibili, è facile vedere come esse siano strumenti semplici, rapidi e potenti, e come, quindi, siano necessarie nelle apliczioni.
> Anche in questo libro, come negli *Elementi*[55], gli enti geometrici *punto* e *vettore*, necessari in Geometria, in Fisica e in Meccanica, compariscono sempre indipendenti dalle loro coordinate rispetto a qualsiasi sistema di riferimento.[56] (Burali-Forti and Marcolongo 1909, viii)

In a book published in French a few years later they developed the same type of argument:

> En réalité les calculs vectoriels ordinaires sont des *tachygraphies des coordonnées cartésiennes*, (...) Mais il est possible d'envisager le calcul vectoriel d'une toute autre manière. Notre calcul diffère substantiellement des calculs ordinaires en ce qu'il *peut opérer directement sur les éléments géométriques et physiques, sans avoir jamais besoin de recourir à aucune coordonnée*. Notre calcul peut ainsi s'appeler *intrinsèque*, ou *absolu*, ou *autonome*.[57] (Burali-Forti and Marcolongo 1912, vii)

Their motivations were therefore quite similar to Leibniz's, since the axiomatic approach was, for them, a way to build an intrinsic algebra, independent of the coordinates. In this sense, it created the theoretical bases for an algebra on geometrical entities. However, the power of generalization and unification was limited to the framework of geometry and physics, unlike what Pincherle had tried to do. Burali-Forti and Marcolongo succeeded in making Grassmann's work (limited to the applications to geometry and physics) better known. They also made the use of vectorial methods more efficient and popular. For instance, they defined the scalar and vector products geometrically and used a vectorial interpretation of the determinant via the formula: $\det(u,v,w) = (u \wedge v).w$; they also used the two products in order to give geometrical interpretations of the different types of linear operator in dimension not more than 3. Spectral theory for geometrical linear transformation was also introduced with the use of the two products

The beginning of the axiomatic theory of vector spaces remained therefore quite limited. Moreover, all the works in this perspective stemmed, more or less directly, from Grassmann. Nevertheless, they were different for essential reasons. Peano's and Burali-Forti's and Marcolongo's works, although they offered a wider viewpoint than Grassmann's theory (due to the axiomatic aspect) remained more limited in their potentiality for generalization because of a dependence on the geometrical model. Using an axiomatic approach seemed to be a choice that the limited range of applications could not justify; this is one of the reasons why these works did not lead to a spread of the use of the axiomatic approach. The situation was quite different concerning Pincherle, but his work did not meet with much more success;

that is to say, that the resistance to the axiomatic approach was very strong at the time.

3.1.4. Hermann Weyl

Another pioneer of the axiomatic approach in linear algebra was Hermann Weyl. In the first edition of his book *Raum - Zeit - Matterie,* in 1918, he gave an axiomatic definition of an affine space based on a vector space ('Lineare Vektor-Mannigfaltigkeit')[58], not limited to three-dimensional space.[59] Weyl's main goal was a presentation of the theory of relativity, using four dimensional-space. He explained in detail his choice for an axiomatic presentation:

> Um die Raumgesetzte in ihrer vollen mathematischen Hammronie zu erfassen, müssen wir von der besonderen Dimensionzahl $n=3$ abstrahieren. Es hat sich nicht nur in der Geometrie, sondern in noch erstaunlicherem Masse in der Physik immer wieder gezeigt, dass, sobald wir die Naturgesetze, von denen die Wirklichkeit beherrscht ist, erst einmal völlig durchdringen, diese sich in mathematischen Beziehungen von der durchsichtigsten Einfachheit und vollendesten Harmonie darstellen. Den Sinn für diese einfachheit und Harmonie, den wir heute in der theoretischen Physik nicht missen können, zu entwickeln, scheint mir eine Hauptaufgabe des mathematischen Unterrichts zu sein; sie ist für uns eine Quelle Hoher Erkenntnisbefriedigung.[60] (Weyl 1918, 20-21)

Following his purpose, he gave two examples of vector spaces which have nothing to do with geometry (a model of the mixture of four gases and an adding-machine) to show the general nature of the concepts he introduced. Then, he showed how the axioms of the affine structure could apply perfectly to $\mathbb{R}^n$ and linear equations, the main result being that:

> Sind $L_1(\xi)$, $L_2(\xi)$, ..., $L_h(\xi)$ h linear unabhängige Linearformen, so bilden die Lösungen ξ der Gleichungen
> $$L_1(\xi) = 0, L_2(\xi) = 0, ..., L_h(\xi) = 0$$
> eine$(n-h)$-dimensionale linear Vektor-Manningfaltigkeit.[61] (ibid., 22)

He also treated the case of non-homogeneous equations, with the concept of affine space. Finally, he explained how the axiomatic theory could have been deduced 'naturally' from the study of linear equations instead of geometry:

> So würde man auch ohne Geometrie von der Theorie der linearen Gleichungen her auf die natürlichste Weise nicht nur zu unsern Axiomen geführt werden, sondern auch zu den weiteren Begriffsbildungen, die wir an sie angeschlossen haben. Ja es wäre sogar in mancher Hinsicht zweckmässig (wie namentlich die Formulierung des Satzes über homogene Gleichungen zeigt), die Theorie der linearen Gleichungen auf axiomatischer Basis in der Weise zu entwickeln, dass man die hier von der Geometrie her gewonnen Axiome an die Spitze stellt. Sie würde dann gültig sein für irgend ein Operationsgebiet, das jenen Axiomen genügt, und nicht bloss fürr die 'Wertsysteme von n Variablen'. [...]
> Ist es nun auf der einen Seite ausserordentlich befriedigend, für die vielerlei Aussagen über den Raum, räumliche Gebilde und räumliche Beziehungen, aus denen die Geometrie besteht, diesen einen Gemeinsamen Erkenntnisgrund angeben zu können, so muss auf andern Seite betont werden, dass dadurch aufs deutlichste hervortritt, wie wenig die Mathematik Anspruch darauf machen kann, das anschauliche Wesen des Raumes zu erfassen: von dem, was den Raum der Anschauung zu dem macht, was er *ist* in seiner ganzen Besonderheit und was er nicht teilt mit 'Zuständen von Rechenmaschinen' und 'Gasgemischen' und 'Lösungsystemen linearer Gleichungen', enthält die Geometrie nichts. Dies 'Begreiflich' zu machen oder ev. zu zeigen, warum

und in welchem Sinne es unbegreiflich ist, bleibt der Metaphysik überlassen.[62] (ibid., 22-23)

This quotation is essential in order to understand in which sense the axiomatic theory of vector spaces is more than just a generalization of geometry. For Weyl, this was a model of space, a drastic formalization of which he felt the limits as much as the power. In this sense, he showed very clearly here how his work differed from that of Grassmann. Indeed, the latter did not reject metaphysics; he integrated it as part of mathematical creation - as opposed to an axiomatic approach which put at a distance the reference to the intuitive geometrical space and rejected metaphysics. Weyl also showed how important the study of linear equations had been in the genesis of a linear model. He was the first to theorize so clearly the linear structure of $\mathbb{R}^n$, which was, for him, the foundation of the theory of linear equations.

3.1.5. Modern Algebra

Apart from Pincherle, the three other examples of axiomatic approach we have studied so far were strongly related to geometry. However, there has been another trend in the progress of axiomatic linear algebra; in the context of what will be called Modern Algebra, the concepts of basis and dimension were mainly studied within field theory. In this context, the question was to characterize an extension-field as the set of all linear combinations of the exponents of a primitive element, having the order of the extension with respect to the original field. The question of generation was therefore essential. We cannot give a detailed historical report of the treatment of this question here, but we will examine two major works: one by Richard Dedekind and the other by Ernst Steinitz, which were among the first to have clearly elucidated this question and played a fundamental role in the development of field theory as well as linear algebra.

Richard Dedekind

In the fourth edition of Gustav Peter Lejeune-Dirichlet's *Vorlesungen über Zahlentheorie* (Dirichlet 1893), Richard Dedekind added a new 11th supplement entitled *Allgemeine Zahlentheorie*, in which one paragraph presented a very general approach to finite dimensional linear algebra (ibid., §164). A field A being given, a system T of n numbers $\omega_1, \omega_2, ..., \omega_n$, is said to be reducible with respect to A, if there are n numbers $a_1, a_2, ..., a_n$, in A, which are not all zero, such that: $a_1\omega_1 + a_2\omega_2 + ... + a_n\omega_n = 0$. This is, of course, the definition of what is now known as linear dependence. Dedekind showed that any subsystem of an irreducible system is irreducible and that if a system contains a reducible subsystem, it is reducible. Then he defined what an algebraic number with respect to A was, and said that it meant that there existed a number n such that the $n+1$ first powers of the algebraic number constituted a reducible system with respect to A. Then he defined the grade ('Grad') of an algebraic number (the minimal value of n), and remarked that any reducible system with at least one non zero number contained an irreducible subsystem, of which the remaining numbers could be expressed uniquely as linear combinations. This led him to the definition of a space Ω ('Schaar'), as the set of all linear combinations of an irreducible set of n numbers. He called these n elements a basis of Ω, and defined the coordinates of an element of Ω as the coefficients of the unique linear combination of the elements of the basis representing the given element. Immediately after that, he presented three properties that he proved to be characteristic of Ω; i.e., of the set of all linear combinations of n independent elements with respect to A.

> I. Die Zahlen in Ω reproduciren sich durch Addition und Subtraction, d.h. die Summen und Differenzen von je zwei solchen Zahlen sind ebenfalls Zahlen in Ω.
> II. Jedes product aus einer Zahl in Ω und einer Zahl in A ist eine zahl in Ω.
> III. Es giebt n von einander unabhängige Zahlen in Ω, aber je n +1 solche zahlen sind von einander abhängig.[63] (Dirichlet 1893, 468)

For the deduction of the properties from the definition, as Dedekind pointed out, only the second part of III needed a proof, and he used a recurrence. Assuming the property for any space has a basis of less than n elements, he took $n+1$ elements, α, α_1, ..., α_n, in a space Ω, with a basis of n elements. The result being obvious if all elements are zero, it could be assumed that α is not zero and that its first coordinate is not zero. Therefore there exist n coefficients c_i such that each $\alpha_i + c_i\alpha$ has zero as first coordinate. The n new numbers are now in a subspace of dimension less than n and Dedekind concluded with the use of recurrence. This result was equivalent to what Grassmann obtained with the exchange theorem. Nevertheless, the approach was quite different. Moreover, this proof does not use the theory of linear equations although it uses representation with coordinates.

To prove that the three properties are characteristic of a space as defined by Dedekind is quite simple. He also immediately deduced that any irreducible system of n numbers is a basis of Ω. Raising the problem of change of basis, he also proved that a system of n numbers is irreducible if and only if the determinant of their coordinates on the original basis is not zero.

The impossibility of finding a system of less than n generators was not explicitly stated, but as was said about Grassmann in 1844, this question could be easily solved, as the definition put the emphasis on the original set of n independent generators.

At the end of the paragraph, Dedekind turned to the case of an extension-field; i.e., when the space is also closed under multiplication. In this context, the main results proved by Dedekind were:

- In an extension-field with a basis of n elements, any number is algebraic and its grade is at most n.
- Such an extension-field B is said to be of n-th grade with respect to A, and therefore any system of n independent elements constitutes a basis of B.
- If θ is algebraic with respect to A and of grade n, then $A(\theta)$ is an extension-field of A of n-th grade and $1, \theta, ..., \theta^{n-1}$, constitute a basis of $A(\theta)$.

In this last supplement, Dedekind also studied the properties of modules, and especially of their bases, but this goes beyond our purpose. In §164, the main results of which have been presented briefly above, Dedekind's approach was very close to a modern presentation of the elementary results of finite dimensional linear algebra. The question of dimension was examined with care, and using original tools as compared with Grassmann's use of the exchange theorem and what had been done in the theory of equations.

Ernst Steinitz

In 1910, Steinitz published his *Algebraische Theorie der Körper*, which marked an important stage in the history of Modern Algebra and was to be a reference work for at least a quarter of a century. In this major work, Steinitz gave a precise definition of linear dependence over a field R, and defined a finite extension of order n:

> Ist R Teilkörper von L, so sagen wir: L ist *in bezug auf R endlich und vom Grade n* - in Zeichen
>
> $$[L:R] = n$$
>
> - , wenn man in L, n in bezug auf R linear unabhängige Elemente angeben kann, während jedes System von mehr als n Elementen aus L in bezug auf R linear abhängig ist.[64] (Steinitz 1910, 199)

This definition was identical to the definition of the number of dimensions of a linear space given by Peano or Burali-Forti and Marcolongo. As for Weyl, any influence was very unlikely to have happened since the Italians' work was practically unknown in Germany. There was no explicit reference to Dedekind either, but there was a natural proximity between the two mathematicians which has to be taken into account.

After this definition, Steinitz showed that in an extension of order n, for any set of n elements, the linear independence is equivalent to the fact that any element of the extension cannot be expressed in more than one way as a linear combination of these n elements. This was quite obviously a consequence of the preceding definition and revealed nothing new with respect to his Italian predecessors. In fact, Steinitz pursued a goal that he had not explicitly stated. For him, a basis of L was a set of elements such that any element of L can be expressed uniquely as a linear combination of them. Steinitz's goal was to prove that any basis has n elements and that any set of n independent elements is a basis of L. To achieve this goal, he needed to prove that the order of an extension could not exceed the number of generators. In other words, he had to connect a result concerning generation to a property of dependence. His definition of a basis introduced the coordinate system, so his proof was set up within the context of n-tuples and linear equations:

> Weiss man hingegen, dass jedes Element β von L sich *wenigstens* auf eine Weise in der Form $\beta = c_1\alpha_1 + \ldots + c_n\alpha_n$ darstellen lässt, so erhält man für iregendwelche $n+1$ Elemente $\beta_0, \beta_1, \ldots, \beta_n$, Darstellungen von der Form
>
> $$\beta_i = c_{i1}\alpha_1 + \ldots + c_{in}\alpha_n \quad (i = 0, \ldots, n)$$
>
> und kann daher nach dem in §1 angeführten Satz über linearer homogene Gleichungen[65] $n+1$ Elemente $d_0, d_1, \ldots, d_n$, die nicht alle 0 sind, in R so bestimmen, dass $d_0\beta_0 + d_1\beta_1 + \ldots + d_n\beta_n = 0$. The extension L wird. Dier erweiterung L ist also endlich und *höchstens* vom Grade n .[66] (ibid., 200)

The 'Exchange Theorem'

After Grassmann's 1862 *Ausdehnungslehre*, this was the first explicit proof of the fact that a set of generators cannot have fewer elements than the number of dimensions, despite the fact that this was implicit in Dedekind's work, and very easily deducible. However, Steinitz and Grassmann's approach and tools were very different. Grassmann knew that a proof could be given with the use of elimination theory, but he preferred the exchange method for explicit reasons given in 1862.

In fact, further on in his theory, Steinitz used something very close to the exchange method, but in the context of transcendental extensions in which algebraic dependence replaces linear dependence: « Let S be a system of elements of an extension-field L, an element a of L is said to be algebraically dependent on S (with respect to R), if a is algebraic with respect to the field $R(S)$. » (ibid., 288). Two systems are said to be equivalent if any element of each system is algebraically dependent on the other system. On the other hand, a system is said to be irreducible if it is not equivalent to any of its subsystems. Transferred into the context of linearity, two equivalent irreducible systems correspond to two bases of the same linear space. After a few preparatory results, Steinitz gave three final theorems which contain, for algebraic dependence, results equivalent to the exchange theorem: the theorem about the completion of an independent system into a basis; the invariance of the number of elements in all the different bases of the same linear space; and properties of the dimension of a subspace of a finite dimensional linear space (ibid., 290-291). This paragraph was presented in such a progressive, deductive way that it was close to an axiomatic definition of a general concept of dependence (which would include algebraic as well as linear dependence) from which the concepts of dimension and basis could be deduced. Therefore, little would have needed to be done in order to translate the preceding results into the linear context, but Steinitz did not mention (at least here) this possibility. About twenty years later, two of his successors, Bartel L. van der Waerden in (1930/31, 1:96) and Emanuel Sperner, setting up Otto Schreier's work in (1931/35)[67], were among the first to use the exchange method (and used this term) in the context of linear dependence. In fact, both of them referred to Steinitz, although we have never found proof that Steinitz did actually use the exchange method in the context of linear dependence. On the other hand, there is no reference to Grassmann, although his work was well-known by German algebraists of the time. Nevertheless, even if an influential relation could be conjectured, in spite of the gap of nearly a century, the two approaches are embedded in very different mathematical and philosophical contexts.

It would be too long here to analyze in detail all the works between 1890 and 1930 in which questions connected to linear algebra were solved in the framework of extension-fields, hypercomplex systems or linear associative algebra, as well as rings, and ideals (framework in which Emmy Noether was the first to give the definition of a module). More detail can be found in Study (1898), Shaw (1907), Cartan (1908), Scorza (1921) or Dickson (1923).

Bartel L. van der Waerden

Van der Waerden's *Moderne Algebra* marked the beginning of a new era in Algebra, and in mathematics in general. His book which was based on lectures given by Emil Artin and Emmy Noether is also the heritage of the works of many German algebraists during the four or five preceding decades (for more details see (van der Waerden 1975)). The great novelty was to give a unified general approach to the new algebra, presented in a textbook accessible to beginners. In this sense, it was used by several students, some of whom became leading mathematicians in the following years: such as most of the members of Bourbaki's group, for instance.

In the first edition, published in two volumes in 1930 - 1931, the results on bases and dimension were presented - with use of the exchange theorem - within the theory of extension-fields (chapter 5). In chapter 15, about linear algebra, the concepts of linear dependence and basis were introduced for modules. In this context, the concept of dimension is more complex. Van der Waerden examined the conditions under which a module had a basis, and established several results about modules with the use of matrix theory and linear equations theory. When he came to vector space, the ring became a field, and he proved the invariance of the number of elements in a basis, specifying previous results in this new context. But he also mentioned that another proof of the invariance was given about extension-fields. After that, he presented a general study of systems of linear equations as an application to module theory. He established the connection between the concept of rank and the problem of dimension with the use of the exchange theorem (which he named after Steinitz). In the following editions, the importance of linear algebra increased and the treatment of vector space became more independent of the general theory of modules.

With this treatise, the concepts of dimension, rank and basis reached their maturity. The study of linear problems became unified within one theoretical approach: vector space theory, whose elementary foundations were presented in a form which is still today the standard in most research and textbooks. Although, from a technical point of view, Grassmann's approach to the concepts of dimension and basis had been rediscovered, and even perfected, his philosophical contribution had been left aside. For this reason, his work is still important, even concerning such elementary concepts as basis and dimension.

In the various works we have briefly analyzed above, the authors tried to give better theoretical foundations to the set of finite-dimensional linear problems with the use of an axiomatic approach, thus offering an alternative to the methods with coordinates. As early as 1888, Peano had mentioned the possibility of infinite-dimensional linear systems, although he only gave the example of polynomials without any applications. Pincherle used several infinite-dimensional linear sets, but he was an exception. Indeed, even if some new problems were modeled via the new axiomatic approach, the main concern was a reorganization of what was already known in other frameworks. In this sense, this trend was part of a more general movement which would lead to a complete reorganization of algebra and mathematics, with the growing importance of the axiomatic presentation of algebraic structures. On another hand, the axiomatic presentation of linear algebra was also a way to give a better theoretical basis to vectorial calculus and, therefore, to introduce a new perspective in geometry and in some parts of physics, in opposition to methods with coordinates. This was a new answer to Leibniz's concerns.

However, in spite of these various works, in different areas, and converging toward a new organization of mathematics, the axiomatic viewpoint did not succeed for quite some time. In particular, in questions about linear differential and integral equations, a generalization to infinite dimensions of methods from the theory of determinants remained widely used until around 1930. This is what we shall analyze in the next section through the constitution of functional analysis.

3.2. Infinite-dimensional linear problems

For a modern mathematician, infinite-dimensional vector spaces cannot be understood without the use of an axiomatic definition, and it is clear that techniques and tools from linear numerical equations cannot be used effectively in linear questions related to functional analysis, to which the axiomatic approach is particularly adapted. Therefore, it would seem natural that in the history of linear algebra there would be a strong relation between the development of functional analysis and the first axiomatic definitions of vector space. But, to a great extent, this is a false idea. This lack of connection is even more surprising due to the fact that the two questions emerged around the same time and co-existed with very little contact for nearly half a century. Indeed, even at the beginning of the 20th century, in spite of some innovative works like those of Baire and Lebesgue, mathematicians were used to representing functions by a series (Fourier's or other types of power series). Therefore, a function was seen as an infinite sequence of coefficients, a natural generalization as an infinite n-tuple. In this context, the first works related to linear questions about functions were a continuation in a generalized approach of the corpus constituted for finite dimensions around linear equations and determinants. This generalization, based on analogies, became more and more sophisticated and integrated results on the convergence of certain types of series. On the other hand, topological considerations on distances and norm led to the integration of a geometrical viewpoint. This allowed the debate about the necessity of synthetic methods versus analytic methods in geometry to be transferred into an infinite-dimensional context, which would eventually lead to the success of the axiomatic approach. However, the first works did not adopt a formal viewpoint; their authors wanted to solve particular differential problems with effective tools. Even if a modern reader can see the first results on what would be called Hilbert's spaces (at first in terms of a series for which the series of squares converge, ℓ^2, then in terms of functions whose square is Lebesgue-integrable, L^2), and then even more sophisticated spaces like the L^p spaces and their dual L^q (such that $p>1$ and $\frac{1}{p} + \frac{1}{q} = 1$), to reach the most general function vector spaces. The analogy with the solving of linear systems and other finite-dimensional problems was the main source of inspiration. Nevertheless, the growing quantity of methods and the increasing difficulty in their rigorous justification, as well as the comparison with geometry, led to the necessity of a unification and therefore of the creation of more formal approaches. One crucial point was the necessity to obtain the most efficient and general type of topological structure. In this sense, it soon became indispensable to overcome the bare generalization of the Euclidean norm with an infinite number of coordinates. Indeed the notion of normed vector space played a central role in the last phase of the constitution of functional analysis. We will now develop these ideas.[68]

3.2.1. Origins

Several 'traditional' problems from physics (vibrating cords, heat equation, movement of planets, etc.) led to linear differential equations. Since the end of the 18th century, these problems had been solved by tools and methods from the theory of determinants, and later, of matrices or bilinear and quadratic forms. General results were at least implicitly known at that time: like the fact that a linear combination of solutions of a homogeneous linear differential equation or system of equations is also a solution of the system, or the fact that the general solution of a differential equation is the sum of a particular solution and the general solution of the homogeneous equation. More sophisticated and specific tools, like the Wronskian, invented· by Josef Wronski and Cauchy, came as a complement to the theoretical corpus attached to the notion of determinant. It was through this analytical approach for differential systems of equations that, at the end of the 19th century, the theoretical corpus described in chapter 4 was to be generalized to infinite dimensions.

As early as 1822, Joseph Fourier, using the representation of functions by a series, was led to the solving of linear systems with an infinite number of unknowns and equations (Fourier 1922, 168 or 212). Considering the state of knowledge about series at that time, he was not able to give a rigorous solution to the problem. However, he drew out the basic principle that influenced many followers. This is known in French as the 'principe des réduites' and it consists in considering the truncated systems, including only the first n equations in the first n variables. Once this system has been solved, the problem is to find the limit of the solution when n tends to infinity. The theoretical difficulty is, on one hand, the existence of the limit and, on the other hand, the adequacy of the limit of the solutions with the solution of the infinite system. Fourier's first attempt remained unknown for nearly half a century. Among those that followed were: Fürstenau (1860) and Kötteritzsch (1870), but George William Hill (1877), Henri Poincaré (1886 et 1900) and Helge von Koch (1891et 1892-3) gave the most significant contributions to this question. They created a generalized theory of determinant dealing with denumerable infinite dimension. However, their problems were not only algebraic; they had to find the right kind of conditions for the solutions of the truncated systems to converge toward a solution of the infinite system. In 1913, Frédéric Riesz made critical comments with respect to these pioneers' works:

> Pour appliquer la méthode classique des déterminants aux systèmes infinis, il fallait imposer des conditions plus ou moins restrictives, et il faut bien avouer que c'est la méthode et non le problème qui exigeait ces restrictions.[69] (Riesz 1913, 42)

However, some attempts were made to question the validity of the methods and the extreme technicality of the determinants. For instance in 1909, Otto Töplitz wrote:

> Die Sätze über die Auflösung von n linearen Gleichungen mit n Unbekanntent pflegt man mit Hilfe der Theorie der Determinanten abzuleiten. Man kann jedoch unter diesen Sätzen solche, deren *Inhalt* der determinantenbegriff garnicht enthält (wenn auch der überliche Beweis nicht davon frei ist), von den übrigen scheiden, deren *Wortlaut* man nicht formulieren könnte, ohne von Determinanten zu reden. [...]
> Im folgenden sollen die Sätze der ersten Classe auf eine besondere Art auf *unendliche* lineare Gleichungssysteme übertragen werden.[70] (Töplitz 1909, 88-89)

Then Töplitz proved in a formal way the equivalence of any finite linear system with a triangular system, in connection with the concept of rank., and without using any determinant. He introduced the notion of '*zeilenfinite*' systems; that is, an infinite system in which any equation uses only a finite number of unknowns. He generalized his preceding theorem to this type of system. Finally, he used this theorem to examine the different cases for the existence and the size of the solution set of any '*zeilenfinite*' system, without using determinants. Töplitz's approach remained isolated. However, it shows the theoretical difficulty in the generalization to infinite linear systems.

3.2.2. *Fredholm Integral Equation*

Infinite linear problems were not only encountered via differential equations, but also via integral equations (obtained in some cases by transformation of partial differential equations). In this framework, between 1904 and 1910, David Hilbert published six papers which played a central role in what became functional analysis. But before we come to these texts, we must examine the contribution by Ivar Fredholm. In a paper dated from 1903, he drew out the bases of new essential ideas and methods which were to inspire many followers, including Hilbert.

Fredholm's question came from a traditional problem of mechanics which had led Niels Abel, and then Joseph Liouville, Carl Neumann and Vito Volterra, to study the following types of integral equations:

(a) $\displaystyle\int_0^1 f(x,y)\,\varphi(y)\,dy = \Psi(x)$;

where f and Ψ are given, and φ is the unknown; or:

(b) $\displaystyle\varphi(x) + \int_0^1 f(x,y)\,\varphi(y)\,dy = \Psi(x)$

(c) $\displaystyle\lambda\,\varphi(x) + \int_0^1 f(x,y)\,\varphi(y)\,dy = \Psi(x)$

Fredholm focused his research on equation (b), when f is a 'finite' (i.e., bounded) integrable function. We will now try to explain his approach in order to show how he used the analogy with the solving of finite systems of linear equations in a very sophisticated manner. He started by stating:

> Il existe une quantité D_f qui joue par rapport à l'équation fonctionnelle (b) le même rôle que joue le déterminant par rapport à un système d'équations linéaires. Pour définir D_f j'introduis la notation abrégée

$$(1)\quad f\begin{pmatrix} x_1, & x_2, & \ldots, & x_n \\ y_1, & y_2, & \ldots, & y_n \end{pmatrix} = \begin{vmatrix} f(x_1,y_1) & f(x_1,y_2) & \ldots & f(x_1,y_n) \\ f(x_2,y_1) & f(x_2,y_2) & \ldots & f(x_2,y_n) \\ \hdotsfor{4} \\ f(x_n,y_1) & f(x_n,y_2) & \ldots & f(x_n,y_n) \end{vmatrix}$$

et je pose

$$(2)\quad D_f = 1 + \int_0^1 f(x,x)\,dx + \frac{1}{2!}\int_0^1\int_0^1 f\begin{pmatrix} x_1, x_2 \\ x_1, x_2 \end{pmatrix} dx_1\,dx_2 + \ldots$$

$$= \sum_{n=0}^{\infty} \frac{1}{n!}\int_0^1\cdots\int_0^1 f\begin{pmatrix} x_1, x_2, \ldots, x_n \\ x_1, x_2, \ldots, x_n \end{pmatrix} dx_1\,dx_2\ldots dx_n.$$

[71] (Fredholm 1903, 367)

At first, he proved the convergence of D_f, using a result about infinite determinants proven by Hadamard. Then he generalized the notion of minor for his type of determinant. With these tools, he established several formulae, which he generalized to the case when f is replaced by λf (this gave a new equation close to type (c) above). In this sense, he showed that the minors in this generalized case were related to the values of the derivatives of order n of $D_{\lambda f}$ with regard to λ. Finally he was able to prove that if D_f was zero, there was necessarily a first minor which was not zero (in modern terms this means that the kernel of any compact operator is finite-dimensional).

Fredholm did not say much about his method. The basic idea, already used by some predecessors, was to study a system with n equations and n variables, like in Fourier's *'principe des réduites'*. Here, one considers n values x_i each in one of the n sub-intervals of length $\frac{1}{n}$ of $[0,1]$, then one substitutes to x in equation (c) successively each value of x_i. On the other hand, the integral in y is replaced by an approximation in terms of a Riemann sum. It is then easy to obtain the n following linear equations with n unknowns:

$$\varphi(x_i) + \frac{\lambda}{n} \sum_{j=1}^{n} f(x_i, x_j)\varphi(x_i) = \Psi(x_i) ; \qquad i=1 \text{ to } n.$$

Therefore, when Ψ is the zero function, $(\frac{-\lambda}{n})$ is the inverse of an eigenvalue of the matrix $(f(x_i, x_j))_{i,j=1 \text{ to } n}$, whose determinant is formally identical to the one introduced (without any explanation) by Fredholm and whose characteristic polynomial is:

$$P_n(\lambda/n) = \sum_{k=0}^{n} (\lambda/n)^k \left(\sum_{s \in Sn, inc.} f \begin{pmatrix} x_{s(1)}, x_{s(2)}, \dots, x_{s(k)} \\ x_{s(1)}, x_{s(2)}, \dots, x_{s(k)} \end{pmatrix} \right)$$

Of course, the next essential difficulty is to make n tend to infinity. It is possible to show that the characteristic polynomial tends toward a whole function in λ which is exactly $D_{\lambda f}$. Fredholm did not explain this passage in detail, but the origins of his method leave no doubt. We will see that Hilbert made explicit use of this method, taking care to justify the passage from the finite to the infinite dimension. Unlike Fourier's method, Fredholm's was not the limit case of a finite problem; it introduced new objects right from the beginning and the analogy reflected the method which used generalized objects. In this sense, the fact that Fredholm gave very little explanation on the origins of his objects made the analogy even more powerful.

Then Fredholm introduced a totally new viewpoint:

En considérant l'équation (b) comme transformant la fonction $\varphi(x)$ en une nouvelle fonction $\psi(x)$ j'écris cette même équation
(7) $S_f\varphi(x) = \psi(x)$,
et je dis que la transformation S_f appartient à la fonction f(x,y).
Les transformations (7) forment un groupe.[72] (Fredholm 1903, 372)

The idea of considering a functional equation to be the transformation of one function into another was already present in some work by Volterra, around 1886. However, Fredholm's viewpoint was more revolutionary, even if he did not make any reference to any kind of functional space. Indeed, the notion of operator on functions was really operational in Fredholm's work; that is what makes his work so important. He noticed the fact that the product of two operators S_f and S_g is another operator S_F and he characterized the function F attached to the product:

$$F(x,y) = g(x,y) + f(x,y) + \int_0^1 g(x,t)f(t,y) \, dt.$$

He clearly stated the solving of equations of type :

$$\varphi(x) + \int_0^1 f(x,y) \, \varphi(y) \, dy = \Psi(x),$$

in terms of the inversion of the operator S_f, which meant finding g such that: $S_f S_g = S_0 = \text{id}$. Within this framework, he was able to establish what is still known as Fredholm's alternative:

> La condition nécessaire et suffisante pour qu'il existe une solution différente de zéro de l'équation :
> $$S_f \varphi(x) = 0$$
> c'est que $D_f = 0$. Si n est l'ordre du premier mineur de D_f qui soit différent de zéro, l'équation donnée possède n solutions linéairement indépendantes.[73] (Fredholm 1903, 375)

Then he showed that equation (b) had a single solution only if $D_f \neq 0$. This solution is $S_g \Psi(x)$, where g is such that $S_f S_g = S_0$. When $D_f = 0$, he showed that a necessary and sufficient condition for the equation to have a solution was:

$$\int_0^1 \Psi_k(x)\Psi(x) = 0 \; ; \; k = 1 \text{ to } n \; ;$$

where Ψ_k are n independent solutions of the homogeneous equation. It is important to note that these relations were not interpreted in geometrical terms of orthogonality and norm. We will see that Schmidt introduced this viewpoint in 1908.

Fredholm's results made a complete analogy with the finite-dimensional case. Finally he made a few more generalizations, but the essential part of his paper is what we have summarized above. Fredholm's contribution was not only important for the results he established, but above all because of the novelty of his approach which introduced a new concept such as an operator on functions, and, made a very powerful use of the analogy with the theory of determinants in finite dimension.

3.2.3. David Hilbert

Hilbert's contribution is much more dense, yet it is essentially analytical and does not compete with Fredholm's creative use of formalism. This may sound surprising for those who know Hilbert mostly for his contribution to the foundations of geometry. His work about Fredholm's equations offers a great variety of methods and new results in varied contexts, which became an inspiration for several mathematicians working on the subject. It would be impossible to summarize here the whole content of the six papers. The starting point is a very detailed presentation of Fredholm's main results in the specific case when the kernel f(x,y) is a symmetrical function. This led Hilbert to give a complete account of symmetrical finite operators in terms of bilinear forms (like Frobenius). Most of the results of this section were already known, but the novelty was in the use of a new method which made the generalization to infinite dimension easier. The main result concerns the possibility of a diagonalization of symmetrical operators in an orthonormal basis of eigenvectors. Hilbert then generalized to infinite dimension. He showed, with a rigorous proof, the convergence of the characteristic polynomial. He studied the case of a symmetrical kernel in great detail and gave several applications. In the fourth text (1906), Hilbert studied infinite quadratic forms and he pointed out their importance in several contexts. His method presented several similarities with the previous study of symmetric operators. Formally, the main difference is that expressions such that $\int_a^b (x(s))^2 ds$ are replaced by $\sum_{n=1}^{\infty} a_n^2$, and something analogous for the scalar product (note that Hilbert never used the term scalar product, and generally very few geometrical terms). However, the analogy is totally implicit and Hilbert made no comment and presented everything twice in detail. Regarding quadratic forms, his main goal was to study the spectrum of an infinite quadratic form. We will not give the details, which are rather more difficult than in finite dimension, since it uses Stieljes's integral for the discontinuous part of the spectrum. Finally, Hilbert used these results to study, in a new manner, Fredholm's equation. He did not limit himself to the symmetric case and he used Fourier series to make the connection with the finite case. He established the main properties. With this new approach, he proved again most of the results he had already established in his first texts with a symmetrical kernel, without explicitly referring to this case. Then he gave new applications.

    Of course, this is only a very brief overview of Hilbert's texts which are very rich. The diversity of the problems and methods shows various ways of using analogy with the finite cases. Similarities are visible, yet not emphasized. Hilbert never tried to make any effort toward a unification of the different context in a more formal, synthetic, approach. Nevertheless, in the introduction to the book published in 1912, in which his six papers were collected, he admitted the necessity for such a unification:

> [...] der systematische Aufabau einer allgemeinen Theorie der linearen Integralgleichungen für die gesamte Analysis , inbesondere für die Theorie der bestimmten Integrale und die Theorie der entwicklung willkülicher Funktionen in unendliche Reihen, ferner für die Theorie der linearen Differentialgleichungen und der analytischen Funktionen sowie für die Potentialtheorie und Variationsrechnung von höchster Bedeutung ist. [...]

And again, when analyzing the method used for solving the functional equations:

> Die Methode [...] besteht darin, dass ich von einem algebraischen Problem, nämlich dem Problem der orthogonalen Transformation einer quadratischen Form von n Variabeln in eine Quadratsumme ausgehe und dann durch strenge Ausführung des Grenzüberganges für n = ∞ zur Lösung des zu behandelnden tranzendenten Problemes gelange.[74] (Hilbert 1912, 2-3)

One can see here the importance and the limitations of Hilbert's work: solving problems of analysis and keeping the analogy with finite dimension explicit. It was to take another fifteen years before the concept of Hilbert space emerged from this approach.

3.2.4. A Topological Approach

Before we come to this point, it is necessary to give an account of another trend of research which played an important role in the constitution of functional analysis and abstract functional spaces. The origins of these works take place in the calculus of variations. In this context, Karl Weierstrass was one of the first to introduce the notion of neighborhood for functions, in an implicit topological approach to functional spaces. In this tradition, Italian mathematicians like Volterra and Pincherle, but also Guilio Ascoli and Cesare Arzelà, studied the properties of what they called functions of lines (i.e. functions of functions) introducing concepts of continuity and derivative. In France, Jacques Hadamard and Maurice Fréchet worked along the same line. In 1906, Fréchet published his doctorate entitled, *Sur quelques points du calcul fonctionnel,* in the introduction to which he presented the essential ideas for an axiomatic approach to topological spaces of functions, showing the novelty of his viewpoint:

> "Nous dirons qu'une *opération fonctionnelle* U est définie dans un ensemble E d'éléments de *nature quelconque* (nombres, courbes, points, etc.) lorsqu'à tout élément A de E correspond une valeur numérique déterminée de U : U(A). La recherche des propriétés de ces opérations constitue l'objet du *Calcul Fonctionnel*.
> *Le présent travail est une première tentative pour établir systématiquement quelques principes fondamentaux du Calcul Fonctionnel et les appliquer ensuite à certains exemples concrets.*[75] (Fréchet 1906, 1)

What may seem so obvious today shows the great modernity of Frechet's approach. He defined different types of abstract sets corresponding to different types of topology. There was no mention, even implicit, of any algebraic structure. In this sense, this text is not quite relevant for our topic. However, its influence is important since the approach in terms of structure (even topological) of functional sets is a decisive step toward the modern notion of functional space. We will refer to other contributions of Fréchet at the end of this paragraph.

We will now examine how the notion of abstract Hilbert space was elaborated from Hilbert's own work, with the influence of ideas along the lines of what we have just described, after a process of generalization and formalization involving several contributors.

3.2.5. Frigyes (Frédéric) Riesz, Ernst Fischer and Erhard Schmidt

A first decisive step was taken in 1907 independently by Riesz and Fischer. Both of them used the new framework of Lebesgue's integral theory in order to generalize Hilbert's method for the study of Fredholm's equation to the case when the square of the kernel is Lebesgue-integrable. Although they used different methods, each of them showed implicitly a result equivalent in modern terms to the isomorphism between ℓ^2 et L^2. Riesz used Fourier series in an analytical approach when Fischer's method, based on the definition of mean-square convergence, was more synthetic and formal. The same year, Riesz wrote a paper in which he analyzed the differences in terms of complementarity between his and Fischer's method:

> L'idée de représenter une fonction par ses constantes de Fourier devait devenir très familière. De cette façon, on parvenait à représenter l'ensemble des fonctions sommables sur un sous-ensemble d'une infinité dénombrable de dimensions. Quel est ce sous-ensemble? Jusqu'à aujourd'hui on ne sait pas le dire.
>
> Or pour une classe plus spéciale, pour le système des fonctions sommables et de carré sommable, la solution ne pose plus tant de difficultés. Pour cette classe, il existe un lien plus intime entre la fonction et sa série de Fourier (…). Pour cette classe de fonctions on peut définir une notion de distance et l'on peut fonder sur cette notion une théorie géométrique des systèmes de fonctions, théorie qui ressemble à la géométrie synthétique.[76] (Riesz 1907c., 1409-1410)

In his idea, the geometry on the infinite-dimensional space of coordinates with Euclidean distance is the analytical form of this synthetic geometry.

Riesz's and Fischer's works are very important because they draw out the 'intimate connection' between two major types of problem solved by Hilbert. The analogy with geometry applies only on the topological structure of the spaces ℓ^2 and L^2, not on their algebraic structure. Nevertheless, the duality analytic/synthetic geometry makes a connection with the similar question regarding vectors and analytic geometry in n-dimensional spaces.

The next essential step was taken by Schmidt who made an explicit use of geometric language. In a text published in 1908, he presented a study of the space ℓ^2 (without using this notation) in which he systematically used geometrical terms and tried to show the power of the analogies. He considered the series (real or complex) square summable, that he just called functions, and introduced the symbols:

$$(A;B) = \sum_{x=1}^{x=n} A(x)B(x) \text{ , and } \|A\|^2 = (A; \overline{A})$$

to which he does not give any name. Then he gave and proved most of the results which are today classical in the geometrical approach of Hilbert spaces, especially concerning the representation on a complete infinite set of orthonormal functions.

It is from Riesz's and Schmidt's contributions that the idea arose of thinking geometrically in Hilbert's spaces, and more generally in any functional space. This change of perspective was essential and has proved to be fruitful. Moreover, it unified the analytic, geometric, as well as algebraic origins of linear algebra. On the other hand, it shows that the development of functional analysis is very important to understand and questions the supposedly 'natural' connection between linear algebra and geometry.

At this stage of the development of what will become functional analysis, the notion of Hilbert space did not exist as such. A structural correspondence had been drawn out between two sets of objects, unified in a dual geometric approach. Further generalizations of Hilbert's work would have to be made, involving new types of spaces, before a formal definition of Hilbert space could be given.

In 1910, Riesz tried to generalize his 1907 result and introduced what we now call the L^p space, p>1. Using Lebesgues's integral theory, and Hölder's and Minkowski's inequalities (well-known at that time), he established the dual relations between L^p and L^q, for $\frac{1}{p} + \frac{1}{q} = 1$. He also defined the notion of weak convergence and generalized a theorem he had proved in L^2, on the basis of a result by Hadamard on square summable series. This theorem, often known today as Riesz's representation theorem states that for any bounded (for the L^p norm) linear operator A in L^p, there exists a function a in L^q, such that :

$$A(f) = \int_a^b a(x)f(x)\,dx, \text{ for any f in } L^p.$$

Moreover, in this text, Riesz used systematically a notation in terms of linear operator, which made him the first one after Fredholm to use such a formal approach. He introduced the important notion of adjoint (or transposed) operator and solved Fredholm's equation through the eigenvalue problem. Not only did he generalized Hilbert's (and his own) work in the case L^p, but he also introduced a more formal approach in the case L^2.

In a later work, from 1916, originally published in Hungarian and then in German under the title, *Über lineare Funktionalgleichungen*, Riesz presented a more general approach to integral equations. Indeed, he gave one of the first axiomatic definitions of the norm and of a bounded linear operator. In this paper he established the main results of what is now known as Riesz-Fredholm's theory of compact operators. Whereas the whole treatise dealt with the set of continuous real functions on [a;b], most of the results were obtained in such a formal manner that they could easily have been generalized to any complete normed function space, if such a notion had been explicitly defined

In the same vein, in 1921, Eduard Helly gave a very formal approach to solve infinite systems of linear equations. He used general finite and infinite-dimensional normed vector spaces. If the notion of norm was defined axiomatically, the structure of vector space remained implicit. Indeed, Helly only used the representation with coordinates.

3.2.6. Stefan Banach, Hans Hahn and Norbert Wiener
During the first three decades of the 20th century, more and more general spaces of functions were introduced in various contexts. The topological structure was formalized axiomatically, but the linear structure remained implicit, transparent. Within two years, however, three mathematicians, Banach, Hahn and Wiener, working independently produced an axiomatic definition of a complete normed vector-space (known today as a Banach space) in a context related to functional analysis, but without making explicit reference to previous axiomatic definitions of vector spaces from the time of Peano to Weyl. Moreover, none of them defined the algebraic structure separately.

Chronologically, Wiener's work was the first to be published, in the proceedings of the International Congress of Mathematicians in 1920 in Strasbourg, when its author was only starting his career. The algebraic part of his definition presented an affine structure with an underlying vector space. This approach differed from that of Banach and Hahn, but was the point of view that was later adopted by Fréchet, and then by Banach himself. On the other hand, Wiener's axioms did not take into account the addition (as if it were implicit); they only concerned the multiplication by positive numbers and did not mention the axiom : $1.\xi = \xi$. The norm was defined with the modern axioms. Wiener gave a few examples, but his paper remained quite theoretical. In 1922, Wiener published two more short papers on the subject. In the first paper, he defined a structure (Vr) (called 'restricted vector system'), for which he added the property of commutativity of addition. In the same year he recognized the prior rights of Banach's work. (Banach had defended his doctoral thesis a few months before the Strasbourg Congress, but his work was only published in 1922, in French). Wiener also admitted that Banach's axioms were given « in a form more immediately adopted to the treatment of the problem in hand » (Wiener 1923). After this date, Wiener turned his interest to other mathematical subjects (see (Wiener 1956, 60-64)).

Hahn's work, published in 1922, was partially inspired by Helly's, mentioned above. Unlike Wiener and Banach, axiomatization was not a goal in itself for Hahn: he wanted, above all, to unify, in a general theory, several problems dealing with functional and differential equations. He started his paper by introducing what he called a linear space ('Linear Raum') of which he gave an axiomatic definition corresponding to the modern notion of normed vector space. The elements were called points (not vectors). The axioms of the algebraic part of the structure were rather similar to our modern ones (nothing is missing and none is redundant). The definition of the norm was identical to Helly's, to which he referred, and is the modern one. Then Hahn defined the notion of completeness ('Vollständigkeit'). After this brief formal start, he gave several examples of linear spaces of functions and operators with applications. Except for the first two definitions, this work was therefore oriented toward applications and did not give many general results. In a paper published in 1927, Hahn gave some accounts of Banach's work and studied Fredholm's equation in a general Banach space. This led him to numerous general results about complete linear subspaces. The most important of these results was certainly the theorem known today as the Hahn-Banach theorem :

> **Satz III.** Sei R_0 ein vollständiger linearer teilraum von R und $f_0(x)$ eine Linearform in R_0 der steigung M. Dann gibt es eine Linearform f(x) in R der steigung M, die auf R_0 mit f_0 überstimmt.[77] (Hahn 1927, 217)

It is also in this text that Hahn defined the notion of dual space, in a topological sense, and established the first results connected to this notion.

Among the three, Banach's work had the most important influence. In the introduction of his doctoral dissertation entitled : *Sur les opérations linéaires dans les ensembles abstraits et leur application aux équations intégrales,* defended in 1920, only partially published (in French) in 1922, Banach wrote:

> L'ouvrage présent a pour but d'établir quelques théorèmes valables pour différents champs fonctionnels, que je spécifie dans la suite. Toutefois, afin de ne pas être obligé à les démontrer isolément pour chaque champ particulier, ce qui serait bien pénible, j'ai choisi une voie différente que voici : je considère d'une façon générale les ensembles d'éléments dont je postule certaines propriétés, j'en déduis des théorèmes et je démontre ensuite de chaque champ fonctionnel particulier que les postulats adoptés sont vrais pour lui.[78] (Banach 1967/79, 2:308)

This quotation shows that Banach was explicitly seeking an axiomatic approach. This may seem new for that time, especially in the field of analysis. However, it is worth noticing that this attitude was quite characteristic of the Polish mathematicians of the time. In his definition of a complete normed vector space,

Banach distinguished three sets of axioms, although he did not give different names to the corresponding structures; in particular, to the algebraic structure on its own. However, after giving the first set of axioms, Banach gave a few examples: the (geometric) vectors, Grassmann's forms, the set of complex numbers and the set of quaternions. Although he did not refer to any preceding axiomatic definitions of vector space, his choice of axioms was closer to Burali-Forti's and Marcolongo's, and was also incomplete and redundant. However, it is very likely that Banach, Wiener and Hahn, each discovered their list of axioms as suggested by Moore (1995).

Most of Banach's treatise was devoted to the establishment of general results on linear operators in his space. In the last part, he showed that various functional sets satisfied his list of axioms: these include the set of continuous functions, L^p spaces, the sets of bounded measurable functions, of functions whose (n-1) first derivatives are absolutely continuous and the n-th derivative is either continuous or L^p, etc. But there was no application, although it seems that some applications were actually present in his original doctorate, which was never published. Therefore, this first text by Banach had a limited influence. It was only with his book, *Théorie des opérateurs linéaires,* published in 1932, that Banach made a decisive contribution to the field of functional analysis. Before we come to this point, we have to give an account of a few other works made between 1920 and 1932.

3.2.7. *Maurice Fréchet*

Fréchet was one of the first mathematicians who used the axiomatic definition of a complete normed vector space in his work. He knew both Wiener's and Banach's contributions and around 1925 he adopted a definition of his own which combined the advantages of the two. From Wiener, he kept the distinction between an affine and a vectorial structure, but he was formally closer to Banach's approach. In relation with his preceding work, Fréchet distinguished the notions of points and vectors and introduced the notion of deviation ('écart') or, distance for points, and a corresponding concept of norms (valid for vectors). He defined axiomatically the notions of abstract vector space and abstract affine space, for which he also gave geometrical definitions. Moreover, he also defined a more general notion of topological affine space (not necessarily normed). Finally, he gave a large number of examples of spaces having the different structures.

In 1928, he published, in a collection edited by Borel and dedicated to the theory of functions, a book entitled: *« Les espaces abstraits et leur théorie générale considérée comme introduction à l'analyse générale »,* in the introduction of which he gave a well documented historical account. The results presented in this book were organized in a general axiomatic and abstract manner :

La méthode qui a été employée avec succès par MM. Volterra et Hilbert consiste à remarquer qu'une fonction (continue, par exemple) peut être déterminée par la connaissance d'une infinité dénombrable de paramètres. Et alors on traitera d'abord le problème comme s'il n'avait à faire intervenir qu'un nombre fini de paramètres, puis on passera à la limite.

Nous croyons que cette méthode a joué un rôle, mais qu'elle a fini son temps. C'est un artifice inutile de substituer à la fonction une suite infinie de nombres qui d'ailleurs, peut être choisie de plusieurs façons[79]. On le voit bien, par exemple dans la théorie des équations intégrales où les solutions de Fredholm ou de Schmidt sont beaucoup plus

simples et plus élégantes que celle de Hilbert, ce qui n'enlève pas à ce dernier le
mérite essentiel d'avoir obtenu par sa méthode un grand nombre de résultats
nouveaux.[80] (Fréchet 1928, 4-5)

This quotation is important in order to understand the respective role of the
different contributions and the reasons why, between 1920 and 1930, Fréchet's or
Banach's viewpoints finally eclipsed other methods (which had been productive and
necessary in their own time, but had now become redundant). It is also important for
our purpose to point out Fréchet's note on the comparison with geometry: the
axiomatic definition of vector spaces is a way to avoid the disadvantage of
coordinates, in infinite dimensional spaces as well as in finite dimensional spaces.

3.2.8. *John von Neumann and Quantum Mechanics*
It is time now to see how the notion of abstract Hilbert space finally emerged. The
final step was taken by John von Neumann in 1927[81], in his attempt to establish
consistent and rigorous mathematical bases for quantum mechanics. After the initial
works, essentially by Max Planck, Albert Einstein and Niels Bohr, two different
approaches to quantum mechanics coexisted : 'the viewpoint of matrices' developed
by Werner Heisenberg, Max Born and Pascual Jordan and 'the viewpoint of waves'
pursued by Erwin Schrödinger. Without going into detail, it is necessary to know,
for our purpose, that these two viewpoints led to quite different models. The first
model used infinite sequences and matrices; the second was based on the notion of
wave function. The essential mathematical problem in the two models was to solve
a functional equation leading to an eigenvalue problem, which could be approached
by infinite systems in the first model (ℓ^2) or by operators on a functional space (L^2)
in the second. Once again the duality analytic / synthetic surfaced. Schrödinger
showed the mathematical equivalence of the two approaches in 1926. Von Neuman
refined his work and sought for deeper connections in order to make the analogy
more consistent. This led him to the general definition of an abstract Hilbert space.

Die in I.3 skizzierte Methode kann darauf heraus, den "diskreten" Raum in
Indexwerte, Z (=1,2,...), mit dem kontinuierlichen Zustandsraum Ω des mechanischen
Systems (Ω is k-dimensional, wenn k die Zahl der klassisch-mechanischen
Freiheitsgrade ist) in Analogie zu setzen. Dass dies nicht ohne einige Gewaltäigkeit
um Formalismus und Mathematik gelingen kann, ist kein wunder : die Räume Z und Ω
sind wirklich sehr verschiden, und jeder Versuch, sie in Beziehung zu setzen, muss
auf grosse Schwierigkeiten stossen.[82] (von Neumann 1932, 19)

Von Neumann analyzed in detail the two methods for solving the eigenvalue
problem and showed their analogies. This led him to the characteristics of an
abstract Hilbert space. He then gave the axiomatic definition of a "Hilbert space" as
a complex infinite-dimensional vector space with an inner-product being complete
and separable (i.e., there exists a countable family of vectors being dense). In his
book, published in 1932, von Neuman gave a full account, in a very formal way, of
the properties of Hilbert spaces. He started his presentation with the study of the
finite-dimensional complex inner product space. The same year, Marshall Stone
published his book, which would remain a reference on linear operators in Hilbert
spaces. After these works, the notion of Hilbert space was extended to spaces which
were not necessarily isomorphic. Finally, it is important for our study to underline

the fact that the emergence of an abstract Hilbert space standardized the study of finite-dimensional inner product space in an axiomatic presentation.

3.2.9. Banach's Decisive Contribution

On the other hand, after seven years devoted to other mathematical subjects, Banach published, in 1929, two papers about linear operators in functional spaces in which he enlarged his initial contribution, taking into account Fréchet's recent work. In particular, the second of these papers presented the most general version of Hahn-Banach's theorem. In 1932 he published a book entitled *'Théorie des opérateurs linéaires'*, which was to influence all functional analysis and would become the reference. This book gave not only a very complete theoretical approach, in an axiomatic presentation, but also numerous applications to most of the classical problems of the field. It also opened new perspectives and tackled new types of questions which would appear to be essential for the further development of functional analysis.

3.2.10. Conclusion

Solving infinite-dimensional linear problems was subject to profound changes between 1880 and 1932, when the modern approach was laid out. As a conclusion we will now draw out the main aspects of this evolution:

i) The natural tendency, consisting in generalizing from what is well-known and works well, led mathematicians to solve infinite-dimensional linear problems through a generalization of methods with coordinates and determinants. This was reinforced by the dominant conception of a function as a series of an infinite but countable number of coefficients. As a consequence, early axiomatic approaches did not meet much success.

ii) Hilbert's various methods, developed about Fredholm's equation, opened up the field of investigations. In this context, successive generalizations and sophistication led progressively to the drawing out of similarities and unified approaches. This process involved mostly methods and tools, the eigenvalue problem being one of the central questions.

iii) Another essential idea was to create a concept of function of functions (Volterra, Pincherle, Fréchet). In order to define concepts such as limit, continuity or derivability of such functions of functions, one needed to reflect on the topological structure of sets of functions, starting with notions of distance and norm (Fréchet and Riesz).

iv) Wiener's, Hahn's, Banach's and Fréchet's contributions to the axiomatic approach were crucial, yet they cannot be fully understood without referring to the previous evolution. In particular, their success was, for a great part, due to the fact that they appeared at the right time to answer questions which had come to the right stage of maturation; especially concerning questions raised in points ii) and iii).

v) The topological aspect was essential in the process that led to the emergence of the concept of functional space. The linear structure remained implicit for a long time. In this sense, the notion of normed vector space appeared before the notion of vector space was clearly conceptualized. Terms such as vectors or points were imported from geometry through analogy, within Hilbert's space, with topological questions connected with notions of scalar product, distance and norm.

vi) In this process, even if only infinite-dimensional spaces were concerned, the process of generalization and unification threw a new light on the presentation of finite-dimensional questions. In particular, the result on abstract Hilbert space led to a more formal presentation of finite-dimensional Euclidean vector spaces and spectral theory.

3.3. Epilogue

At the beginning of the nineteen-thirties, the axiomatic presentation of vector spaces became predominant in two major fields of mathematics: functional analysis and algebra, represented by Banach's and van der Waerden's books. Von Neumann's contribution was representative of a third field, however, in which the axiomatic approach to vector spaces was essential. This was due to the fact that Hilbert.'s space played a specific role in functional analysis, especially through its connection with quantum mechanics. Therefore, in this field, not only mathematicians, but also physicists were involved. Nevertheless, the algebraic and analytic aspects of the axiomatic theory of vector spaces remained for a time disconnected until, in 1941, Israil Gelfand introduced the notion of Banach algebra.

The main question that we still have to examine concerns the question of the diffusion of the theory among mathematicians and its transposition into teaching. In the early thirties, Germany was further ahead than any other countries, but the rise of Hitler rapidly changed the situation, due to the massive immigration of mathematicians to, not only other parts of Europe, but especially to the United States of America. In the US, German mathematicians had a strong influence on the development of modern algebra and functional analysis.[83]. Garrett Birkhoff's and Saunders MacLane's stay in Germany, on several occasions during the thirties, influenced their research as well as their view on the teaching at university level. In 1941, they published the first edition of their *Survey of Modern Algebra*, based on their courses given since 1936.[84] This was one of the first university textbooks presenting modern algebra at this level. Linear algebra played a central part in this book: vector spaces and linear transformations were presented in an axiomatic manner.[85] The next year, in 1942, Paul R. Halmos published his book entitled, *'finite-dimensional vector spaces'*. Unlike Birkhoff and MacLane, Halmos' goal was not essentially algebraic: rather he intended to introduce students to Hilbert's abstract space, insisting on the geometrical aspects of functional analysis:

> My purpose is to emphasize the simple geometric notions common to many parts of
> mathematics and its applications, and to do this in a language which gives away the
> trade secrets and tells the student what is in the back of the minds of people proving
> theorems about integral equations and Banach spaces. (Halmos 1942, i).

These two books had a great influence on university teaching within the US, as well as in many other countries. They were re-edited several times.[86]

In France, the thirties marked the beginning of a fundamental change; young mathematicians, including those who would form the early Bourbaki, had to face the conservatism of their elders, symbolized by Edouard Goursat's textbook, the most popular, if not unique, reference for the *'licence' of mathematics* (Masters degree).

During the First World War, many young mathematicians had died,[87] therefore there was a generation gap that created a strong mental opposition between the younger and the older mathematicians in the universities. The elders had suffered from isolation from Germany and were not aware of the new algebra. For example, a book like van der Waerden's *Moderne Algebra* was totally ignored by older mathematicians, but it was a total revelation for young students of the *'Ecole Normale Supérieure'* who discovered, in this book, a new way of doing mathematics. When they became university assistants, some of them felt the necessity to renew the teaching. In fact, the Bourbaki's adventure started from a conversation between Henri Cartan and André Weil, both of whom were teaching in Strasbourg (see Weil 1992, 104-106). At the end of the thirties, in provincial universities (Paris was still in the hands of older mathematicians) some younger professors introduced modern mathematics in the teaching of the *'licence'*. For instance, in 1939-40, when Strasbourg fell under German authority, the university staff was sent to Clermont-Ferrand. In this context, young lecturers had much more freedom than anywhere else. That year, Henri Cartan taught a completely new course in the *'licence'*, starting with an axiomatic presentation of algebraic and topological structures, including the theory of vector spaces. When Cartan got appointed to the *'Ecole Normale Supérieure'* the following year, André de Possel (another early Bourbaki) replaced him in Clermont-Ferrand and taught a similar course. Still in Clermont-Ferrand, in 1940-41, Charles Ehresmann taught a course in cinematic, in which he gave an axiomatic presentation of finite-dimensional vector and affine spaces and linear transformations. Until 1954, in France, such experiments were few and far between. Bourbaki was founded in 1934: the volume about linear and multilinear algebra appeared in 1947. One strong principle of Bourbaki was to adopt the most general presentation and come to particular cases in a second phase. Concerning linear algebra, this led them to put forward the concept of module over a ring, a position that many mathematicians, especially abroad, refuted. Nevertheless, Bourbaki was not the only one to promote the axiomatic theory of vector spaces. In 1935, Gaston Julia gave a series of conferences at the Sorbonne about the mathematical bases for Quantum Mechanics. These conferences, addressed to physicists as well as mathematicians, were built from an elementary standpoint. Therefore, Julia presented in detail the theory of finite dimensional Euclidean vector spaces: first in a classical way, with coordinates and determinants, then in a manner inspired by Töplitz (1909) and finally in an axiomatic approach, insisting on the positive aspects of the last one. The rest of the conferences were devoted to the study of the most general Hilbert space. The text of the conferences was published in two volumes by the Sorbonne from the notes of two students of the *'Ecole Normale Supérieure'*. A few years later, at the request of George Bruhat, André Lichnérovicz wrote a textbook of elementary mathematics for physicists in the spirit of the famous book by Courant and Hilbert[88]. Bruhat died in captivity during the war, and the book was only published in 1947 under the title, *Algèbre et Analyse Linéaires*. The first chapter about linear equations starts with these words:

> En traitant des problèmes concrets, apparemment fort éloignés les uns des autres, les mathématiciens et les physiciens constatent souvent qu'ils sont ramenés à des calculs linéaires formellement identiques. Une notion importante va nous aider à comprendre les raisons de ce phénomène : la notion d'espace vectoriel.[89] (Lichnérovicz 1947, 7)

This is followed by an axiomatic presentation of the theory of vector spaces. This book, much more accessible than Bourbaki's, was essential for the diffusion of the axiomatic theory of vector spaces in France.

What is striking in the preceding quotation, as well as in Halmos', is that, as soon as the late forties, it seemed obvious that all linear problems bore so much similarity that the axiomatic viewpoint imposed itself naturally. It meant that the idea that the axiomatic structural approach was the best way to model linear situations had rapidly become the standard approach between 1930 and 1950. It in fact became a dogma that nobody questioned any more

A decisive step in the renewal of French university teaching was taken in 1954. That year, Georges Valiron, who was in charge of the mathematical certificate for the *'licence'* , became seriously ill just before the beginning of the term. In the emergency, the dean of the faculty consulted Henri Cartan, who was then an eminent figure of the *'Ecole Normale Supérieure'*. This was how Gustave Choquet, then a young protégé of Bourbaki, acquired the responsibility of the mathematical teaching for the *'licence'*. He imposed a radical change with a modern approach; i.e., he introduced the students with the axiomatic structural approach. Rapidly, the Parisian revolution spread to all of France. In 1956, modern mathematics officially entered the national curriculum of the *'classes préparatoires'*, the most prestigious part of the French scientific teaching. In this reform, the theory of vector spaces played a central role as the part of new algebra that would renew the teaching of geometry. As a consequence of this change, the teaching of vector analysis, already present in courses like cinematic from a geometrical viewpoint (the vectors were directed line-segments attached to an origin), became more formal and was preceded by an introduction to general finite-dimensional vector spaces in an axiomatic presentation. In secondary schools, the reform was more difficult to implement: the innovators had to convince the national inspectors of the consistency of their project. This was the goal of the Lichnérovicz commission. On the other hand, the teachers had to be trained. With this objective in mind, the national association of mathematics teachers (A.P.M.E.P., *Association des Professeurs de Mathématiques de l'Enseignement Public*) published, in 1962, a series of books based on conferences given by André and Germaine Revuz in Paris, as part of an experimental project. In 1963, the second volume was dedicated to the theory of finite-dimensional vector spaces. Partly based on Bourbaki's ideas, this book became a reference for the teaching of linear algebra in France. We will not retrace here in detail the history of the reform of modern mathematics in French secondary schools. It is sufficient to know for our purpose that, in 1969, the national curriculum imposed the teaching of the axiomatic theory of finite-dimensional vector-spaces from the first year of the *'Lycée'* (age 15), introducing a fundamental change in the teaching of geometry, which became a bare application to linear algebra. This was preceded by an animated debate among mathematicians. Two books, published the same year, 1963, in the same collection, reflect the type of questions at stake at that time : *'Algèbre linéaire et géométrie'* by Jean Dieudonné and *'De l'enseignement de la géométrie'* by Choquet. Both of them agreed that the teaching of geometry in secondary school should be changed, but Choquet was opposed to Dieudonné. Indeed Choquet defended the idea that the structure of vector spaces should not be given too early; he tried to promote an axiomatic approach to geometry which was more intuitive and did not refer explicitly to the axioms of vector spaces. Dieudonné wanted students to get used, as early as possible, to the ideas that were essential for the mathematics at stake in the *'licence'*. In this sense, the model of geometrical space, as the Euclidean three-dimensional vector space, was the best preparation for Hilbert space and general functional spaces which were essential in the *'licence'*. The reform implemented in 1969 was essentially based on Dieudonné's ideas. It lasted until the beginning of the 80s. We will not try here to analyze the reasons for its failure. Anyway, the counter-reform progressively evacuated all axiomatic approaches to linear algebra from secondary schools; even Cartesian geometry gradually became reduced to a minimum. As a consequence, in the first year of university, what used to be only a generalization to infinite-dimensional vector spaces, became a first course introducing linear algebra. Moreover, the linear algebra course, representing about a third of the curriculum of the first year in a science university, has been, for the last 15 years, the first teaching of an axiomatic structure. On the other hand, students have very little practice of analytical geometry. At first, university teachers, who may have not always realized the radical changes, in secondary school teaching, had difficulties understanding their students' problems in linear algebra. Then, most universities started reducing the formal aspect of their teaching. Most of them tried to limit the first semester to direct applications to geometry, sometimes not even introducing the general concepts. More recently, it became more and more fashionable to introduce linear algebra through the solving of systems of linear equations and applications to matrix calculus.

In some cases, this was inspired by the anglo-american tradition. Nevertheless, linear algebra remains a difficult course, not only feared by students but also by teachers, especially those who become frustrated by students' incomprehension of ideas which to them seem so elementary (see the second part of this book).

4. CONCLUSION

Most of the influential mathematicians today have been trained at a time when it seemed natural to model linear problems in terms of the axiomatic theory of vector spaces. They are convinced of the usefulness of axiomatic structures, which have been the bases of their mathematical culture. When teaching linear algebra and facing students' misunderstanding of what seems so elementary to them, they feel totally disarmed and are often unable to understand the inconsistent answers from their students. Generally, the main reproaches made by teachers concern the lack of practice and competence in formal language, elementary logic, set theory and geometry.

The historical analysis we have just completed shows that the modern axiomatic viewpoint is the ultimate achievement of a long process which involved several various parts of mathematics. In this sense, the concept of vector space, so elementary in terms of structure, encapsulates, in a very elaborate product, the result of a long and complex process of generalization and unification. Yet the inevitable risk of reduction that masks the complexity of the modeling process in the treatment of linear problems is the price that has to be paid for the efficiency of the axiomatic structural approach. Because of this, a knowledge of the historical development provides the teacher and the researcher with a field of investigation from which they can better understand the students' difficulties and, more generally, from which they can put some "meat" to the "bare bones" of axiomatic approach.

From a didactical point of view, it is important to be aware of the unifying and generalizing nature of such a theory as that of vector spaces. The recurrent historical resistance to the axiomatic viewpoint shows that the fact that the efficiency cannot be seen immediately while solving one specific problem is a reason to avoid investing in a formal approach. In fact, between 1880 and 1930, the question of using, or not using an axiomatic approach was linked with the organization of the knowledge, not with the efficiency in problem solving. The analogy between the finite and the infinite cases led to a generalization of the finite cases, but this evolution changed the perception of the finite cases. To think of a function as a single object, and not as an infinite collections of values, is a profound change that brought a new light on the duality analytic/synthetic, even in finite dimensions. The axiomatic approach imposed itself as the best way to unify the whole set of linear problems in a formal setting.

On the other hand, the historical background is fundamental in order to understand the complexity of the connections between linear algebra and geometry; from the parallelogram of velocity and forces until Hilbert spaces, including Leibniz's and Grassmann's contributions. The important role played by the study of systems of linear equations in the history of the theory of vector spaces is also a point that deserves more concern than some historical works may have suggested. Many of the elementary concepts of linear algebra took shape in this framework, even if the use of the theory of determinants made the context not always very easy to analyze. For instance, such a concept as the rank has a crucial role in the study of linear systems, while it is nearly of no interest in geometry (if limited to three dimensions).

We would, of course, not promote a teaching that would try to recreate, at best, the historical conditions in the teaching of linear algebra. Nevertheless, we claim that the knowledge of the historical development is a necessity in order to have an epistemological control, either when analyzing existing teaching projects or when experimenting new ways of teaching linear algebra. The various works presented in the next part of this book will show how this epistemological control is essential in some aspects of the didactical analyses.

5. NOTES

[1] Who knows, for instance, that the first calculations using determinants started at least around 1750 (and even 1679, according to a letter by Leibniz not published until 1850)? Determinants were, at that time, practical tools for solving systems of linear equations. The rules for their calculation were expressed with long and complicated sentences without any formalism - a far cry from the modern concept of determinant in today's multilinear algebra.

[2] Cramer's paradox comes from two (apparently contradictory) results which were well-known and believed to be true at that time:
1) Two algebraic curves ('lignes courbes' in Euler's vocabulary) intersect in as many points as the product of their order. There are cases when all these points are real, finite and distinct.
2) Any n-th order algebraic curve is defined, by a n-th degree polynomial equation in two variables. Such a polynomial equation has $\frac{(n+1)(n+2)}{2}$ coefficients, one being arbitrary, therefore $\frac{(n+1)(n+2)}{2}-1$ $=\frac{n(n+3)}{2}$ points are necessary and enough to entirely determine an n-th order algebraic curve.

The paradox is visible for $n \geq 3$, indeed then $n^2 \geq \dfrac{n(n+3)}{2}$, and so it appears that two n-th order algebraic curves may have in common as much or even more points as is enough to entirely determine one of them.

[3] One will see that it is not possible to determine the two unknowns x and y, as if one eliminates x then the other unknown y disappears by itself and one gets an identical equation, from which it is not possible to determine anything. The reason for this accident is quite obvious as the second equation can be changed into $6x - 4y = 10$, which being simply the first one doubled, is thus not different.

[4] Thus when one says that in order to determine three unknowns, it is enough to have three equations, one must add the following restriction: that is, the three equations differ from each other in such a way that none is already comprised in the others.

[5] Euler does not only mean that the two equations are identical to each other but that each of them is an identity, in other words equations that are true for any value of the unknowns.

[6]
$5x + 7y - 4z + 3v - 24 = 0$,
$2x - 3y + 5z - 6v - 20 = 0$,
$x + 13y - 14z + 15v + 16 = 0$,
$3x + 10y - 9z + 9v - 4 = 0$,
would be worth only two, since if one extracts from the third the value:
$x = -13y + 14z - 15v - 16$
and then substitutes this value in the second, in order to have:
$y = \dfrac{33z - 3v - 52}{29}$ et $x = \dfrac{-23z + 33v + 212}{29}$,
these two values of x and y can be substituted into the first and fourth equations; that will lead to two identical equations (see preceding note), so that the quantities z and v will remain undetermined.

[7] When one asserts that in order to determine n unknown quantities, it is enough to have n equations expressing their mutual relations, it is necessary to add the restriction that all the equations are different from one another, or that none of them is contained in the others.

[8] When two lines of the fourth order meet in 16 points, because 14 points, if they lead to equations which are all different, are sufficient to determine a line of this order, these 16 points will always be such that three or more of the equations resulting from them are already comprised in the others. Because of this, these 16 points do not determine anything more than do 13, or 12 or even less points and in order to determine the line entirely, one must add to these 16 points one or two others.

[9] To be precise, Cramer does not use a double index but letters with a single index, the letter referring to the unknown (a for x, b for y, etc.) and the index to the equation.

[10] Cramer makes the calculations for $n=2$ et 3, implicitly supposing that the main determinant differs from zero, then he says that this can be easily generalized to greater values of n.

[11] The term of determinant was introduced by Cauchy in a memoir presented to the French Academy in 1812 and published in 1815. This was also the first work with a theoretical approach in which more than merely a rule for calculation was given. Indeed, Cauchy analyzes the determinant from the perspective of a function in n variables « taking only two opposite values through a permutation of its variables ».

[12] The first results in this direction were developed by several mathematicians; for a conjoint overview, see (Muir 1960, 1:227-235 ; 2:14-17 , 50-52 , 85-86 , 3:84-85 , 227-235 2:85-86 , 86-92).
[13] Several particular solutions
$A_1^{(\chi)}, ..., A_n^{(\chi)}, (\chi = 1, ..., k)$
will be said to be independent or different, when $c_1 A_\alpha^{(\chi)} + ... + c_k A_\alpha^{(\chi)}$ cannot be zero for $\alpha = 1$,

..., n, without c_1, ..., c_k being all zero, in other words, when the k linear forms
$A_1^{(\chi)} u_1 + ... + A_n^{(\chi)} u_n$ are independent.

[14] He does not use this term or any other in fact, but he shows that if m is the maximal order of nonzero minors, all the solutions can be written as a linear combination of any set of m independent solutions

[15] When in a determinant all the sub-determinants (minors) of $(m+1)$th order vanish, but not all minors of m-th order, I call m the *rank* of the determinant.

[16] The principles of analytic geometry were set out independently by René Descartes in one appendix of his famous 'Discours de la méthode' entitled 'Géométrie', published in 1637 and by Pierre de Fermat in a memoir entitled *Ad Locos Planos et Solidos Isagoge* dated from 1643 (but of which the main results had certainly been established between 1636 and 1638). It was only published after Fermat's death in 1679.

[17] Any problem in geometry can easily be reduced to such terms that a knowledge of the lengths of certain lines is sufficient for its construction. Just as arithmetic consists of only four or five operations, namely, addition, subtraction, multiplication, division and the extraction of roots, which may be considered a kind of division, so in geometry, to find required lines it is merely necessary to add or subtract other lines; or else, taking one line which I shall call unity in order to relate it as closely as possible to numbers, and which can in general be chosen arbitrarily, and having given two other lines, to find a fourth line which shall be to one of the given lines as the other is to unity (which is the same as multiplication); or, again, to find a fourth line which is to one of the given lines as unity is to the other (which is equivalent to division); or, finally, to find one, two, or several mean proportionals between unity and some other line (which is the same as extracting the square root, cube root, etc., of the given line. And I shall not hesitate to introduce these arithmetical terms into geometry, for the sake of greater clearness. (translation from the Dover edition by Smith & Latham, 1954).

[18] According to Greek tradition, until the 17th century, geometrical problems were separated into three categories:
- plane problems (using only straight lines and circles)
- solid problems (using only conics)
- linear problems (using other types of 'lines', i.e. curves).
Thus linear meant something very different from what it means now.

[19] Beware of not taking this term in its modern sense. Here it means 'relative to the line' as a new type of geometrical magnitude in opposition to numbers, the usual magnitudes in algebra.

[20] I am still not satisfied with Algebra, because it does not give the shortest methods or the most beautiful constructions in Geometry. That is why I believe that we need still another analysis which is distinctly geometrical or linear (see footnote above), and which will express *situation* directly, as Algebra expresses magnitude directly. And I believe that I have found the way and that we could represent figures and even machines and movements by characters, as algebra represents numbers or magnitudes. (Translation from Crowe 1967, 3)

[21] I have discovered certain elements of a new characteristic which is entirely different from algebra and which will have great advantages in representing to the mind, exactly and in a way faithful to its nature, even without figures, everything which depends on sense perception. Algebra is the characteristic for undetermined numbers or magnitudes only, but it does not express situation, angles, and motion directly. Hence it is often difficult to analyze the properties of a figure by calculation, and still more difficult to find very convenient geometrical demonstrations and constructions even when the algebraic calculation is completed. But this new characteristic, which follows the visual figures, cannot fail to give the solution, the construction and the geometric demonstration all at the same time, and in a natural way and in one analysis, that is, through determined procedure.
(Translation from Crowe 1967, 3)

[22] For a detailed account of Leibniz's contribution see (Echeveria and Parmentier 1995) which offers a selection of translated (into French) and annotated texts by Leibniz, among which many had never been published before, on the subject.

[23] Following the publication of Leibniz's letter in 1833, the German mathematical society *Jablonowski Gesellschaft* , organized a contest to reward a work able to realize Leibniz' unaccomplished program. Hermann Grassmann, of whom we will speak in detail in a following section, finally got the prize in 1845.

[24] These works are given here in chronological order, nevertheless most of them did not became known at the time they were first produced, essentially due to the lack of fame of their authors. For more details on this section of the history of mathematics see: (Cauchy 1849), (Cartan 1908), (Budon 1933), (Crowe 1967) and (Artigue and Deledicq 1992).

[25] Given any number = v of points $A,B,C, ..., N$ with coefficients $a,b,c,...n$ where the sum of the coefficients does not equal zero, there can always be found one (and only one) point S - the centroid - which point has the property that if one draws parallel lines (pointing in any direction) through the given points and the point S and if these lines intersect some plane in the points $A',B',C', ..., S'$, then one always has:
$a.AA' + b.BB' + ... + n.NN' = (a + b + c + ... + n)$. SS',
and consequently, if the plane goes through S itself, then,
$a.AA' + b.BB' + ... + n.NN' = 0$. (Translation from Crowe 1967, 49)

[26] Grassmann, who rediscovered in his *Ausdehnungslehre* from 1844, independently from Möbius, the principles of the barycentric calculus, will show that in the case when the sum of coefficients equals zero, the sum of the weighted points must be considered as a 'segment' (equivalent to what we now call a vector), of which he gives the characteristics. Grassmann's approach is wider than Möbius and the barycentric calculus is only one aspect of his theory. His constant use of a dialectical process between intuition and formalism allows him to overcome the geometrical intuition and to obtain this result as a consequence of the regularity of an algebraic structure (see section on Grassmann).

[27] *Fundamental theorem* : In the equipollences, terms are transposed, substituted, added, subtracted, multiplied, divided, etc., in short, all the algebraic operations are performed which would be legitimate if one were dealing with equations, and the resulting equipollences are always exact. As we said in 5°, non-linear equipollences can only be referred to figures in a single plane.
(translation from Crowe 1967, 53).

[28] Maxwell's equations are a good example of the simplifications made possible by quaternions or geometrical calculus compared to the long and meaningless calculations that one had to compute with the analytical approach using Cartesian coordinates.

[29] The principle of permanence had been stated by the English School of Algebra, in particular by George Peacock It stated that any new algebra, in order to be acceptable, should have the well-known properties of arithmetic. These properties were in fact only partially explicit but corresponded to the structure of commutative field. The fact that complex numbers satisfy the principle had made it even stronger. Only the quaternions came to break this principle.

[30] For more detail on this point see: (Novy 1973), (Pycior 1979 and 1981) or (Richards 1980).

[31] Due to a lack of success, no second part would ever be published.

[32] Grassmann had studied philology at the University of Berlin. He knew many languages including very rare dialects. He did very significant work in linguistics, for which he got immediate recognition, unlike in mathematics or physics. His translation from Sanskrit into German verse of the Hindu *Rig Veda* is still referred to today.

[33] Theory of tides.

*) Vergleiche: J.G. Grassmanns Raumlehre Theil II. pag. 194 und dessen Trigonometrie p.10.

[34] The initial incentive was provided by the consideration of negatives in geometry; I was used to regarding the displacements AB and BA as opposite magnitudes. From this, it follows that: if A,B,C are points of a straight line, then AB+BC=AC is always true, whether AB and BC are directed similarly or oppositely; that is, even if C lies between A and B. In the latter case, AB and BC are not interpreted merely as lengths, but rather their directions are simultaneously retained as well, according to which they are precisely oppositely oriented. Thus the distinction can be drawn between the sum of lengths and the sum of such displacements, in which the directions were taken into account. From this there followed the demand to establish this latter concept of a sum, not only for the case that the displacements were similarly or oppositely directed, but also for all other cases. This can most easily be accomplished if the law AB+BC=AC is imposed even when A,B,C do not lie on a single straight line. While I was pursuing the concept of the product in geometry as it had been established by my father*) [cf. J.G. Grassmann, *Raumlehre*, part II, p.194 and *Trigonometrie*, p.10 (Berlin: G. Reimer, 1824 and 1835)], I concluded that not only rectangles, but also parallelograms in general, may be regarded as products of an adjacent pair of their sides, provided one again interprets the product, not as the product of their lengths, but as that of the two displacements with their directions taken into account. (translation from Kannenberg 1995, 9)

[35] Friedrich Schleiermacher, *Dialektik*, Berlin: G. Reimer, 1839. This compilation was written by Ludwig Jonas, a student of Schleiermacher, at his request, from his notes.

[36] The position of geometry relative to the theory of forms depends on the relation of space perception to pure thought. Although just now we said that perception confronts thought as something independently given, it is not thereby asserted that space perception emerges only from the consideration of solid objects; rather, it is that fundamental perception imparted to us by the openness of our senses to the sensible world, which adheres to us as closely as body to soul. It is the same with time and with the perception of motion based on time and space, wherefore one could count the pure theory of motion (phorometry) among the mathematical sciences with as much justice as geometry. (Translation from Kannenberg 1995, 24-25).

[37] Now we characterize a method of treatment as scientific if the reader is thereby on the one hand led necessarily to recognize the individual truths, and on the other is placed in a position from which he can survey each point in the broader sweep of the development. (Translation from Kannenberg 1995, 30).

[38] Grassmann wanted his theory to be independent from the three other branches of mathematics (see introduction). Therefore, in the 1844 version of the Ausdehnungslehre, he did not use the concept of number as a primitive concept but deduced it from the division of collinear displacements. On the other hand, at the beginning of his work such a statement as « a displacement is a third of another one » is impossible. Instead, collinear displacements are said to belong to the same type of evolution. For more quantitative appreciation of collinearity, Grassmann uses the rather obscure notion of « corresponding parts of a series of displacements ».

[39] The development presented in the previous paragraph prepares us for an independent description of the higher order system. So far this has been presented as dependent on knowing the fundamental methods of evolution by which it was generated. As we will show, we can remove this dependence since the same system of m-th order is generable by any m methods of evolution belonging to it that are mutually independent (in the sense of §16), that is, that are included in no system of lower order (than the m-th). (Translation from Kannenberg 1995, 58-59).
[40] This term is not used by Grassmann; it was introduced by German algebraists around 1930, in reference to Steinitz; this point will be discussed further on.

[41] And since one can continue this process, it follows that one can always generate the same system by any m mutually independent methods of evolution [...]. (Translation from Kannenberg 1995, 60).

[42] If there were a set of $n<m$ generators, any set of m generators could be replaced by a subset of n generators, but this contradicts the assumption of independence of the generators.

[43] In some textbooks this result is known as Grassmann's formula. In the Ausdehnungslehre, Grassmann deduces it from the concept of *outer multiplication ('äuseres Produkt')*, which is, with the *regressive product ('angewandte Produkt')*, the most original of Grassmann's contributions.

* Das Zeichnen $^>_<$ zusammengesetzt aus $>$ und $<$ soll ungleich bedeuten. (*Grassmann's footnote*)

[44] The change in the position of the adjective 'algebraisch', compared to the identical proposition in Nr. 9, is somewhat troublesome. Is it a misprint, or did Grassmann really think that the two different ways of putting the sentence were equivalent ?

45

The Theory of Extension
in a complete and more rigorous form
[...] (A foreword of 8 pages, not reproduced here, foillows).
First section.
The elementary connections of extensive magnitudes
Chapter 1.
Addition, subtraction, multiples and fractions of extensive magnitudes.

1. Explanation. I say that a magnitude a is derived from the magnitudes $b, c \ldots$ by the numbers $\beta, \gamma, \ldots,$ if
$$a = \beta b + \gamma c + \ldots$$
where $\beta, \gamma, \ldots$ are real numbers, rational or irrational, different from zero or not. I also say that, a is in this case numerically derived from $b, c, \ldots$

2. Explanation. Moreover, I say that two or more extensive magnitudes $a, b, c, \ldots$ stand in a numerical relation, or that the set of extensive magnitudes $a, b, c, \ldots$ is subject to a numerical relation, if any one of them can be numerically derived from the others, thus if one can write for instance
$$a = \beta b + \gamma c + \ldots$$
where $\beta, \gamma, \ldots$ are real numbers. In case the set contains only one magnitude a, we will say that the set is subject to a numerical relation, only if $a = 0$.
[...]

3. Explanation. I call unit any magnitude, that allows to derive numerically a series of magnitudes, and I call this unit, a primitive unit, if it cannot be derived numerically from any other unit. I call the unit of numbers, that is one, the absolute unit and the other I call them relative units. Zero can never serve as a unit.

4. Explanation. I call system of units any set of magnitudes, standing in no numerical relation to one another, and that can allow to derive numerically other magnitudes by means of arbitrary numbers.
[...]

5. Explanation. I call extensive magnitude any expression that is numerically derived from a system of units (yet not reduced to the absolute unit), and I call the numbers that belong to the units the derivation numbers of this magnitude [...].

6. Explanation. To add two extensive magnitudes, derived from the same system of units, means to add their derivation numbers belonging to the same unit, that is
$$\Sigma \, \alpha e + \Sigma \, \beta e = \Sigma \, (\alpha + \beta) \, e.$$

7. Explanation. To subtract an extensive magnitude to another one derived from the same system of units means to subtract their derivation numbers belonging to the same unit, that is
$$\Sigma \, \alpha e - \Sigma \, \beta e = \Sigma \, (\alpha - \beta) \, e.$$
Remark. With respect to the meaning of parenthesis, I use the convention that a polynomial or product of several factors written with no parentheses is equivalent to the expression written with parentheses, in which all parentheses start at the beginning, thus
$$a + b + c = (a + b) + c, \quad abc = (ab)c$$
and so on.

8. *For the extensive magnitudes a, b, c, the following fundamental forms hold:*
1) $a + b = b + a,$
2) $a + (b + c) = a + b + c,$

3) $a + b - b = a$,
4) $a - b + b = a$.
Proof.
[...]
9. *For the extensive magnitudes all the laws of the algebraic addition and subtraction hold.*
Proof. Indeed these laws can, as it is known, be deduced from the four fundamental forms of Nr. 8.
10. Explanation. To multiply an extensive magnitude by a number means to multiply each of its derivation number by that number, that is
$\Sigma \, \alpha e \, . \, \beta = \beta \, . \, \Sigma \, \alpha e = \Sigma \, (\alpha \beta) e$.
11. Explanation. To divide an extensive magnitude by a number not equal to zero means to divide each of its derivation number by that number, that is
$$\Sigma \, \alpha e : \beta = \Sigma \, (\frac{\alpha}{\beta}) \, e \, .$$
12. *For the multiplication and the division of the extensive magnitudes (a, b) by the numbers (α, β) the following fundamental forms hold:*
1) $a \beta = \beta a$,
2) $a \beta \gamma = a \, (\beta \gamma)$,
3) $(a + b) \gamma = a \gamma + b \gamma$,
4) $a \, (\beta + \gamma) = a \beta + a \gamma$,
5) $a \, . \, 1 = a$,
6) $a \beta = 0$
if and only if either $a = 0$, or $\beta = 0$,
7) $a : \beta = a \frac{1}{\beta}$, *if* $\beta \begin{smallmatrix} > \\ < \end{smallmatrix} 0.$ *

proof.
[...]
13. *For the multiplication and division of extensive magnitudes by numbers all the algebraic laws of multiplication and division hold.*
Proof. For from the fundamental forms (1 to 6) of the foregoing proposition follow, in a well-known way, all the algebraic laws of multiplication, and from the form (7) of the same proposition, division is reduced, as in algebra, to multiplication. Thus the algebraic laws of division remain true for the division of extensive magnitudes by numbers.
[...]
(unpublished translation by Kannenberg, reproduced here by courtesy of the author)

[46] What Peano knew of Grassmann's work in detail is not clear. It is possible that he never read the full version of 1844, but only the summary that Grassmann had published in *Grunerts Archiv* in 1845.

[47] For the coincidence of two equal and similar systems, A, B, C, D, ... and A', B', C', D', ... in space of three dimensions, in which the points D, E, ... and D', E', ... lie on opposite sides of the planes ABC and A'B'C', it will be necessary, we must conclude from analogy, that we should be able to let one system make a half revolution in a space of four dimensions. But since such a space cannot be thought, so is also coincidence in this case impossible. (translation from Smith 1929, 526)

[48] Peano uses the symbols '=' to designate an equivalence, '<' for an implication and '∩' for the conjunction 'and'.

[49] **72.** There exist systems of entities for which the following definitions are given:
1. The equality of two entities **a** and **b** of the system is defined: It means that a proposition, written as **a** = **b**, is defined, which expresses a condition between the two entities of the system, satisfied by some pairs of entities and not satisfied by other pairs, and that satisfies the logic equations:
$(a = b) = (b = a)$, $(a = b) \cap (b = c) < (a = c)$[49]
2. The *sum* of two entities **a** + **b** is defined: It means that an entity, written **a+b**, is defined that belongs again to the given system and that satisfies the conditions:
$(a = b) < (a+c = b+c)$, $a + b = b + a$, $a + (b + c) = (a + b) + c$.
And the common value of the two sides of the last equality will be written $a + b + c$.

3. Let **a** be an entity of the system, and m a positive integer, with the expression m **a** we designate the sum of m entities equal to **a**. It is easy to see, given **a**, **b**... entities of the system, m, n, ... positive integers, that:

$(a = b) < (m\,a = m\,b)$; $m(a+b) = m\,a + m\,b$; $(m + n)a = m\,a + n\,a$; $m(n\,a) = (mn)a$; $1a = a$.

We suppose that a meaning is given to the expression m **a**, for any real number m, so that the preceding equations are still satisfied. The entity m **a** will be called the *product* of the (real) number m by the entity **a**.

4. Finally we suppose that there exists in the system an entity, that we call *zero entity*, and designate by 0, such that, for any entity **a**, the product of the number 0 by the entity **a** will always make the entity 0, or $0a = 0$.

If we give to the expression **a** - **b** the meaning of **a** + (-1)**b**, it follows that:

$$a - a = 0 \; ; \; a + 0 = a \,.$$

DEF. *The systems of entities for which the definitions 1, 2, 3, 4 are given in such a way that the given conditions are satisfied, are called* linear systems.

[50] DEF. *The number of dimensions of a linear system is the maximum number of entities independent of each other that can be found in the system.*

[51] At first, let us note that any analytical function in one variable is characterized by the values given to a generally infinite but denumerable number of parameters. It is thus possible to consider the classes of functions that contain all linear combinations of their elements, for instance; the whole set of functions being regular in the neighborhood of a given point as spaces having generally an infinite but denumerable number of dimensions. The distributive operators applicable to the functions of a same class are, therefore, a natural generalization of what 'linear transformations' are in finite-dimensional linear spaces; and this concept, especially because it is attached to a simple and meaningful notation, will induce in a synthetic mode, and with use of a constant analogy with geometry, several relations of composition, of decomposition, of classification in groups, or of transformations of these operators.

[52] Between the given elements α_1, α_2, ..., α_n, there exists r independent linear relations, and no more, $(r<n)$; the set of elements $\alpha = a_1\,\alpha_1 + a_2\,\alpha_2 + ... + a_n\,\alpha_n$ contains n-r independent elements, and no more, and is therefore $(n\text{-}r)$-dimensional.

[53] Regarding purely formal considerations, it will not be necessary to specify whether the series converges or diverges, since it will only be useful in order to group in one element the whole succession of numbers a_0, a_1, ..., a_n, ... But regarding the applications to the theory of functions, it will be necessary to take into account the fact that such a series converges.

[54] This term designates the linear operators on geometrical space.

[55] This is a reference to a previous book by the same authors : *Elementi di calcolo vettoriale con numerose applicazioni alla geometria, alla meccanica e alla fisica-matematica*, Bologna: N. Zanichelli, 1909.

[56] Vectorial homographies are included in the more general concept of linear transformations, linked to theory of determinants and matrices through the coordinates. But we have decided as a main goal to present the homographies as *absolute entities*, and not only as a stenography for coordinates. In this way, the absolute geometrical characteristic of homographies is highlighted and it is easy to see how they were, therefore, useful for applications.

Furthermore, in this book, as in the *Elements*, geometrical entities such as points and vectors - useful in geometry, physics and mechanics - will always appear as independent of their coordinates, relative to any system of reference.

[57] In reality, ordinary vector calculus is a *stenography for Cartesian coordinates* [...], but it is possible to see vector calculus in a totally different way. Our algebra differs quite substantially from ordinary algebra in the sense that it can *operate directly on geometrical and physical entities, without ever needing to use any coordinates*. Our algebra can be said to be intrinsic, or absolute, or autonomous.

[58] He does not make any reference to Peano, Pincherle or Burali-Forti and Marcolongo. In a footnote, he says that Grassmann, in his 1844 *Ausdehnungslehre*, made a systematic treatment of affine geometry in dimension more than three.

[59] Weyl introduces the dimension as an axiom : « there are n linearly independent vectors, but every $n+1$ are dependent on one another ». Nevertheless, he adds that this result could have been deduced from the theory of equations, but without saying how. Therefore, his approach, like Peano's or Burali-Forti's and Marcolongo's, remains ambiguous on the concept of dimension.

[60] To recognize the perfect mathematical harmony underlying the laws of space, we must discard the particular dimensional number $n=3$. Not only in geometry, but to a still more astonishing degree in physics, has it become more and more evident that as soon as we have succeeded in unraveling fully the natural laws which govern reality, we find them to be expressible by mathematical relations of surpassing simplicity and architectonic perfection. It seems to me to be one of the chief objects of mathematical instruction to develop the faculty of perceiving this simplicity and harmony, which we cannot fail to observe in the theoretical physics of the present day. It gives us deep satisfaction in our quest for knowledge. (translation by Henry L. Brose in (Weyl 1952, 23))

[61] if $L_1(\xi)$, $L_2(\xi)$, ..., $L_h(\xi)$ are h linearly independent linear forms the solutions ξ of the equations
$$L_1(\xi) = 0, L_2(\xi) = 0, ..., L_h(\xi) = 0$$
form an $(n-h)$-dimensional linear vector manifold. (translation in Weyl 1952, 25)

[62] In this way we should not only have arrived quite naturally at our axioms without the help of geometry by using the theory of linear equations, but we should also have reached the wider conceptions which we have linked up with them. In some ways, indeed, it would appear expedient (as is shown by the above formulation of the theorem concerning homogeneous equations) to build up the theory of linear equations upon an axiomatic basis by starting from the axioms which have here been derived from geometry. A theory developed along these lines would then hold for any domain of operations, for which these axioms are fulfilled, and not only for a « system of values in n variables ». [...]
Now, if on the one hand it is very satisfactory to be able to give a common ground in the theory of knowledge for the many varieties of statements concerning space, spatial configurations, and spatial relations which taken together, constitute geometry, it must on the other hand be emphasized that this demonstrates very clearly with what little right mathematics may claim to expose the intuitional nature of space. Geometry contains no trace of that which makes the space of intuition what it *is* in virtue of its own entirely distinctive qualities which are not shared by 'states of addition-machine' and 'gas-mixtures' and 'systems of solutions of linear equations'. it is left to metaphysics to make this 'comprehensible' or indeed to show why and in what sense it is incomprehensible. (translation in Weyl 1952, 25-26)

[63] I. The numbers in Ω reproduce themselves through addition and subtraction; i.e., the sum and the difference of any two numbers are in Ω.
II. Any product of a number in Ω by a number in A is a number of Ω.
III. There are n independent numbers in Ω, but any $n+1$ such numbers are dependent.

[64] Let R be a subfield of L. L will be said to be *finite with respect to R and of order n* - symbolized as
$$[L:R] = n$$
- if there are in L, n elements linearly independent over R, while any set of more than n elements of L are linearly dependent over R.

[65] This theorem proves that a system of less than n homogeneous linear equations with n unknowns has a non-zero solution.

[66] If one knows that every element β of L can be expressed in *at least* one way as
$$\beta = c_1\alpha_1 + ... + c_n\alpha_n$$
then for any $(n+1)$ elements $\beta_0, \beta_1, ..., \beta_n$, there are expressions like :

$$\beta_i = c_{i1}\alpha_1 + \ldots + c_{in}\alpha_n \quad (i = 0, \ldots, n)$$

and therefore, according to the theorem of §1 about homogeneous linear equations [see preceding footnote], $(n+1)$ elements d_0, d_1, ..., d_n, not all being 0, can be found in R, such that $d_0\beta_0 + d_1\beta_1 + \ldots + d_n\beta_n = 0$. The extension L is thus finite and its order is n at most.

[67] To be found on page 20 of the English translation mentioned in bibliography.

[68] For complementary information on this subject, see Hellinger and (Töplitz 1923), (Pincherle 1912), (Hadamard 1912), (Bernkopf 1966 and 1968), (Monna 1973) and (Dieudonné 1981).

[69] To be able to apply the classical method of determinants to infinite systems, it was necessary to impose some more or less restrictive conditions, and I must admit that the method required these restrictions, not the problem itself.

[70] The theorems about the solutions of n linear equations with n unknowns are traditionally deduced from the theory of determinants. However, it is possible to distinguish among them, those for which the *content* does not use anything in relation to the concept of determinant (even if their usual proof does), from those which cannot be expressed, without the help of determinants. [...]
In the following, we will enlarge, in a particular manner, the theorems of the first class, to the infinite systems of linear equations.

[71] There exists a quantity D_f which plays, with regard to the functional equation (b), the same role as the determinant plays with regard to a system of linear equations
In order to define D_f, I introduce the following notation:

$$(1) \quad f\begin{pmatrix} x_1, & x_2, & \ldots, & x_n \\ y_1, & y_2, & \ldots, & y_n \end{pmatrix} = \begin{vmatrix} f(x_1,y_1) & f(x_1,y_2) & \ldots & f(x_1,y_n) \\ f(x_2,y_1) & f(x_2,y_2) & \ldots & f(x_2,y_n) \\ \ldots & \ldots & \ldots & \ldots \\ f(x_n,y_1) & f(x_n,y_2) & \ldots & f(x_n,y_n) \end{vmatrix}$$

And I state

$$
\begin{aligned}
(2) \quad D_f &= 1 + \int_0^1 f(x,x)\,dx + \frac{1}{2!}\int_0^1\int_0^1 f\begin{pmatrix} x_1, x_2 \\ x_1, x_2 \end{pmatrix} dx_1\,dx_2 + \ldots \\
&= \sum_{n=0}^{\infty} \frac{1}{n!}\int_0^1 \ldots \int_0^1 f\begin{pmatrix} x_1, x_2, \ldots, x_n \\ x_1, x_2, \ldots, x_n \end{pmatrix} dx_1\,dx_2 \ldots dx_n.
\end{aligned}
$$

[72] By considering equation (b) as a transformation of the function $\varphi(x)$ into a new function $\psi(x)$ I can write this same equation
(7) $S_f\varphi(x) = \psi(x)$,
and I can say that the transformation S_f belongs to the function $f(x,y)$.
The transformations (7) *constitute a group.*
[73] The necessary and sufficient condition for the following equation:
$$S_f\varphi(x) = 0$$
to have a non zero solution, is that $D_f = 0$. If n is the order of the first non zero minor of D_f, the equation has exactly n linearly independent solutions.

[74] [...] the systematic construction of a general theory of integral equations for the whole of analysis, especially for the theory of definite integral and the theory of the representation of arbitrary functions in infinite series, besides for the theory of linear differential equations and analytic functions, as well as for potential theory and calculus of variations, is of the highest importance [...]
[The method] consists in starting from an algebraic problem, namely the problem of the orthogonal transformation of a quadratic form in n variables into a sum of squares, and by rigorous passage to the limit for $n = \infty$, succeed in solving the considered transcendental problem.

[75] We will say that a *functional operator* U is defined in a set E of element of *any nature* (numbers, curves, points, etc.) when to each elements A of E corresponds a given numerical value of U: U(A). The search of properties of these operators is the object of *Functional Calculus*.
Our present work is a first attempt to establish in a systematic way some fundamental principles of Functional Calculus and to apply them to some concrete examples.

[76] The idea to represent a function by its Fourier series was to become quite familiar. In this way, it was possible to represent the set of summable functions on a subset with a countable infinity of dimensions. What is this subset? Until today nobody can say.
Yet for a more specific class, the set of summable functions whose square is also summable, the solution presents less difficulty. For this class, there exists a more intimate connection between a function and its Fourier series [...] For this class of functions, it is possible to define a notion of distance that will be the basis for a geometrical theory of systems of functions, a theory which has the characteristics of synthetic geometry.

[77] **Theorem III.** Let R_0 be a complete linear subspace of R and $f_0(x)$ a linear form in R_0 with bound M. then it exists a linear form $f(x)$ in R with bound M, which coincides with f_0 in R_0.

[78] This book pursues the goal to establish some theorems, valid for various functional fields, that I will specify later. However, in order to avoid isolated proofs for each field, which would be quite painful, I have chosen a different approach; that is, I consider, in a general way, the sets of elements for which I postulate certain properties, and from which I deduce some theorems, and then I prove that the chosen postulates are true for each particular functional field.

[79] Notre critique doit être étendue dans le même sens et avec les mêmes limitations que celle qui s'adresse à un emploi abusif des coordonnées dans les questions qui relèvent de géométrie pure (c'est une note de Fréchet).

[80] The method which MM. Volterra and Hilbert employed consists in remarking that a function (continuous for instance) can be represented by a countable infinity of parameters. Therefore, the problem can be solved as if it had only a finite number of parameters, and then by a passage to the limit. We think that this method played an important role but is now outmoded. Replacing the function by an infinite sequence of numbers, which can be chosen in several manners is an unnecessary artefact (*note*: our reproach must be extended in the same sense and with the same limitations than when applying to the abusive use of coordinates in problems referring to pure geometry). This is easy to see, for instance, in the theory of integral equations, for which Fredholm's or Schmidt's solutions are much more simple and elegant than Hilbert's, even if it is greatly to his credit that a lot of new results have been established with his method.

[81] Seven years after the first axiomatic definition of normed vector spaces was given.

[82] The quotation is from a book published in 1932, but most of the results were already in the 1928 paper.

[83] See. MacLane (1981).

[84] For a historical report on the writing of this book, see (Birkhoff and MacLane 1992)

[85] Note that the notion of module (vector space on a ring) was only marginal.

[86] They are still references for today's students in some universities.

[87] The students of the *'Ecole Normale Supérieure'* were then officers and many were in front rows.

[88] This book originally published in German in 1927, was translated into English and revised in 1953. The vectors are presented in an analytic form through coordinates.

[89] While dealing with concrete problems, apparently far distant from each other, mathematicians and physicists often realize that they end up doing linear computations formally identical. One very important notion will help us to understand the reasons underlying this phenomenon: the notion of vector space.

REFERENCES - PART I

Abreviations : *Acta. Math.* (Acta Mathematica), *Ann. Maths* (Annals of Mathematics); *Arch. Hist.* (Archive for History of Exact Sciences), *Bull. SMF* (Bulletin de la Société Mathématique de France), *C.R. Acad. Sc.* (Comptes-Rendus de l'Académie des Sciences de Paris), *Fund. Math.* (Fundamenta Mathematicae), *Hist. Math.* (Historia Mathematica), *J. Crelle* (Journal für die reine und angewandte Mathematik), *Göttingen* (Nachrichten der Königlichen Gesellschafte der Wissenschaft zu Göttingen), *Math. Ann.* (Mathematische Annalen), *Mon. Phys. Math.* (Monatshefte für Mathematik und Physik), *Stud. Math.* (Studia Mathematica), *Palermo* (Rendiconti del Circolo Matematico di Palermo), *Trans. London* (Philosophical Transactions of the Royal Society of London).

Except when specified, the translations of historical texts have been made by Dorier.
Original references are given along with reedited versions, including in complete works collections (designated by C.W.).

Jean Robert Argand, Essai sur une manière de représenter les quantités imaginaires dans les constructions géométriques, *Ann. Maths* **5** (1806), 33-147; reed., J. Hoüel ed., Paris: Gauthier-Villars, 1874; Paris: Blanchard, 1971.

Michèle Artigue and André Deledicq, *Quatre étapes dans l'histoire des nombres complexes: quelques commentaires épistémologiques et didactiques,* Cahier DIDIREM , n° 15, Paris: IREM de Paris VII, 1992.

Stefan Banach, Sur les opérations linéaires dans les ensembles abstraits et leur application aux équations intégrales, *Fund. Math.* **3** (1922), 133-181, or [C.W., 2:306-348].

Stefan Banach, Sur les fonctionnelles linéaires, en deux parties, *Stud. Math.* **1** (1929), 211-216, 223-239, or [C.W., 2: 375-395].

Stefan Banach, *Théorie des opérateurs linéaires*, Warsaw: Funduszu Kultury Narodowej, 1932, or [C.W., 2:19-217].

Stefan Banach, *Œuvres*, 2 vols., Varsaw: Editions Scientifiques de Pologne, 1967/79.

Giusto Bellavitis, Sopra alcune applicazioni di un nuovo metodo di geometria analitica, *Il Poligrafo Giornale di Scienze, Lettre ed Arti. Verona* **13** (1833), 53-61.

Giusto Bellavitis, Saggio di applicazioni di un nuovo metodo di geometria analitica (Calcolo delle equipollenze), *Annali delle Scienze del Regno Lombaro-Veneto. Padova* **5** (1835), 244-259.

Giusto Bellavitis, *Spozisione del metodo della Equipollenze,* Modena, 1854, reed. *Exposition de la méthode des équipollences,* trad. C.-A. Laisant, Paris: Gauthier-Villars, 1874.

Michael Bernkopf, The development of function spaces with particular references to their origins in integral equation theory, *Arch. Hist.* **3(1)** (1966), 1-96.

Garrett Birkhoff and Saunders MacLane, *A Survey of Modern Algebra*, New-York: MacMillan, 1941; reed., 1948, 1967.

Garrett Birkhoff and Saunders MacLane, A Survey of Modern Algebra : the fifteen anniversary of its publication, *The Mathematical Intelligencer* **14** (1992), 26-31.

Nicolas Bourbaki, *Eléments de mathématique, Livre II, chap. 2: 'Algèbre linéaire'*, Paris: Hermann, 1947.

J. Budon, Sur la représentation géométrique des nombres imaginaires (Analyse de quelques mémoires parus de 1795 à 1820), *Bulletin des Sciences Mathématiques* (2e série) **57** (1933), 175-200 and 220-232.

Adrien Quentin Buée, Mémoire sur les quantités imaginaires, *Trans. London* **96** (1805), 23-88.

Cesare Burali-Forti and Roberto Marcolongo, *Omografie vettoriali con applicazioni alle derivate rispetto ad un punto e alla fisica matematica*, Turin: G. B. Pretrini di Giovani Gallazio, 1909.

Cesare Burali-Forti and Roberto Marcolongo, *Analyse vectorielle générale I, Transformations linéaires*, translated by. P. Baridon, Pavia: Mattei, 1912.

Elie Cartan, Nombres complexes (version élargie de l'article allemand de E. Study, 1898), *Encyclopedie des sciences mathématiques pures et appliquées*, tome I, vol. 1, fasc. 3, Article I-5, 1908, pp. 329-468; or [C.W., 1(2):107-247].

Elie Cartan, *Leçons sur les invariants intégraux,* Paris: Hermann, 1922.

Elie Cartan, *Œuvres Complètes,* 3 parts-6 vols., Paris: Gauthier-Villars, 1953.

Augustin-Louis Cauchy, Mémoire sur les fonctions qui ne peuvent obtenir que deux valeurs égales et de signes contraires par suite de transpositions opérées entre les variables qu'elles renferment, *Journal de l'Ecole Polytechnique* **10** (1815), or [C.W., 1(1^e série):91-169].

Augustin-Louis Cauchy, Sur les quantités géométriques et sur une méthode nouvelle pour la résolution des équations algébriques de degré quelconque, *C.R. Acad. Sc.* **29** (1849), 250-258, or [C.W., 11(1^e série):152-160].

Augustin-Louis Cauchy, *Œuvres complètes*, 2 séries, 26 vol., Paris: Gauthier-Villars, 1882-1956.

Arthur Cayley, Chapters in the Analytical Geometry of (n) Dimensions, *Cambridge Mathematical Journal* **4** (1843), 119-127, or [C.W., 1:55-62].

Arthur Cayley, Sur quelques résultats de géométrie de position, *J. Crelle* **31** (1846), 213-227, or [C.W., 1:317-328].

Arthur Cayley, A Memoir on the Theory of Matrices, *Trans. London* **148** (1858), 17-37, or [C.W., 2:475-496].

Arthur Cayley, *Collected Mathematical Papers*, 13 vols., Cambridge: University Press, 1889.

Gustave Choquet, *Certificat de calcul différentiel et intégral*, les cours de la Sorbonne, Paris: CDU, 1955-56.

Gustave Choquet, L'enseignement de la géométrie, Paris: Hermann, 1964.

Richard Courant and David Hilbert, *Methoden der mathematischen Physik,* 2 vols., Berlin: Springer, 1924; reed., 1930, 1937; *Methods of mathematical physics,* trad. Courant, New-York: Intersciences Publishers, 1953; reed. 1955.

Gabriel Cramer, *Introduction à l'analyse des Courbes algébriques*, Genève: Cramer et Philibert, 1750.

Michael J. Crowe, *A History of Vector Analysis: The Evolution of the Idea of a Vectorial System,* Notre Dame: University Press, 1967, reed., New-York: Dover, 1985.

René Descartes, *Le Discours de la méthode pour bien conduire sa raison et chercher la vérité dans les sciences*, Leyden : Jan Maire, 1637; reprint ed., *La Géométrie*, ed. M. Leclerc et J-C. Juhel, Nantes: Edition de L'AREFPPI, 1984; *The Geometry of René Descartes*, trad. D.E. Smith et M. L. Lathan, New York: Dover Publications, 1954.

Leonard Eugene Dickson, *Algebras et their arithmetics,* Chicago: University Press, 1923; reed., New-York: Dover, 1960.

Jean Dieudonné, *Algèbre linéaire et géométrie élémentaire*, Paris: Hermann, 1964.

Jean Dieudonné, L'algèbre linéaire dans les mathématiques modernes, *Bulletin de l'APMEP* **253** (1966), 315-329.

Jean Dieudonné, *History of Functional Analysis (Mathematics Study 49)*, Amsterdam: North-Holland , 1981.

Gustav Peter Lejeune Dirichlet, *Vorlesungen über Zahlentheorie*, Braunschweig: Vieweg, 1863. Réédition avec des supplméents de Richard Dedekind, 1871 (ed.1), 1879 (ed.2), 1893 (ed.4); reed. of ed.4, New-York: Chelsea Publishing Company, 1968.

Jean-Luc Dorier, *Analyse historique de l'émergence des concepts élémentaires d'algèbre linéaire,* Cahier DIDIREM, n°7, Paris: IREM de Paris VII, 1990.

Jean-Luc Dorier, L'émergence du concept de rang dans l'étude des systèmes d'équations linéaires, *Cahiers du séminaire d'histoire des mathématiques* (2e série) **3** (1993), 159-190.

Jean-Luc Dorier, A general outline of the genesis of vector space theory, *Hist. Math.* **22(3)** (1995), 227-261.

Jean-Luc Dorier, Basis and Dimension: From Grassmann to van der Waerden, in (G. Schubring, ed.) *Hermann Günther Grassmann (1809-1877): Visionary mathematician, Scientist and Neohumanist Scholar* , Dordrecht: Kluwer Academic Publisher, 1996.

Jean-Luc Dorier, L'"Ausdehnungslehre' de Grassmann : Une étape clef dans la théorisation du linéaire, in (D. Flament ed.), *Le nombre une hydre à n visages - entre nombre complexe et vecteur -,* Paris: editions de la Maison des Sciences de l'Homme, 1997.

Javier Echeveria and Marc Parmentier (eds.), *G.W. Leibniz - La caractéristique géométrique,* Paris: Librairie Philosophique J. Vrin, 1995.

Charles Ehresmann, *Cours de cinématique,* polycopié de l'université de Clermont-Ferrand , 1940-41.

Ferdinand G. M. Eisenstein, Allgemeine Untersuchungen über die Formen dritten Grades mit drei Variabeln, welche des Kreistheilung ihre Entstehung verdanken (en deux parties), *J. Crelle* **28** (1844), 289-374 et **29** (1845), 19-53.

Leonhard Euler, Sur une contradiction apparente dans la doctrine des lignes courbes, *Mémoires de l'Académie des Sciences de Berlin* **4** (1750), 219-223, or [C.W., 26:33-45].

Leonhard Euler, Problema algebricum ob affectiones prorsus singulares memorabile, *Novi Comentarii Academiae Scientiarum Petropolitanae* **15**, (1770), 75-106, or [C.W., 6:287-315].

Leonhard Euler, *Opera omnia*, 3 series - 57 vols., Lausanne: Teubner - Orell Füssli - Turicini, 1911-76.

Pierre de Fermat, *Ad locos planos et solidos isagoge*, Toulouse, 1643; reed., in *Varia opera mathematica*, 3 vols., ed. P. Tannery et C. Henri, Paris: Gauthier-Villars, 1891-1912, 85-96.

Desmond Fearnley-Sander, Hermann Grassmann and the Creation of Linear Algebra, *American Mathematical Monthly* **86** (1979), 809-817.

Desmond Fearnley-Sander, Hermann Grassmann and the Prehistory of Universal Algebra, *American Mathematical Monthly* **89** (1982), 161-166.

Ernst Fischer, Sur la convergence en moyenne, *C.R. Acad. Sc.***144** (1907), 1022-1024.

Dominique Flament, La 'lineale Ausdehnungslehre' (1844) de Hermann Günther Grassmann, in *1830-1930 : A Century of Geometry: Epistemology, History and Mathematics*, Lecture notes in Physics vol. 402, Berlin/New-York /Paris: Springer, 1992, pp. 205-221.

Dominique Flament, *Hermann Günther Grassmann, La Science de la grandeur extensive, la lineale Ausdehnungslehre*, Paris: Blanchard, 1994. Prefaced French trad. of [51].

Joseph Fourier, *Théorie analytique de la chaleur*, Paris: Firmin Didot Père et Fils, 1822; reed., Paris: J. Gabay, 1988.

Maurice Fréchet, Sur les opérations linéaires, en deux parties, *Transactions of the American Mathematical Society* **5** (1904), 493-499; **6** (1905), 134-140.

Maurice Fréchet, Sur quelques points du calcul fonctionnel, *Palermo***22** (1906), 1-72.

Maurice Fréchet, Sur les ensembles de fonctions et les opérations linéaires, *C.R. Acad. Sc.***144** (1907), 1414-1416.

Maurice Fréchet, Les espaces vectoriels abstraits, *Bulletin of the Calcutta Mathematical Society* **16(1)** (1925-26), 51-62.

Maurice Fréchet, Les espaces abstraits topologiquement affines, *Acta Math.* **47** (1926), 25-52.

Maurice Fréchet, *Les espaces abstraits et leur théorie considérée comme introduction à l'analyse générale*, Gauthiers-Villars, 1928; reed, 1951.

Ivar Fredholm, Sur une classe d'équations fonctionnelles, *Acta Math.* **27** (1903), 365-390.

Georg Ferdinand Frobenius, Über das Pfaffsche Problem, *J. Crelle* **82** (1875), 230-315, or [C.W., 1: 249-334].

Georg Ferdinand Frobenius, Note sur la théorie des formes quadratiques à un nombre quelconque de variables, *C.R. Acad. Sc.* **85** (1877), 131-133, or [C.W., 1: 340-342].

Georg Ferdinand Frobenius, Über homogene totale Differentialgleichungen, *J. Crelle* **86** (1879), -19, or [C.W., 1:434-453].

Georg Ferdinand Frobenius, Theorie der linearen Formen mit ganzen Coefficienten, *J. Crelle* **86** (1879), 146-208, or [C.W., 1: 503-565].

Georg Ferdinand Frobenius, Zur Theorie der linearen Gleichungen, *J. Crelle* **129** (1905), 175-180, or [C.W., 3 : 349-354]

Georg Ferdinand Frobenius, Über den Rang einer Matrix, *Sitzungsberichte der Königlich Preussischen Akademie der Wissenschaft zu Berlin* (1911), 20-29 et 128-129, or [C.W., 3: 479-488].

Georg Ferdinand Frobenius,*Gesammelte Abhandlungen*, 3 vols., ed. J-P. Serre, Berlin/Heidelberg/New-York: Springer, 1968.

Carl Friedrich Gauss, *Disquisitiones arithmeticae*, Leipzig, 1801, or [C.W.,1]; reed., *Recherches Arithmétiques*, trad. A.C.M. Poulet-Delisle, Paris: Courcier, 1803; Paris: Blanchard, 1979.

Carl Friedrich Gauss, Theoria residuorum biquadraticorum - Commentatio secunda, lu à Göttingen le 23 Avril, 1831, imprimé dans [C.W., 2:69-178].

Carl Friedrich Gauss, *Werke*, 12 vol., Leipzig: Teubner, 1863-1929.

Israil Gelfand , Normierte Ringe, *Matematicheskii Sbornik* **9** (1941), 3-23.

Hermann Grassmann, *Die lineale Ausdehnungslehre*, Leipzig: Otto Wigand , 1844, or [53, 1:1-139]. *Quotations in English are taken from Kannenberg's translation (Grassmann 1995).*

Hermann Grassmann, *Die Ausdehnungslehre, vollständig und in strenger Form*, Berlin: Th. Chr. Fr. Enslin, 1862, or [53, 2:1-383].

Hermann Grassmann, *Gesammelte mathematische und physikalische Werke*, 3 vols., ed. F. Engel, Leipzig: Teubner, 1894-1911; reprint ed., New-York/London: Johnson Reprint Corporation, 1972.

Hermann Grassmann, *A new branch of mathematics. The Ausdehnungslehre of 1844 and other works, trans. Lloyd C. Kannenberg,* 1995, Chicago, La Salle: Open court.

Hans Hahn, Bemerkung zu den Untersuchungen des Herrn M. Fréchet: sur quelques points de calcul fonctionnel, *Mon. Phys. Math.* **19** (1908), 247-257.

Hans Hahn, Über Folgen linearer Operationen, *Mon. Phys. Math.* **32** (1922), 1-88.

Hans Hahn, Über linearer Gleichungssysteme in linearer Räumen, *J. Crelle* **157** (1927), 214-229.

Paul, R. Halmos, *Finite Dimensional Vector Spaces,* Princeton: University Press, 1942; reed. 1947, reed., 1948, 1949, 1951, 1953, 1955.

William Rowan Hamilton, On Quaternions or a New System of Imaginaries in Algebra, *Philosophical Magazine* **25** (1844), 489-495.

William Rowan Hamilton, *Elements of Quaternions*, 2 vols., Dublin, 1866; reed., New-York: Chelsea Publishing Company, 1969.

Ernst Hellinger et Otto Töplitz, Integralgleichungen und Gleichungen mit unendlichvielen Unbekannten, in *Encyklopädie der mathematischen Wissenschaften,* Leipzig: Teubner, II-C-13, pp. 1135-1601, 1923-27; reed. as a book, New-York: Chelsea, 1953.

Ernst Hellinger, Hilberts Arbeiten über Integralgleichungen und unendliche Gleichungssystem, in *Hilberts Gesammlte Abdhandlungen,* 3 vols., Berlin, 1935; reed. New-York: Chelsea, 1965, 3:94-140.

Eduard Helly, Über Systeme linearer Gleichungen mit unendlich vielen Unbekannten, *Mon. Phys. Math.* **31** (1921), 60-91.

David Hilbert, Grundzüge einer allgemeinen Theorie der linearen Integralgleichungen, *Göttingen* (1904), 49-91.

David Hilbert, Anwendung der Theorie auf lineare Differentialgleichungen, *Göttingen* (1904), 213-259.

David Hilbert, Anwendung der Theorie auf Probleme der Funktionentheorie, *Göttingen* (1905), 307-338.

David Hilbert, Theorie der Funktionen von unendlich vielen Variabeln, *Göttingen* (1906), 157-227.

David Hilbert, Neue Begründung und Erweiterung der Theorie der Integralgleichungen, *Göttingen* (1906), 439-480.

David Hilbert, Anwendung der Theorie auf verschiedene Probleme der Analysis Geometrie und Gastheorie, *Göttingen* (1910), 355-417.

David Hilbert, *Grundzüge einer Allgemeinen Theorie der linearen Integralgleichungen,* Leipzig/Berlin: Teubner, 1912; reed. 1924; New-York: Chelsea, 1953.

Gaston Julia, *Introduction mathématique aux théories quantiques, première partie : espaces vectoriels à n dimensions. Opérateurs linéaires. Matrices,* redigé par J. Dufresnoy, Cahiers scientifiques, fasc. 16, Paris: Gauthier-Villars, 1935; red. 1949, 1955; *deuxième partie,* redigée par R. Marrot, Cahiers scientifiques, fasc. 19, Paris: Gauthier-Villars, 1936; reed. corrigée et augmentée, 1955.

Helge von Koch, Sur une application des déterminants infinis à la théorie des équations différentielles linéaires, *Acta Math.* **15** (1891), 53-63.

Helge von Koch, Sur les déterminants infinis et les équations différentielles linéaires, *Acta Math.* **16** (1892-93), 217-295.

Joseph Louis Lagrange, *Œuvres complètes,* 14 vols., Paris: Gauthier-Villars, 1867-1892.

Gottfried Wilhelm Leibniz, lettre à Christian Huyghens - Hanover ce 8 de Sept. 1679, *Christi. Hugenii aliorumque seculi XVII. virorum celebrium exercitationes mathematicae et philosophicae,* ed. Uylenbroek, Hagen: Hagae comitum, 1833, 2:6-12 (this is the reference that one can find in [Grassmann C.W., 1:415-420] in which most of the letter and the whole essay are reproduced) or [C.W. 2: 17-25].

Gottfried Wilhelm Leibniz, *Leibnizens Mathematische Schriften,* ed. C. I. Gerhardt, 2 vols., Berlin: Julius Pressner, 1850; reed., *Œuvres Mathématiques de Leibniz,* Paris: Librairie de A. Frank Editeur, 1853.

Albert C. Lewis, H. Grassmann's 1844 Ausdehnungslehre and Schleiermacher's Dialektik, *Annals of Science* **34** (1977), 103-162.

Albert C. Lewis, Justus Grassmann's School Programs as Mathematical Antecedents of Hermann Grassmann's 1844 Ausdehnungslehre, in *Epistemological and Social Problems of the Sciences in the Early Nineteenth Century,* ed. H. Jahnke et M. Otte, Dordrecht: D. Reidel Publishing Company, 1981, pp. 255-267.

André Lichnérovicz, *Algèbre et analyse linéaires,* Paris: Masson, 1947.

Saunders Mac Lane, History of abstract algebra: origin, rise and declaine of a movment, in *American mathematical heritage : algebra and applied mathematics,* Texas Tech Uninversity, Mathematical Series n°13, 1981, pp. 3-35.

August Ferdinand Möbius, *Der Barycentrische Calcul,* Leipzig: Johan Ambrosius Barth, 1827, or [C.W., 1:1-388].

August Ferdinand Möbius, *Die Elemente der Mechanik des Himmels,* Leipzig: Weidemannsche Buchhet lung, 1843, or [C.W., 4:1-318].

August Ferdinand Möbius, *Gesammelte Werke,* ed. R. Baltzer, 4 vols., Leipzig: S. Hirtzel KG, 1915; reed., Wiesbaden: Dr. Martin Sändig oHG, 1967.

A. F. Monna, *Functional analysis in historical perspective,* Utrecht: Oosthoeck Publishing Company, 1973.

Gregory H. Moore, The axiomatization of linear algebra, *Historia Mathematica* **22(3)** (1995), 262-303.

C. V. Mourey, *La vraie théorie des quantités négatives et prétendues imaginaires,* Paris, 1828; reed., Paris: Mallet-Bachelier, 1861.

Thomas Muir, *The Theory of Determinants in the Historical Order of Development*, 4 vols., London: MacMillan, 1890-1923; reed., New-York: Dover Publications, 1960.

John von Neumann, Mathematische Begründung der Quantenmechanik, *Göttingen* (1927), 1-57.

John von Neumann, *Mathematische Grundlagen der Quantenmechanik*, Berlin: Springer, 1932; reed. New-York: Dover, 1943; *Les fondements mathématiques de la mécanique quantique,* trad. A. Proca, Paris: librairie Alcan, 1946; reed. Paris: Gabay 1988; trad. ang. Robert T. Beyer, Princeton: University Press, 1955.

Lubos Novy, *Origins of Modern Algebra*, trad. J. Tauer, Leyden: The Netherlands Noordohoff International Pub., 1973.

Michael Otte, The ideas of Grassmann in the context of the mathematical and philosophical tradition since Leibniz, *Hist. Math.* **16** (1989), 1-35.

Giuseppe Peano, *Calcolo geometrico secundo l'Ausdehnungslehre di H. Grassmann e precedutto dalle operazioni della logica deduttiva*, Torino: Fratelli Bocca Editori, 1888.

Giuseppe Peano, Intégration par séries des équations différentielles linéaires, *Math. Ann.* **32** (1888), 450-456.

Salvatore Pincherle and Ugo Amaldi, *Le operazioni distributive*, Bologne: Zanichelli, 1901.

Henri Poincaré, Sur les déterminants d'ordre infini, *Bull. SMF* **14** (1886), 77-90.

René de Possel, *Cours de calcul différentiel et intégral*, polycopié de l'université de Clermont-Ferrand, 1940-41.

Helena M. Pycior, Benjamin Peirce's Linear Associative Algebra, *Isis* **70** (1979), 537-551.

Helena M. Pycior, George Peacock and the British Origins of Symbolical Algebra, *Hist. Math.* **8** (1981), 23-45.

André and Germaine Revuz, *Le cours de l'APMEP, II. Espaces vectoriels*, Paris: APMEP, 1963.

Joan L. Richards, The Art and the Science of British Algebra: A Study in the Perception of Mathematical Truth, *Hist. Math.* **7** (1980), 343-365.

Fredéric Riesz, Sur les systèmes orthogonaux de fonctions, *C.R. Acad. Sc.* **144** (1907a), 615-619; or [C.W. 1: 378-381].

Fredéric Riesz, Sur les systèmes orthogonaux de fonctions et l'équation de Fredholm, *C.R. Acad. Sc.* **144** (1907b), 734-736; or [C.W. 1: 382-385].

Fredéric Riesz, Sur une espèce de géométrie analytique des systèmes de fonctions sommables, *C.R. Acad. Sc.* **144** (1907c), 1409-1411; or [C.W. 1: 386-388].

Fredéric Riesz, Über orthogonale Funktionensystem, *Göttingen* (1907d), 116-122; or [C.W. 1: 389-395].

Fredéric Riesz, Untersuchungen über Systeme integrierbarer Funktionen, *Math. Ann.* **69** (1910), 449-497; or [C.W. 1: 441-489].

Fredéric Riesz, *Les systèmes d'équations linéaires en une infinité d'inconnues*, Paris: Gauthier-Villars, 1913, or [C.W., 2:829-1016].

Fredéric Riesz, Über lineare Funktionalgleichungen (trad. d'un texte hongrois de 1916), *Acta Math.* **41** (1918), 71-98; or [C.W. 2: 1053-1080].

Fredéric Riesz, *Œuvres Complètes*, 2 vols., Paris : Gauthier-Villars, 1960.

Erhard Schmidt, Zur theorie der linearer und nichtlinearer Integralgleichungen, 2 parts, *Math. Ann.* **63** (1907), 433-467; **64** (1907), 161-174.

Erhard Schmidt, Über die Auflösung linearer Gleichungen mit unendlich vielen Unbekannten, *Palermo*20 (1908), 53-77.

Otto Schreier and Emanuel Sperner, 'Einführung in der analytische Geometrie und Algebra', *Hamburger Mathematische Einzelschriften,* in two parts, **10** (1931) and **19** (1935); English Trans., *Modern Algebra and Matrix Theory*, trans. Martin Davis and Melvin Hausner , New-York: Chelsea Publishing Company, 1951.

Gert Schubring (ed.), *Hermann Günther Grassmann (1809-1877): Visionary mathematician, Scientist and Neohumanist Scholar* , Dordrecht: Kluwer Academic Publisher, 1996.

James Byrnie Shaw, *Synopsis of linear associative algebra, a report on its natural development and results reached up to the present time*, Washington: Carnegie institution, 1907.

David Eugene Smith, *A source book in mathematics,* New-York: McGraw Hill Book Co., 1929, reed. New-York: Dover 1959. *(Pages are given according to the Dover edition)*

Henry John Stanley Smith, On Systems of Linear Indeterminates, Equations, and Congruences, *Philosophical Transactions of the Royal Society of London* **151** (1861), 293-326, or in *Collected Mathematical Papers*, reprint ed., New-York: Chelsea Publishing Company, 1965, 1:367-409.

Ernst Steinitz, Algebraische Theorie der Körper, *J. Crelle* **137** (1910), 167-309; reprint ed., ed. H. Hasse et R. Baer, Berlin/Leipzig: De Gruyter, 1930.

Marshall Harvey Stone, *Linear transformations in Hilbert space and their applications to analysis,* New-York: American Mathematical Society Colloquium Publications, vol. XV, 1932; reed., 1958.

Otto Töplitz, Über die Auflösung unendlichvieler linearer Gleichungen mit unendlichvielen Unbekannten, *Palermo*28 (1909), 88-96.

Bartel. L. van der Waerden, *Moderne Algebra*, 2 vols., Berlin: Springer, 1930-31.

Bartel L. van den Waerden, On the sources of my book Moderne Algebra, *Hist. Math.* **2** (1975), 31-40.

John Wallis, *Opera mathematica*, 3 vols., Oxford: Oxianae, 1693.

John Warren, *A Treatise on the Geometrical Representation of the Square Roots of Negative Quantities*, Cambridge, 1828; reed. *Trans. London* **119** (1829), 241-254.

André Weil, *Souvenir d'apprentissage*, Bâle/Boston/Berlin: Birkhäuser, 1991.

Caspar Wessel, *Om Directionens Analytiske Betegning,* (lu à l'Academie Royale du Danemark en 1797), Copenhague, 1798; reed. *Nye Samling af det Kongelige Danske Videnskabernes Selskabs Skrifter* (2) **5** (1799), 496-518; *Essai sur la représentaion analytique de la direction*, trad. H.G. Zeuthen, Copenhagen/Paris: H. Valentiner et T. N. Thiele Editeurs, 1897.

Hermann Weyl, *Raum-Zeit-Materie*, Berlin: Springer, 1918; reed., *Space-Time-Matter*, trad. L. Brose, New York: Dover, 1952.

Norbert Wiener, On the theory of sets of points in terms of continuous transformations, *Compte-rendu du congrès des mathématiciens*, Strasbourg: 1920, 312-315; or [C.W., 1: 281-284].

Norbert Wiener, The group of linear continuum, *Proceedings of the London Mathematical Society* **20** (1922), 329-346; or [C.W., 1: 285-302]

Norbert Wiener, Limit in terms of continuous transformation, *Bull. SMF* **50** (1922), 119-134; or [C.W., 1: 303-318].

Nobert Wiener, Note on a paper of M. Banach, *Fund. Math.* **4** (1923), 136-143; or [C.W., 3: 676-683].

Norbert Wiener, *I am a mathematician. The later life of a prodigy*, New-York: Doubleday, 1956.

Norbert Wiener, *Collected works with commentaries*, 4 vol., ed. P. Masani, Cambridge (MA) /London: The MIT Press, 1976.

PART II

TEACHING
AND
LEARNING
ISSUES

JEAN-LUC DORIER, ALINE ROBERT,
JACQUELINE ROBINET AND MARC ROGALSKI

PART II - CHAPTER 1
THE OBSTACLE OF FORMALISM
IN LINEAR ALGEBRA

A Variety of Studies From 1987 Until 1995

1. INTRODUCTION

In the first four chapters of this second part, we will study the teaching of linear algebra as it is introduced in first year university studies in France, when students are generally 18 to 20 years of age.

In France, the teaching of linear algebra was entirely remodeled with the 'reform of modern mathematics' in the sixties. At that time, the influence of Bourbaki and a few others led to the idea - which was based on a very democratic concern - that geometry could be more easily accessible to students if it were founded on the axioms of the structure of affine spaces. Therefore the axiomatic theory of finite-dimensional vector spaces was taught in the first year of secondary school (age 15). The fate of this reform and the reaction it aroused are well known. Therefore, from the beginning of the eighties, the reform of the teaching of mathematics in French secondary schools gradually led to the removal of any subject related to modern algebra. Moreover, the teaching of geometry focused on the study of transformations on elementary figures and analytic geometry is now barely taught at secondary level. On the other hand, formal theories became unpopular and students entering university nowadays have very little practice of any formal mathematical subject. This situation has created a total change in the background of students, for whom the teaching of linear algebra represents the first contact with such a 'modern' approach. Because of this, the teaching at first year university level changed and became less theoretical. In many universities[1], it was decided to prepare the students for the teaching of linear algebra by a preparatory course in Cartesian geometry or/and by a course in logic and set theory. Yet, in secondary school, students still learn the bases of vector geometry and the solving of systems of linear equations by Gaussian elimination. Therefore, they have some knowledge on which the teaching of linear algebra can be based. For the moment, at the universities, the idea of teaching students the elementary axiomatic theory of vector spaces within the first

DORIER J.-L. (ed.), The Teaching of Linear Algebra in Question, 85—124.
©2000 *Kluwer Academic Publishers. Printed in the Netherlands.*

two years of scientific studies has not be seriously questioned, and the teaching of linear algebra, in France, remains quite formal.

In this context, Robert and Robinet (1989) showed that the main criticisms made by students toward linear algebra concern the use of formalism, the overwhelming amount of new definitions and the lack of connection with what they already know in mathematics. It is quite clear that many students have the feeling of having landed on a new planet and are not able to find their way in this new world. On the other hand, teachers usually complain about their students' erratic use of the basic tools of logic or set theory. They also complain that students have no skills in elementary Cartesian geometry and consequently cannot use intuition to built geometrical representations of the basic concepts of the theory of vector spaces. These complaints correspond to a certain reality, but the few attempts at remediation - with previous teaching in Cartesian geometry and/or logic and set theory - did not seem to improve the situation substantially. Indeed, many works have shown that difficulties with logic or set theory cannot be interpreted without taking into account the specific context in which these tools are used. Dorier (1990) tested, with statistical tools, the correlation between the difficulties with the use of the formal definition of linear independence and the difficulties with the use of the mathematical implication, in different contexts. Although these two types of difficulties seemed at first closely connected, the results showed clearly that no systematic correlation could be made. This means that students' difficulties with the formal aspect of the theory of vector space are not just a general problem with formalism, but mostly a difficulty of understanding the specific use of formalism in the theory of vector spaces, and the interpretation of the formal concepts in relation with more intuitive contexts like geometry or systems of linear equations in which they historically emerged. We will analyze this point in more detail here.

We have developed a research program on the learning and teaching of linear algebra in first year of science university. This work, which started in the late 80s, includes not only the elaboration and evaluation of experimental teaching, but also an epistemological reflection, built upon a dialectic process on the historical analysis of the genesis of the concepts of linear algebra (as presented in the first part of this book) and on a didactical analysis of the teaching and the difficulties of the students.

In the chapter 1, we show, through various diagnostic studies conducted between 1987 and 1994, that students' difficulties in linear algebra reveal a single massive obstacle appearing throughout all successive generations and for nearly all modes of teaching: what we naively call *the formalism obstacle*. Students recognize it themselves (as indicated in a survey conducted in 1987). Teachers are also aware of it, but the research we present in this chapter specifies the nature of the difficulties encountered. Thus, we exemplify not only the commonplace difficulties due to formal manipulations, but we also clarify how the lack of prior knowledge in logic and elementary set theory contribute to the production of errors in linear algebra itself.

In the first part, Dorier has amply characterized the nature of the concepts, explaining that linear algebra is the final result of the vast undertaking of formalization; thereby unifying and making economical and analogous solutions possible to all the problems stemming from linearity. This does not, however, remove the necessity of mastering a certain degree of formalism in order to gain access to these solutions: for instance, one must work on equations while momentarily setting aside what they represent yet know when to return to them; one must replace explicit transformations by tables of numbers; play on different bases without explicit representations; reason either on vectors or on their coordinates without confusing them, etc. But this also requires, in various problems, to identify a structure and to judge what making the structure explicit brings as positive but also negative effects. In particular, it requires the ability to establish equalities of sets, to distinguish between objects, subobjects, homomorphisms and the subsets that are attached to them, etc.

It is through this maze that we will lead the reader, extracting as much as possible from student productions in a wide variety of contexts.

Yet, in a certain way, all mathematics is characterized by a certain degree of formalism. Two interdependent particularities make this teaching task especially delicate in linear algebra, in our opinion. The generalizing and unifying nature of this theory, analyzed from a historical and epistemological point of view in the first part of this book, makes it difficult to introduce to students with particular well-chosen problems. For example, problems have yet to be found where certain notions from the beginning of linear algebra would come into play implicitly; by their very nature, they have not yet become part of the student's toolbox. Consequently, passing from the elementary algebra and geometry knowledge learned in secondary school to linear algebra is particularly problematic.

Thus we introduce, in chapter 2, the notion of 'level of conceptualization' to help define steps in the acquisition of mathematical notions. Under this notion, the teacher's task is to facilitate the student's passage from one level of conceptualization to the following one, as set out in the official syllabus. We maintain, therefore, that depending on the notions to be taught and the official syllabuses, the pedagogical means available to the teacher differ, with, in the case of linear algebra, the added difficulty mentioned above of the epistemological specificity of the theory. Yet a rapid study of the organization in current secondary education of this passage to linear algebra, reveals, in fact, how far removed institutions are from coping with the difficulties, inherent in the learning of linear algebra, even if certain notions are addressed implicitly. The problems remain to be resolved at the first year university level.

In this context, chapter 3 will present a teaching project for linear algebra in first year university studies, taking into account the syllabuses and students' prior knowledge, from the diagnosis of the first problems and our hypotheses on the specificity of the theory. One of the didactical tools recommended is recourse to the meta level of teaching: a reflection on mathematics and learning mathematics, within well-adapted and complex scenarios.

In chapter 4, we will return more specifically to this meta lever, though it is not reserved solely for linear algebra. In particular, we will show certain of its effects through three examples, and we will also clarify specific difficulties that we have encountered in our current research, particularly in terms of evaluation of these effects. This will be followed by our future perspectives.

We will conclude with a global view on our research and the problems encountered.

2. FIRST DIAGNOSIS, A 1987 STUDY

In October and November 1987, a study was conducted to determine students' knowledge and ideas on linear algebra after their first year of university studies during which they had studied the following notions: vector spaces and subspaces, linear transformations, systems of linear equations, linear forms, matrices, and determinants. To accomplish this, 379 second-year students completed a questionnaire before beginning any new instruction in linear algebra. Their responses provided a description of their knowledge and ideas on linear algebra (Robert et Robinet 1989).

2.1. The Questionnaire

NAME:
SOME QUESTIONS ABOUT LINEAR ALGEBRA:

1. True or false (justify your answer): v and u being two linear transformations of a vector space over itself :
$$v \circ u = 0 \Leftrightarrow Im(u) = Ker(v).$$

2. Give some examples of vector spaces you know?

3. Let A_1, A_2, A_3 be three points in the plane with coordinates (a_1,b_1), (a_2,b_2), (a_3,b_3), find the coordinates of the points M, N and P, if they exist, such that A_1 is the midpoint of [MN], A_2, of [NP], A_3, of [PM]. The coordinates of M, N and P will be denoted respectively (x_1,y_1), (x_2,y_2) and (x_3,y_3).

4. Find, if they exist, the solutions of the following system:

$$\begin{cases} x_1+x_2=a_1 \\ x_2+x_3=a_2 \\ x_3+x_4=a_3 \\ x_1+x_4=a_4 \end{cases}$$

5. Let E be the vector space of all polynomials, with real coefficients and a degree not more than 2, and f the linear mapping over E such that any P of E has an image Q = f(P) such that :
$Q(X) = (2X+1) P(X) - (X^2 -1) P'(X)$
Give the matrix of f in the basis $(1, X, X^2)$.

6. What would you say to a student entering first year of university in order to describe what linear algebra consists of?

7. What are, in your opinion, the main difficulties in the learning of linear algebra?

2.2. *Rapid Analysis of the Contents of the Questionnaire*

Question 1 is very formal: general linear transformations and the inclusion and equality between sets are perhaps still problematic for students who have only just begun to encounter this language. In addition, they must master the notions of linear transformation, image, and kernel. This question tests the students' detailed comprehension of the corresponding notions of linear algebra, and it must not be expected that students construct many counter-examples, even though the question was purposely left vague enough to allow for invention. Nevertheless, they could quite easily respond correctly by referring to the fact that an implication has no a priori reason to be an equivalence.

On the other hand, question 3 is very practical: after writing an equation easily within the reach of a high-school junior (and therefore should not have been an obstacle for a second-year university student), students had only to solve a system of linear equations of Cramer's type with three equations and three unknowns. This skill is required of science-oriented high-school seniors and good results can be expected from students having completed a year of linear algebra at the university level.

Question 4 focuses on knowledge typically acquired during the first year of university studies. It requires solving a 4x4 system of linear equations of rank 3 with solutions if and only if $(a_1-a_2) = (a_4-a_3)$. This exercise demands, therefore, having mastered the solution algorithms with proper understanding of their meaning.

Question 5 also tests the notion of linear transformation, not in a formal manner as in question 1, but rather numerically since a basis of the vector space is given.

Questions 6 and 7 were designed to shed light on how students represent linear algebra. Question 6 was formulated in such a way as to rule out all the mathematical definitions that students could have provided if they had only been asked to explain what linear algebra was.

Question 2 provides more precise information on students' referents on vector spaces; those cited should tell us about their knowledge (are the examples cited truly a vector space structure?) and on their representations (are the vector spaces necessarily finite-dimensional ?).

2.3. *Relativization of the Results Obtained*

This questionnaire was given to students in three sections of second-year university science programs (mathematics and physics) in three different universities (Paris 7, Lille, Paris 6), under very different conditions: students at Paris 7 filled out the questionnaire anonymously during the first week of classes while the students at Lille and Paris 6 filled it out before any teaching of linear algebra, but several weeks after the beginning of classes, and they signed their questionnaires. Students at Paris 7 responded, therefore, after two months of vacation (not necessarily spent studying) with no review of mathematical notions. In addition, since they did not sign their questionnaires, they were undoubtedly not especially motivated to undertake time-consuming calculations

2.4. Analysis of the Questionnaire

2.4.1.   Question 1

The responses were classified into 5 categories:
A: correct responses with correct explanations; B: correct responses without explanation or incorrect explanations; C: incorrect responses and/or nonsensical responses; D: incorrect responses because the inclusion had been assimilated to equality; E: no response.

	A	B	C	D	E	Total
Paris 7	11%	7%	23%	16%	43%	100%
	(16)	(11)	(33)	(24)	(62)	(146)
Paris 6	6%	14%	42%	6%	32%	100%
	(3)	(7)	(21)	(3)	(16)	(50)
Lille	24%	4%	32%	26%	14%	100%
	(43)	(8)	(59)	(48)	(25)	(183)
Total	16%	7%	30%	20%	27%	100%
	(62)	(26)	(113)	(75)	(103)	(379)

The correct responses (A+B), with or without explanation, came from only one-third of the students who responded; similarly, slightly more than one-third of those who responded confused inclusion and equality. Here is an example:

If $v(u(x)) = 0 \Rightarrow v(y) = 0$, *then* $y \in \text{Ker}(v)$ *thus* $u(x) \in \text{Ker}(v)$ *but* $u(x) = y$, *therefore* $\exists\, x \in E \mid u(x) = y$ *thus* $y \in \text{Im}(u)$, *finally* $v \circ u = 0 \Rightarrow \text{Im}(u) = \text{Ker}(v)$.

The correct conclusion was $\text{Ker}(v) \subset \text{Im}(u)$; the whimsical use of formalized language is noteworthy.

The other incorrect responses show both the lack of appropriation of the notions in question and the more or less inadequate mastery of set theory language; furthermore, these two deficiencies feed off each other, as the two examples below demonstrate:

False, because $v \circ u = 0 \Rightarrow \text{Im}(u) \in \text{Ker}(v)$ *and not* $\text{Im}(u) = \text{Ker}(v)$.

Here, there is either confusion between the symbols, or the student translates $v \circ u = 0$ as « the set of the u(x) belongs to Ker v », therefore, in a condensed form, $\text{Im}(u) \in \text{Ker}(v)$ (the set Im(u) is treated as an element).

v *and* u *being 2 linear transformations,* $v \circ u = 0$ *by definition so V vector space*
$\text{Ker } v = \{ \vec{e} \in V, v(\vec{e}) = 0 \}$
$\text{Im } u = \{ \vec{e} \in V, \exists\, \vec{u} \in V \text{ such that: } u(\vec{u}) = \vec{e} \}$
$u(e_1) = e_2 \Rightarrow v(u(e_1)) = v(e_2)$
$v(e_2) = 0 \qquad v(u(e_1)) = 0$
$v \circ u\, (e_1) = 0$
$\text{Im}(u) = u(e_1) = e_2 \text{ and } \text{Ker}(v) = v(e_2) = 0 \Rightarrow \text{Im } u = \text{Ker } v.$

We have reproduced what the student wrote on paper, faithfully respecting the presentation.

It is clear in this example that Im(u) is treated as an element: the student calculates v(Im(u)) to find 0; she should conclude that Im(u) ∈ Ker(v). This shows why the obstacle of formalism is evoked: there is confusion between elements and sets because of an analogy in the form of the expressions.

> *False, there is a counter-example:* $v = 2x + 1$ *and* $u = 3x + 2$
> $Ker(v) \Rightarrow x = -1/2$ *and* $Im(u) \Rightarrow y = 3x + 2$.

Here there is great confusion between vector, linear transformation (more affine), and vector subspaces treated as elements.

2.4.2. Question 2

Secondary school teaching yields a certain familiarity with $\mathbb{R}$, $\mathbb{R}^2$, and $\mathbb{R}^3$ vector spaces through the teaching of geometry (affine and analytic). We would expect to find this frequently cited.

	$\mathbb{R}...\mathbb{R}^3$	$\mathbb{R}^n$	$\mathbb{C}$	Func-tions	Polyno-mials	Matrices	Non-sense
Paris7	52% (76)	24% (35)	24% (35)	12% (18)	27% (39)	5% (7)	11% (16)
Paris 6	32% (16)	34% (17)	26% (13)		28% (14)		4% (2)
Lille	41% (75)	39% (71)	14% (26)	22% (40)	51% (93)	14% (25)	4% (7)
Total	44% (167)	32% (123)	20% (74)	15% (58)	39% (146)	8% (32)	7% (25)

Polynomial spaces are also quite frequently cited, which is not very surprising because they are the principal source of examples and exercises in first year university linear algebra.

Vector spaces of infinite dimension (continuous functions on a real interval, series, etc.) are seldom cited; an explanation may be that vector spaces are given as examples and do not appear (or appear very infrequently) in the usual exercises given in the first year. Curiously, matrix spaces, which appear in many exercises in the second year, are seldom cited. This reflects the caution exercised by teachers who are reluctant to introduce the matrices associated with systems and linear transformations and matrices as vectors in the same year.

2.4.3. Question 3

For this question, responses showing modeling problems were expected but in fact there were very few. Either the question was not answered at all or the modeling was correct. The answers could have been sorted according to the procedures used (Gaussian method or elimination) but the methods used were rarely purely one or the other: the Gaussian and elimination methods were most often intertwined.

	Correct answers	False answers	No answer
Paris 7	55% (80)	39%(57)	6% (9)
Paris 6	62% (31)	24% (12)	14% (7)
Lille	81% (149)	12% (22)	7% (12)
Total	69% (260)	24% (91)	7% (28)

Nearly 70% of the students responded correctly and this result would undoubtedly have been better if the papers from Paris 7 students had been signed.

2.4.4. *Question 4*

	Correct answers	False answers	No answer
Paris 7	11% (16)	68%(99)	21% (31)
Paris 6	18% (9)	70% (35)	12% (6)
Lille	39% (71)	49% (90)	12% (22)
Total	25% (96)	59%(224)	16% (59)

The number of incorrect responses was even higher than appears here, which led to questioning the teachers. Certain replying that, for them, 'solving a system' meant « finding the conditions of existence of solutions and the numbers of such independent conditions. » We therefore included in the correct responses those that gave the condition and 'no solution', if this answer was not filled out, and 'an infinity of solutions', if it was filled out. Different solution methods were found: prior calculation of the determinant, the Gaussian method, the process of elimination, a mixture of these two methods, representing the set of solutions with parametric as well as Cartesian equations, etc. Examining the papers showed that the method used was correlated to the teaching in the first year. Very often the method they initially started with did not lead anywhere. The following are a few typical examples.

- The student writes : *The system is equivalent to*:
$$\begin{cases} a_1-a_2+a_3-a_4=0 \\ x_1+x_2=a_1 \\ x_3+x_4=a_3 \end{cases}$$

and does not know what to do next.
- The student writes : *The system implies:* $a_1-a_2+a_3-a_4=0$
and does not know what to do with this result.

- The student writes: *The system implies:*
$$\begin{cases} x_4=a_4-x_1 \\ x_3=a_3-a_4+x_1 \\ x_2=a_2-a_3+a_4-x_1 \end{cases}$$

and does not know what to do with this result.

Furthermore, many calculation errors are often associated with methodological errors.

All this explains that, even allowing for a loose interpretation of correct responses, only one-fourth of the students succeeded in this type of exercise.

2.4.5. *Question 5*

It should be remembered that asking students to write the matrix of a given linear transformation in a given basis is a classic first year exercise.

Firstly, the correct responses were classified together, then two groups were formed of erroneous answers (i) those that displayed a calculation error, and (ii) those that showed incomprehension, an example of latter being:

$$\begin{pmatrix} 3A+C & 0 & 0 \\ 0 & A+2B+C & 0 \\ 0 & 0 & B+C \end{pmatrix}.$$

It is sometimes difficult to classify the answers, such as in the following matrix:

$$\begin{pmatrix} 3 & 1 & 0 \\ 0 & 1 & 2 \\ 0 & 1 & 1 \end{pmatrix}$$

which seems to be the consequence of a pardonable calculation error, whereas the detailed calculation in certain student papers showed that the column 3, 0, 0 was obtained by evaluating $Q(1) = (2x1+1)x1 + 0 = 3$ (and it is remarkable that the two other columns of the matrix were correct and obtained by evaluating $f(x)$ et $f(x^2)$).

	correct	false	Non-sense	No answer
Paris 7	40% (59)	22% (32)	13% (19)	25% (36)
Paris 6	24% (12)	10% (5)	22% (11)	44% (22)
Lille	40% (73)	11% (20)	27% (49)	22% (41)
Total	38%(144)	15% (57)	21% (79)	26% (99)

In conclusion, we can say that half of the students understood how the matrix of a transformation was calculated. However, nearly one-third of them did not know how to calculate it and, among these students, a large number had not yet assimilated the fact that the coefficients of the matrix are scalars rather than elements of the vector space.

2.4.6. *Question 6*

The vast majority of the responses described mathematical contents; only 13% of the responses for Paris 7, 8% for Paris 6, and 4% for Lille involved statements on linear algebra. In particular, students never mentioned the capacity of linear algebra to study a variety of problems in geometry. It would seem that many of the concepts in linear algebra remain in a concept-object state; the only time when students recognized a tool-like character in it was for solving systems of linear equations. It can be inferred, although this remains to be demonstrated, that the teaching normally done in the first year is very axiomatic and that the utility of linear algebra is not at all emphasized, to such an extent that students do not manage to bring these notions to mind.

2.4.7. Question 7

The learning difficulty mentioned by a large number of students in all three groups was 'abstraction' (perhaps what we call the formal character of algebra). Here are a few examples of student responses.

- *The fact that it is abstract, no intuition is possible, no verification is possible.*

- *It is the sometimes abstract aspect of algebra that bothers me the most, in other words, when reasoning dominates calculation.*

- *This field is very theoretical and we don't see what it all means.*

Other responses emphasized the difficulty linked to a particular notion.

- *The great difficulty is based on vector spaces. I can't seem to grasp them fully.*

- *Undoubtedly understanding sets that are called 'vector spaces' and what they represent.*

A fairly large number of responses primarily involved new definitions and new theorems to understand and learn.

- *The multitude of properties and theorems to learn.*

- *The number of definitions.*

- *The difficulties come from having to know how to regroup what is to be learnt according to the type of problem, which is not easy, given the large number and the variety of results to know.*

Other responses brought out the problems linked to calculations, for example:

- *The difficulties? Calculating quickly with no errors.*

- *Calculation errors are frequent because matrix calculation is generally rather daunting.*

- *The calculations can be long and difficult.*

The last class of responses alluded to the difficulties that could perhaps be attributed to proofs.

For example:

- *Most difficult are the problems requiring a proof.*

- *Reasoning and proofs or rigor and precision or logic and quantifiers.*

Among those students who expressed themselves on this question, there were approximately 40% who found that the discipline was very abstract and that they found it difficult to make use of these notions. Generally speaking it seems that linear algebra does not organize previously acquired knowledge (at least partial knowledge). Students' knowledge is perhaps too limited for them to succeed in abstracting, from examples that they know, the structure of vector space. In particular, their knowledge of analytic geometry never seems to be reinvested. Do they know it poorly or not at all?

Nearly one-third of the students (who answered the question) gave answers which expressed difficulties due to language, formalism, and the abundance of new definitions and theorems. This is undoubtedly amplified by the simultaneity of the introduction of set theory language, the use of quantifiers, and a large number of new definitions and theorems.

2.4.8. Conclusion

In the linear algebra skills that were tested, we have shown that fewer than one-fourth of the students know how to manipulate the notions of image and kernel of linear transformations and that fewer than half the students know how to solve a system of linear equations (4x4) where the numerical calculation is simple. Nearly one-third of the students do not know how to calculate the matrix of a linear transformation, when the vector space is different to $\mathbb{R}$, $\mathbb{R}^2$ or $\mathbb{R}^3$, even though it is isomorphic to one of those.

Fewer than one-third of the students cite $\mathbb{R}^n$ as an example of a vector space, yet Harel's work (1989a and b and 1990) shows that this is a necessary step in the construction of the notion of vector space; there is consequently work to be done in the acquisition of this notion.

For a majority of the students, linear algebra is no more than a catalogue of very abstract notions that they represent with great difficulty. In addition, they are submerged under an avalanche of new words, new symbols, new definitions, and new theorems.

3. A SECOND SURVEY MADE IN 1990 : IS FORMALISM A DIDACTICAL OBSTACLE RELATED TO THE LACK OF PREREQUISITES ?

3.1.   Introduction

The research presented here is based on the analysis of results of the tests given throughout the year, in the field of linear algebra, to students in their first year at a French university.

Our main goals were :
a)   To better determine what constitutes the teaching of linear algebra; especially by analyzing the types of tasks proposed to students, within the questions given by the teachers in the tests.
b)   For a standard section of first year science students with a fairly standard teaching of linear algebra, we wanted to determine the methods, procedures and mistakes of students in relation to the tasks proposed to them, relatively to their individual previous abilities in basic logic and algebra notions.
c)   Being then able to formulate a diagnosis on the different effects of this teaching, we may propose some hypotheses for its possible change.

3.2.   The Methodology and the Hypotheses

We analyzed copies of eight different tests.

We first took eighty-four copies from a pretest on basic notions in logic and algebra. This had been given to students in their first weeks at university, before any specific teaching in these fields. The evaluation of this test gave us the individual level of acquisition for what we thought may be prerequisites for linear algebra.

For the analysis itself, we used a methodology introduced by Boschet and Robert (1985) in their work on the acquisition of real analysis notions in first year at a science university.

Among the questions on the test, we categorized four main types : *quantification* (QA), *implication and equivalence* (EQ), *numerical algebra* (AN) and *algebraic structure* (AS). For the first two types we distinguished three different levels in the tasks induced by the questions. The first one was a purely formal setting, but seen from the outside, since the students were asked to say whether a proposition expressed in formalized language was true or false (QA1 and EQ1). The second one was formal as well, but the task this time was internal, since it asked for the negation of a formalized proposition. Only the QA-type of questions appeared at this level (QA2). The last level corresponded to an interplay between the formal setting and another setting, according to the meaning introduced by Douady (1986). The questions, this time, consisted in requiring a translation of a proposition from a formalized language into an every-day or a graphic formulation, or vice-versa (QA3 and EQ3).

So we obtained seven different types of questions which we could associate with seven different bodies, or 'blocks', of knowledge.

The hypothesis we made, and which is induced by Piaget's work, is, in brief, that the acquisition of new knowledge is usually made possible by the disequilibration of old knowledge followed by its reorganization through complex cognitive mechanisms of assimilation/accomodation. The necessary disequilibration is not usually part of the explicit teaching, and the process described above is then, of course, unconscious. Douady however showed that (at least for primary school pupils) didactical situations in which a notion may be seen in at least two different settings, and in which the pupils have different levels of ability, is suitable to start this dialectical process.

After Boschet and Robert (1985), we think that former knowledge, efficiently applicable in different settings, may be a better guarantee for the acquisition of a new notion. More precisely, we may raise such questions as, Will a student who is very good at formal logic (EQ1, QA1 and QA2), but not very good at dealing with the interaction between formal and natural languages, learn linear algebra less well than a student who is globally of the same level in logic, but who has more homogeneous abilities? For every prerequisite (block of knowledge), is there a minimal threshold of acquisition beyond which the probability of success is much higher ?

To be able to answer these questions, we have defined three different states for every block: full (2), half-full (1), empty (0), according to a mark given to the questions related to it. We have also considered the parameter B, giving the number of empty blocks which measure the number of gaps in previous abilities.

We then obtained nine different variables (the global mark of the test, the seven blocks, and the number of empty blocks) from which to evaluate the level of acquisition of basic logic and algebra notions for each student. A statistical study of the results of the tested population led us to build a new block Q, with the QA_i, to summarize the level of different abilities in questions dealing with quantification.

We finally kept only the following nine variables which we considered to be the most significant: the global mark N, the number of empty blocks B and seven variables (being 1 or 0) for the blocks : EQ1(2), EQ3(2), EQ3(0), Q(2), Q(0), AN(2), AS(2). This seemed to be, with minimal loss, the best way to keep information compact enough and suitable for our further purpose.

Nevertheless, in addition to the specific methodology developed here, some restrictions about the test itself have to be considered to make this evaluation more valid, which is of course only a partial way of considering the contents as well as the level of acquisition of prerequisites in   basic logic and algebra. Indeed, if the questions about logic seemed to be suitable, although necessarily incomplete, the ones about algebra appeared to be less satisfying; i.e., numerical questions were a bit too imprecise to give a good evaluation, and the ones about structure were too 'cultural' to give a real idea of the level of acquisition. By the latter we mean that asking someone to give an example of a group is not enough to evaluate his knowledge about groups.

The seven other tests were : four 'ordinary' two-weekly tests, the mid-term and the final exam, and a special mid-course true/false-test. Except for the last one, all these tests included questions on real analysis subjects.

For each of these tests, we made an a priori analysis which included an explanation of the tasks induced by the questions and the different procedures that could possibly be developed by students. We then gave the statistical results, with marks given to every question and codes to identify special procedures, which we gathered in a table, whose arrays represent the students. We also obtained a global mark for each test. We divided every sample into three categories, according to these marks, from which we managed to balance the distribution numerically.

We analyzed, in this order, thirty-nine papers from the first ordinary test (T1), seventy-four from the mid-term exam (E1), forty-six from T2, fifty-eight from the true/false test (TF), fifty-eight from T3, fifty from T4 and seventy-three from the final exam (E2). The difference in numbers of papers analyzed for each test can be explained by the facts that (a), apart from the mid-term and final exams none of these tests were compulsory and (b), that we were only able to photocopy the tests during the short time that they were in the hands of the correctors. Unfortunately, in the end, we got only nineteen complete sets of papers of the eight tests. Each paper analyzed corresponded to a student whose pretest we had already analyzed so that the students analyzed at each test formed a sample of the main population analyzed for the pretest.

For each test, we made a short analysis of the new data obtained from the pretest with the new sample. We compared the mean-value of the marks, their standard deviation, and the percentages of students having EQ1(2), EQ3(2), EQ3(0), Q(2), Q(O), AN(2), AS(2), B=0, B≤1; B≤2, with the equivalent data for the whole population. In each case, we noticed only little variations for which there was always rather obvious explanations. The samples imposed on us under these material circumstances were deemed representative enough of the whole population to validate our general conclusions

Finally, we analyzed, more precisely, the results of all the tests (including the pretest) for the nineteen students whose eight papers we had. We made several factor analyses of some of the different characters defined on the sample, although the small number of students did not allow us to make a real statistical analysis. Nevertheless, we got quite a lot of information on every student, which would not have been possible with too many students. Moreover, we took the results as they appeared in a real teaching situation, with all its complexity. This kind of analysis for linear algebra has not been made before, as far as we know, in France. So we claim that our work was a necessary first step before carrying out a statistical analysis involving many more students. To be able to look at the correlations between the different components of knowledge in linear algebra over a large statistical population (a few hundred), we have to be more familiar with the contents of the teaching, the different tasks and procedures involved in linear algebra, and we must be able to draw some hypotheses that will help us build tests accordingly. We hope that the kind of analysis we propose meets these aims.

3.3.   The Results

3.3.1.   Global Analysis of the Contents of the Teaching

In most French universities, first year students in science classes follow a two hour a week course over one semester, which represents more or less a fourth of their mathematical course for the full year. The course usually starts with the axiomatic definition of a vector space and finishes with the results about diagonalization of matrices. This is, of course, an average estimation. In fact, since linear algebra has completely disappeared from secondary teaching, even for geometry, a new tendency consists (in the first year of university) of teaching a bit less abstract linear algebra and a bit more linear algebra for geometry.

The abstract part of this teaching is usually feared by students because of its esoteric nature, and by teachers because of the bare obviousness of most of the reasoning, which leaves them without arguments faced with their students' incomprehension.

On the other hand, our historical study (see the first part of this book) has confirmed our belief that linear algebra is a simplifying and unifying concept. For this reason, there is no problem (except a few, far too complicated for students) for which linear algebra is an absolute necessity. Besides, even if linear notions give a more elaborated or a more general answer to a problem, it is often too subtle for students to realize, because they already have enough problems in using concepts, which they are not familiar with, to be able to have a critical look at their work.

This nature, quite specific to linear algebra, leads to a dichotomous attitude in teaching, which is reflected in two different kinds of problems.

The problems of the first kind present applications of linear notions to questions about polynomials, functions or series, etc. They include interplay between different settings and change of point of view. Most of them are, at the same time, real problems and good illustrations of the simplification and generalization offered by finding a solution through linear algebra but only to someone, however, who has, first of all, no difficulty in using linear algebra notions and, secondly, is quite familiar with the subject involved. For instance, most of the problems of interpolation with polynomials have very elegant solutions using the theory of vector spaces. These solutions can be extended to more general situations, but one needs to do quite a lot of calculations to see the simplification given. Besides, in those problems, one usually needs to obtain a lot of results before being able to reach the stage of applying linear algebra. So, if such problems are given to students, one may have to deal with the following two difficulties :

1. The use of linear algebra will be only an effect of the didactical contract, since it is not absolutely necessary in order to solve the problem and the students cannot appreciate the simplification it provides. Students will follow the process of resolution induced by the questions even if they see a solution not using linear algebra.

2. The first questions necessary to approach the linear questions may need so many abilities in different fields that only a few students will manage to answer the questions dealing with the notions of linear algebra. The evaluation of the result of such problems is then more on these questions than on linear algebra.

In the second kind of problems, linear concepts are used in a formal setting without interplay with any other setting. Those might be either very formal and difficult questions about subtle notions (such as supplementary spaces) or, on the contrary, mechanical use of algorithms (such as the search of eigenvalues and eigenvectors of a matrix). In the first case they give useful results, although very hard to obtain; in the second case, they are only training for calculation and easily evaluated contents for tests! These problems do not use 'real' vector space, but very general ones, mostly $\mathbb{R}^n$.

In our analysis, tests T1, E1 and T2, are of the first kind.

E1 is a typical example. The goal of the problem was to obtain Gregory's formula, which gives a polynomial in terms of the values of the P(n+1) - P(n) (n=0 to deg(P)). There is a very attractive solution through the study of the operator $D : P \rightarrow Q$ s.t. $Q(X)=P(X+1)-P(X)$. Yet it is quite long and difficult. It is really only useful then for theoretical reasons or if you need to calculate quite a lot of polynomials. In fact, most of the students didn't succeed in proving all the steps leading to the formula, mostly through lack of technical ability in algebraic calculations. But when they were asked, in the last question, to find the polynomials of degree less than three, whose values in 0, 1, 2 and 3 were given, and although Gregory's formula had been given, they used a direct method and solved a system of four linear equations with the four coefficients of the polynomial as unknowns!

In T1, the question was to find the polynomials of degree less than four whose values in 0 and 1, as well as the ones of the derivatives, were given. The solution induced by the test, was to first find the polynomials whose all four known values were 0, and then to deduce the general solution by addition of any solution, for instance, the one of the third degree. Of course, the first question is obvious, since such polynomials can be divided both by X^2 and $(X-1)^2$, but most of the students did not realize that, and again solved a system of four equations with five unknowns! As they had to solve another system to find the solution to the third degree, they ended with more calculations, plus a theoretical proof, than if they had directly solved the system of four linear equations given by the conditions.

In T2, the questions preparing the linear solution were so technical (they used polynomials with two variables) that hardly any students had a chance to answer any question about linear algebra.

It is clear that there is a real difficulty here. We think that such problems should be introduced by explicit use of the 'meta lever'[2] and that the 'technical' points they raise in the field of algebraic calculation or logical reasoning should not be underestimated.

TF, T3, T4 and E2 are of the second kind. T3, T4 and E2 are mostly applications of numerical algorithms about the search for eigenvalues and eigenvectors, diagonalization or reduction to a triangle form of matrices. But in T3 and E2, we also find some very theoretical questions. The true-false test is typically about abstract notions, although nearly all of them refer to $|R^3$. It would be too long here to develop all the results to this quite specific test; in brief it shows that formal questions about basic notions of linear algebra such as linear independence, generating subsets, supplementary, etc. bring to light some sharp misunderstandings on the part of the students.

Generally, one of the most obvious difficulties for students in all tasks about linear algebra is to be able to keep control of what they are doing. This arises from the confusion between variables and parameters in the resolution of linear systems and leads to one of the best illustrations of it. For example, in E2, students were asked to find an orthogonal basis of eigenvectors. After they found three eigenvectors of two different eigenvalues, they proved, giving all the details, that they were independent, without shortening the proof to the independence of the only two of same eigenvalue, and then that they were orthogonal.

3.3.2. The Main Statistical Results

The factorial analysis on the seven series of marks (excluding those of the prestest) for the nineteen students reveals two sets of tests : T1, E2 and T4 on one side and T2, TF and T3 on the other side; E2 being just between those two groups.

This is a different distribution to the above. These two groups separate the calculating tasks from the more conceptual tasks. Indeed, in the first group of tests there are quite a lot of resolutions of linear systems, asked explicitly (T1 and E2), or appearing as the suitable ways in which to solve questions (determinations of polynomials, eigenvalues or eigenvectors). T4 consists mainly in the use of algorithms for calculations with matrices. T1 and E2 also use algebraic calculation notions for polynomials or integrals. On the other hand, T2 and T3, and mostly TF, deal with more conceptual problems. The final exam seems to be quite a well-balanced compromise of the two.

This separation is given by the second factorial axis of the analysis; the first one separates the globally successful students from the ones who failed. The distribution of students on the first factorial plan is quite harmoniously spread out, which seems to suggest that both numerical and conceptual abilities are useful, but independent, in order to succeed in linear algebra. For instance, it shows that students can globally succeed in linear algebra without having a good conceptual basis. For example, they can find the triangle form of a matrix without having a good knowledge of the concept of supplementary subspace, although it is a basic notion for the theory of matrices' reduction. This points out a contradiction in the teaching of linear algebra. The choice made in most French universities' curricula to teach linear algebra from the definition of a vector space to the diagonalization of matrices all in one year induces a suppression of the teaching of basic concepts in favor of the benefits of more easily taught and evaluated notions such as the reduction of matrices. This is, of course, the effect of the difficulties and the failure encountered in the teaching of abstract notions.

3.3.3. Correlations With the Pretest

The correlations with the pretest are globally quite strong. The main correlation appears with the number of empty blocks. This confirms our hypothesis about the existence of a minimal threshold in the acquisition of previous abilities beyond which chances of success in linear algebra are greater. The AN and AS blocks are not very correlated and, among the blocks related to logic, Q is the most correlated of all. These results seem to show that a certain level of previous abilities in basic logic, mainly abilities in the use of quantification, is required to reach a minimal success in linear algebra.

But some results of the more detailed correlations are a bit surprising.

For instance, in the true/false test, there were the three following propositions:

1. Let U, V and W be three vectors in $\mathbb{R}^3$: *If they are two by two non-collinear, are they independent ?*

2.1. Let U, V and W be three vectors in $\mathbb{R}^3$, *and f a linear operator in* $\mathbb{R}^3$: *If U, V and W are independent, are f(U), f(V) and f(W) independent?*

2.2. Let U, V and W be three vectors in $\mathbb{R}^3$, *and f a linear operator in* $\mathbb{R}^3$: *If f(U), f(V) and f(W) are independent, are U, V and W independent?*

These questions were generally incorrectly answered. In the three cases, students used the formal definition of linear independence and tried different combinations with the hypotheses and the conclusions, leading to apparently erratic proofs that teachers usually reject without further comment.

For instance, to the first question, a majority of students answered 'yes', giving proofs that are examples of the difficulty they have in treating linear (in)dependence globally. Indeed, many students are inclined to treat the question of linear (in)dependence by successive approximations starting with two vectors, and then introducing the others one by one. We will say that they have a local approach to a global question. Certainly, in many cases, at least if it is well controlled, this approach may be correct and actually quite efficient, yet it is a source of mistakes in several situations. The students have built themselves what Vergnaud (1990) calls *théorèmes-en-acte* (i.e., rules of action or theorems which are valid in some restricted situations but create mistakes when abusively generalized to more general cases). Here is a non-exhaustive list of *théorèmes-en-acte* connected with the local approach of linear (in)dependence that we have noticed in students' activities (see also Ousmann 1996):

- If U and V are independent of W, then U, V and W are globally independent
- If U_1 is not a linear combination of $U_2, U_3,..., U_k$, then $U_1, U_2,..., U_k$ are independent
- If U_1, V_1 and V_2 are independent and if U_2, V_1 and V_2 are independent, then U_1, U_2, V_1 and V_2 are independent.[3]

Our historical analysis confirms the fact that there is an epistemological difficulty in treating the concept of linear (in)dependence as a global property (remember the distinction made by Euler with three equations). It follows that special care must be taken in the teaching regarding this point. For instance, the exercise above can be discussed with the students. Moreover, for the teachers, knowing the type of *théorèmes-en-acte* that students may have built would help them to understand the students' mistakes and therefore to correct them more efficiently.

To questions 2.1 and 2.2, many students answered respectively 'yes' and 'no', despite coming close to writing the correct proof for the correct answers. Here is a reconstructed proof that reflects the difficulties of the students:

If $\alpha U + \beta V + \gamma W = 0$ *then* $f(\alpha U + \beta V + \gamma W) = 0$
so f being a linear operator: $\alpha f(U) + \beta f(V) + \gamma f(W) = 0$,
now as U, V and W are independent, then $\alpha = \beta = \gamma = 0$,
so f(U), f(V) and f(W) are independent.

In our first analysis, we concluded that this type of answer revealed a bad use of the mathematical implication, characterized by the confusion between hypothesis and conclusion. But the results of our statistical analysis showed that the correlation was insignificant; in some cases it was even negative. This shows that if a certain level of ability in logic is necessary to understand the formalism of the theory of vector spaces, general knowledge, rather than specific competence, is needed. Furthermore, if some difficulties in linear algebra are due to formalism, they are specific to linear algebra and have to be overcome essentially in this context.

On the other hand, some teachers may argue that, in general, students have many difficulties with proof and rigor. Several experiments that we have made with students show that if they have connected the formal concepts with more intuitive conceptions, they are in fact able to build very rigorous proofs. In the case of the preceding exercise, for instance, after the test, if you ask the students to illustrate the result with a specific example, let us say in geometry, they usually realize very quickly that there is something wrong. It does not mean that they are able to correct their wrong statement, but they know it is not correct. Therefore, one main issue in the teaching of linear algebra is to give our students better ways of connecting the formal objects of the theory with their previous conceptions in order to have a better intuitively-based learning. This implies not only giving examples, but also to show how all these examples are connected and what is the role of the formal concepts with regard to the mathematical activity involved.

For instance, in his work, Ousman (1996) gave a test to students in their final year of *lycée* (just before entering university). Through this test, he wanted to analyze the conception of students on dependence in the context of linear equations, and in geometry, before the teaching of the theory of vector spaces. He gave several examples of systems of linear equations and asked the students whether the equations were independent or not. Of course he noticed mistakes due to a local approach but the answers showed also that the students justified their answers by solving the system. Therefore, they very rarely gave a justification in terms of linear combinations but rather, most of time, in terms of equations vanishing or unknowns remaining undetermined. Their conception of (in)dependence was, like Euler's, that of inclusive dependence and not linear dependence. Yet this is not surprising, since these students, like Euler and the mathematicians of his time, were only concerned with the solving of the system, therefore inclusive dependence was more natural and more relevant for them.

However, the formal concept is the only means by which one can comprehend all the different types of 'vectors' in the same manner with regard to their linearity. In other words, students must be aware of the unifying and generalizing nature of the formal concept. In our research, we used what we called the *meta lever*. Therefore we built teaching situations leading students to reflect on the epistemological nature of the concepts, with explicit reference to their previous knowledge (see chapter IV of this book). In this approach, the historical analysis is a source of inspiration as well as a means of control. Nevertheless, these activities must not merely involve a lecture by the teacher or a reconstruction of the historical development, they must reconstruct an epistemologically controlled genesis, taking into account the specific constraints of the teaching context.

For instance, with regard to linear (in)dependence, French students entering university normally have a good practice of Gaussian elimination for solving systems of linear equations. It is therefore possible in the beginning of the teaching of linear algebra to make them reflect on this technique not only as a tool, but also as a means of investigation of the properties of the systems of linear equations. This does not conform to the historical development, since the study of linear equations was, historically, mostly held within the theory of determinants. Yet, Gaussian elimination is a much less technical tool and a better way for showing the connection between inclusive dependence and linear dependence, since identical equations (in the case when the equations are dependent) are obtained by successive linear combinations of the initial equations. Moreover, this is a context in which such a question as « What is the relation between the size of the set of solutions of a homogeneous system and the number of relations of dependence between the equations? » can be investigated with the students as a first intuitive approach for the concept of rank. This will be discussed in Rogalski's presentation of the experimental teaching in chapter 3 of this book.

4. THE LAST SURVEYS FROM 1991 TO 1994: THE OBSTACLE STILL HAS NOT BEEN OVERCOME[4]

4.1.   The Present State in an Experimental Teaching Method from Responses on Four Tests.

In 1992 and 1993, we gathered the tests of the students in a first-year university class, an experimental section at the University of Lille 1 (Rogalski 1991) and the third chapter of the present book). These were written exams given, either at the beginning of linear algebra instruction (February), or at the end of instruction (April or June). Examination of these tests highlighted the more or less persistent errors of the students as regards the rank of vector systems, while at the same time revealing a satisfactory understanding of vector subspaces.

A first test was analyzed for the years 1992 and 1993, although students had only studied the Gaussian method for systems of linear equations, analytic geometry in $\mathbb{R}^3$, the rank of a set of vectors, and the double definition of the subspaces of $\mathbb{R}^n$ (equations and parametric representation). In 1993, the teacher modified the instruction given to students in accordance with the results of the study from the preceding year. In addition, for 1993, the instruction included questioning on the same subjects at the end of the learning period. The notions at play comprised: the rank of a set of vectors, the representation of subspaces by generators, and parametric or implicit equations.

We also analyzed two tests on the notion of subspace and, more particularly, on the problem of the intersection of subspaces. One is a graded test dating from February 1992; the other took place the following year at the end of the learning period, in April 1993.

In both cases, we looked at students' success but focused particularly on the procedures students put into practice; even though two successive groups of students took the test, thus limiting the scope of the comparisons between the two groups.

4.2. Ranking

- In February 1992, the question that interested us was:
Let $X = (1, 0, 0, -1)$; $Y = (-2, -1, 1, 0)$; $Z = (1, 1, -1, 1)$; $T = (0, 0, 1, 0)$; $V_a = (a, a-1, 1, a-2)$ *and* $W = (1, -1, 1, 2)$ *be vectors in* $\mathbb{R}^4$.
What is the rank of the following sets of vectors:

(X,Y),   (X, Y, Z),   (X, Y, Z, T),   (X, Y, Z, V_a),   (X,Y, Z, V_1)?

- In February 1993, the question that interested us was:

Let: a = (0, -1, 1, 0) ;   b = (2, 1, 1,0) ;   c = (0, 0, 3, 1) ;   d = (2, 0, -1, -1);
e = (1, 0, 1, 1) *and* f = (1, 0, 0, 1) *be vectors in* $\mathbb{R}^4$.

What is the rank of the set of vectors {a, b, c, f}?

For the 1992 test, the major point of interest was determining the rank of (X, Y, Z, T) only, a question comparable to that given in 1993. Students' procedures in 1992 were the following:

1) One notes that rank(X, Y, Z, T) = rank(X, Y, T) (for example, using the result just found: rank (X, Y, Z) = 2).

1-1) There are three vectors, two of which are linearly independent (X and Y), therefore, to find the rank (3 or 2), it is only necessary to verify if the vectors are linearly independent, which requires writing the system a X + b Y + c T = 0 (1).

1-1-1) It is found that a=b=c=0 and therefore the vectors are linearly independent (procedure coded LI, for linear independence).

1-1-2) By applying the Gaussian method to the rows, the system of equations (1) is found to be equivalent to a system of three linearly independent equations and the conclusion is drawn (wrongly because the students had not yet acquired the knowledge necessary to explain this result) that the rank is 3 (coded GR, for Gaussian method on Rows).

1-2) The following method is used (coded 10):

$$\begin{cases} X = (1,\ 0, 0, -1) \\ Y = (-2, -1, 1,\ 0) \\ T = (0,\ 0, 1,\ 0) \end{cases} \text{ has the same rank as: } \begin{cases} X & (1, 0,\ 0, -1) \\ -2X - Y & (0, 1, -1,\ 2) \\ T & (0, 0, 1,\ 0) \end{cases}$$

It was also noted that many students did not write the linear combinations that they used to come to the zero triangle, and that the result was given in a somewhat magic fashion by the number of non-zero real numbers found on the diagonal.

2) The vectors are examined for linear independence by writing the system (3) a X + b Y + c Z +d T = 0. Knowing already, from the preceding question, that X, Y, Z are dependent, it is clear that X, Y, Z, T are also dependent. However, since the solutions for (3) have the form a = b = c and d = 0, this should lead to the conclusion that X, Y, T are independent and therefore that rank(X, Y, Z, T) = 3. But this procedure was not one used by the students. All those who attempted this procedure showed, by using the Gaussian method, that system (3) was equivalent to a system of three linearly independent equations and concluded (wrongly because the students had not yet acquired the knowledge necessary to explain this result) that the rank was 3 (coded GR).

3) The existence of a, b, c such that T = a X + b Y + c Z (4) is sought and since it can be shown that this system has no solution (some students did not take Z into account), it is enough to draw the conclusion (coded LC, for Linear Combinations).

4) The method called '0s and 1s' is used on 4x4 matrix (coded 10). One must transform the system so as to obtain 0 and 1, in the matrix, placed on different lines.

Certain procedures (GR) were strongly linked to a confusion between systems of equations and sets of vectors as in the following example:

X	1	0	0	-1
Y+ 2 X	0	-1	1	-2
X + Z + Y	0	0	0	0

The student wrote « solving by the Gaussian method gives one row of 0 in the last equation of the system so it can be said that the rank of vectors {X, Y, Z} is 2 ».

In 1993, the same procedures were found as in the preceding year, except that three students declared that they were going to carry out a Gaussian method on the coordinates of vectors written in columns, but then applied the method on the rows. This appears to be a sign of the confusion between systems of equations and sets of vectors.

The additional error encountered in 1993 was an error in logic:

« a, b, c, are independent iff there exist $\lambda_1, \lambda_2, \lambda_3, \lambda_4$, such that:

$\lambda_1 a + \lambda_2 b + \lambda_3 c + \lambda_4 f = 0$. »

(X,Y,Z,T)	10 (correct)	GR (false)	GR (correct)	LI (correct)	tot. pop.
Populations	30	44	20	28	113

{a, b, c, f}	10 (correct)	GR (false)	GR (correct)	LI (correct)	tot. pop.
Populations	6	7	71	27	113

The improvement of the scores from one year to the next should be noted, as well as the move toward a method: the vector coordinates are written in columns, the Gaussian method is applied to rows, and a conclusion is drawn. However, only 44% of the 1992 students and 35% of the 1993 students justify this method by solving a system of equations. It seems to be becoming more automatic over the years. The improvement in scores cannot be entirely credited to the teaching techniques because, on the one hand, the students were not the same and, on the other hand, vectors a, b, c, f were linearly independent, making interpretation of the triangulated table much easier.

4.3. Vector Subspaces

In all that follows the vector subspace generated by X, Y... is noted lin(X,Y...).

4.3.1. Questions on the Equality or Inclusion of Subspaces
- In February 1992, this question followed the first question analyzed above :
Prove, without further calculation, if the following statements are true or false
1) $Z \in$ lin {X, Z} ;
2) $Z \in$ lin {X, Y, V_a} ;
3) $T \in$ lin {X, Y, Z} ;
4) lin {X, Y} = lin{Z,V_0}
5) lin {X, Y, T, V_1} = $\mathbb{R}^4$.

- In February 1993, this question followed the one about rank:
Determine the following sets:
a) lin{a, b, c} ∩ lin{a, e}
b) lin{a, e} + lin{a, c, e}] ∩ lin{a, b, c}.

We will now compare question 4) from the February 1992 test and question a) from the February 1993 test.
For lin {X, Y} = lin {Z, V_0}:
If the student noticed that Z = - X - Y and V_0 = X - Z, the inclusion is easily obtained: lin {X, Y} ⊃ lin{Z, V_0}. To prove the equality, either the other inclusion must be shown; lin{Z, V_0} ⊃ lin {X, Y} (for example, by using the fact that Y = - 2 Z -V_0 and X = Z + V_0, or that rank (Z, V_0) = rank(X, Y) and then using the fact that if a subspace is included in another and has the same dimension, one is equal to the other; which would be the same as showing that Z and V_0 are not proportional).
Some students confused inclusion and equality (coded Inc); others did not know how to use the rank to prove an equality.

lin{X,Y}=lin{Z,V_0}	Inc	Correct answer	No answer	Total
Populations	35 (32%)[5]	40 (37%)	5	113

More than half of the students used inclusion and one out of three proved the equality.
For question a) in 1993, the procedures used were the following:
- By finding that the rank of {a, b, c, e} is 4, it is deduced that e does not belong to lin{a, b, c}. Then, with more or less correct reasoning a final conclusion is drawn. (coded p).
- The fact that a vector belongs to the intersection is expressed as: xa+yb+zc = x'a+y'e, which leads to x=x', y=0, z=0 and y'=0, which can allow to conclude that the intersection is included in lin{a}(coded q), however the students sometimes concluded to the equality. (coded q').
- The theorem on the dimension of the intersection is used, but this use was always incorrect (coded s).

lin{a,b,c}∩lin{a, e}	Proc p	Proc q,q'	Proc s	Corr. ans.	Total
Populations	31	21	16	43 (48%)[6]	113

The frequency of the procedures cannot be compared because the 'no calculation' restriction, in 1992, blocked the q procedures. On the other hand, we can compare the rates of success by noting that, in 1992, the students were more confused over the question of inclusion-equality These questions combine two difficulties: the one involving set theory; the other involving the properties of subspaces - both poorly mastered at the beginning stages of learning linear algebra.

4.3.2. Questions on the Representation of Subspaces
In February 1992, the question was:

Let Fλ be the subspace of the solutions in $\mathbb{R}^4$ of the following system:

$$\begin{cases} x + 2\,y - z + t = 0 \\ \qquad y + z \quad = 0 \\ x + \lambda y + \lambda z + t = 0 \end{cases}$$

Give some parametric equations representing Fλ. Find a basis for:
F$\lambda \cap$ lin{X,W}.

For the first part, the use of the Gaussian method was unanimous among those who completed the problem. There were a few calculation errors and $\lambda = -2$ or $\lambda \neq -2$ were at times forgotten, but these errors were not considered serious. The interpretation of the result, on the other hand, was at times not easy and somehow subjective; for example:

For $\lambda \neq -2$, x + t = 0, it is a plane x = -a and t = a
and F$\lambda \cap$ lin{X, W} = lin {(a, -a), (a, 2a)}.

The steps of the solution were correct, and so apparently was the final answer, but the student completely veered off course and gave vectors of $\mathbb{R}^2$ to generate a subspace of $\mathbb{R}^4$. This shows that it is easy to obtain parametric equations by using the Gaussian method, but that it does not prevent a loss of meaning.

For the second part, students most often proceeded by simultaneously using the two parametric representations (the problem encouraged this procedure) and did not know what to do with the systems thus obtained (especially when they gave the same letters for the parameters in the representation of the two subspaces). None of the students successfully completed the study of the intersection with this procedure (coded parapara).

Example 1
F$_{-2}$: (3 b - a, -b, b, a)
a' X + b' W = (3 b - a, -b, b, a) $\Rightarrow$ b' = b et a' = 2 b - a.
Thus (2 b - a) X + b W = F$_{-2}$ thus F$_{-2} \in$ lin{X, W},
but <(1, 2, - 1, 1) ; (0 ,1, 1, 0)> is a basis of F$_{-2}$
thus <(1, 2,- 1, 1) ; (0 ,1, 1, 0)> is a basis of F$_{-2} \cap$ lin{X, W}.

It may be that the student did not know how to interpret b' = b and a' = 2 b - a (the quantifiers are missing and we see the difficulty as one related to insufficient mastery of logic and set theory) and this student slipped up since, instead of giving <(3, -1, 1, 0) , (-1, 0, 0, 1)> as a basis of F$_{-2}$, which was expected given his parametric representation, he gave (1, 2, -1,1) and (0 ,1, 1, 0) which were the coefficients of the two equations defining F$_{-2}$.

In class, teachers recommended taking the parametric equations for a subspace and the Cartesian equations for the other, with the result that certain students attempted to look for the Cartesian equations of lin{X, W}. None of them succeeded in writing them correctly. The most frequent procedure (coded P) consisted in giving the following equations: x - t = 0 and x -2y + z + 2t = 0, which were obtained by taking the coordinates of X and W as coefficients of the unknowns.

Example 2

$F_\lambda \cap \lin\{X, W\}$, *one must solve the following system:*

$$\begin{cases} x + 2\,y - z + t = 0 \\ y + z = 0 \\ (\lambda + 2)\,z = 0 \\ 2\,x - y + z + 2\,t = 0 \\ x - t = 0 \end{cases}$$

The student wrote:

1	2	-1	1	0	V_1
0	1	1	0	0	V_2
0	0	$\lambda+2$	0	0	$V_3 - V_1 + V_2$
1	0	0	-1	0	W
1	-1	1	2	0	X

This student used the Gaussian method correctly on this system and concluded that:

$F_\lambda \cap \lin\{X, W\} = \{V_1, V_2, X - V_1\}$.

The mix-up after double parametric representation and the P procedure were the most frequent errors, but other errors arose which were mostly related to a poorly understood procedure for handling double parameters.

$F_1 \cap \lin\{X, W\}$	Parapara-correct	P	Other mistakes	No answer	%mistakes/ nb answers
Populations	20	13(14%)[7]	61	17	77%

In February 93, the question was :

Give a parametric representation of $\lin\{a, b, c, d\}$, *and a system of linear equations that defines this subspace.*

Even though this question was not identical to the question in the 1992 exercise, it is comparable because certain students used a P-type procedure. They wrote:

a	0	-1	1	0
b	2	1	1	0
c	0	0	3	1
d	2	0	-1	-1

whose equations are:

$$\begin{cases} -y + z = 0 \\ 2\,x + y + z = 0 \\ 3\,z + t = 0 \\ 2\,x - z - t = 0 \end{cases}$$

It is then possible to extract the parametric representation: $y = -x$, $z = -x$, $t = 3x$. There are variants to this procedure: first, transforming the initial table by triangulating it by the Gaussian method and then taking the numbers from the table thus obtained as coefficients of the equations. The same methods are used but with the vector coordinates written in columns.

We also found x a + y b + z c = 0 or d as an equation (coded T), or even a + b - c - d = 0 which is a relation of dependence between the vectors, giving a = k, b = 1, c = m, d =k+l-m as parametric equations (coded V). In addition, we found: "v belongs to lin{a,b,c,d}, if and only if, v = k (a + b - c - d)" (coded V').

The correct procedures stem from the following reasoning: v belongs to lin{a,b,c,d}, if and only if, v = ka + lb +mc or v = ka + lb + mc+ nd, but although the parametric equations were given, the implicit equations sometimes were not (coded C).

lin{a,b,c,d}	P	T	V / V'	C	No answer	Tot. Pop.
Populations	39	16	18	20	31	113

It should be noted that the 'deviant' P procedure was the most popular – chosen by 47% of those who attempted to solve the problem. The representation of subspaces defined by a system of generators was far from being attained, not only for implicit, but also for parametric equations.

The April 1993 test will allow us to judge the progress in acquiring this notion with the following exercise:

One considers in $\mathbb{R}^5$ *the following vectors*: a = (1,0,-1,1,2), b = (0,3,1,-1,1), c = (-1,-1,1,1,0).

(a) *Give a parametric representation of* lin{a,b,c}.
(b) *Give linear equation(s) defining* lin{a,b,c}.
(c) *Determine* lin{a,b} ∩ lin{b,c}.

The task in a) and b) was slightly different from the February task because the equation question was separated into two parts by, in particular, first asking for parametric equations. The majority of students resorted to the C procedure to obtain the parametric equations and then proceeded, by elimination of the parameters, to obtain the implicit equations (using the Gaussian method or by floundering toward the answer).

We noted in parentheses the C procedures evolving toward correct results (not necessarily brought to term or free of calculation errors).

lin {a,b,c}	P	C (correct parametric equations)	Implicit correct equations	Other false ans.	Other cor. ans.	No ans.	Total pop.
Populations	4	89 (85)	53	6	1	6	114

Clear progress can be noted in the procedures between February and April: 'deviant' procedures have sharply declined and are no longer more than 9% of the procedures used (89% in the first test) learning has taken place. A few problems remain in coordinating parametric equations and implicit equations; only 38% of the students gave all the equations correctly. The number of students who revealed their incomprehension, either by deviant procedures or by an incorrect number of equations, was 56; i.e., 52% of those who had attempted to answer the question. Thus, slightly more than half of the students succeeded in mastering the delicate relation between the possible representations of vector subspaces in a simple case.

It should be noted that no algorithm exists for easily finding the implicit equations from the parametric equations. There is also the difficulty, analyzed below, in terms of logic and set theory. Indeed, an equation is a constraint that determines a subset in a set; the system representing the intersection of all subsets.

Later, this knowledge was tested on a more complex problem of subspace intersection (in February 1992 and in April 1993).

In February 1992, an exam, that we will study in more detail below, proposed the following. Of interest here are the procedures used to solve the first question of the linear algebra exercise.

Let E_1, E_2, E_3 represent the subspaces of $\mathbb{R}^4$ defined by the equations:
E_1: 3x-2y-z+t=0 E_2 : x+y+2z-t=0 E_3 : 5x+3z-t=0

(a) Compare E_3 and $E_1 \cap E_2$.

We found 3 distinct solution strategies:

(S1) Apply the Gaussian method to the first two equations, or to all three; affirm that the third equation is a linear combination of the other two (or that the intersection of the three subspaces is equal to the intersection of the first two, or that the intersection of the first two is equal to the intersection of the third with one of the first two). Then conclude to the correct (or opposite) inclusion, or the equality of the two subspaces.

(S2) Find the parametric representation of the intersection of the first two subspaces then find the parametric representation of the third (correctly or incorrectly) and compare the two parametric representations.

(S3) Find the parametric representation of the intersection of the first two subspaces then check if the generators of the intersection of the first two subspaces belong to the third.

E_3 et $E_1 \cap E_2$	S1	S2	S3	success	failure	tot. pop.
Populations	66	11	6	27	92	124

The main causes of failure are analyzed below. There were, in total, 70% failures but 64% started out on the right track.

In April 1993, the question was :

Let E_1, E_2, E_3 be the subspaces of $\mathbb{R}^4$ defined by the equations:
e_1: 3x-2y-z+t=0 e_2 : x+y+2z-t=0 e_3 : 5x+3z-t=0

compare E_3 and $E_1 \cap E_2$.

The same question was asked as in February 1992, but denoting the subspaces and the equations differently (to avoid confusion between equations and subspaces). These were not the same students as in February 1992 but the teacher was the same, and therefore it is meaningful to compare the procedures.

The same procedures, S1, S2, S3, were found, as well as the following:

(S0) Some wrote that the equation of the intersection of two subspaces is the sum or the difference of the equations of each subspace; there is therefore only one equation for the intersection, leading to a false conclusion. This error had been identified in 1992, but it was not retained in the analysis; indeed, the teacher thought that this error was due to students' lack of experience with the subject and that it would quickly disappear. This test, done at the end of the learning period, showed that this was not at all the case and that this was a persistent error; coming undoubtedly from difficulties linked to the use of logic and set theory.

(S1') Directly notice that $e_1 + 2 e_2 = e_3$ and end as in S1.

E_3 and $E_1 \cap E_2$	S1	S1'	S2	S3	S0	others	False ans.	Correct answer	No ans.	total
pop.	3	6	10	1	24	44	19	27	23	114

Under 'others', we classified answers which started with no indication of a specific strategy toward which the student was working ($e_1 + 2 e_2 = e_3$ only (34%[8]), parametric representation of the intersection alone (9%), or the Gaussian method on two or three equations (19%)).

To sum up, there was a total of 70% failures and 76% reasonable starts, showing very little progress in comparison with the results from the students of the preceding year. There were still as many failures due to poor comprehension of inclusion (21%) and an error of logic where students took $e_1 + e_2 = 0$ or $e_1 - e_2 = 0$ as an equation of the intersection (26%) - there was an automatic translation of the connotation of 'intersection' (with respect to subsets) into the connotation of 'or' (with respect to the equations).

In conclusion, it is possible to say that the notions of rank of a set of generators and generated subspace are more or less acquired at the end of the learning phase; that the relation between parametric equations and implicit equations is understood by most of the students at a basic level, but that it can be used only by a third of the students as soon as it must be used in relation with notions of logic connected to inclusion or intersection. Further analysis of the February 1992 exam and a small-group work session will give us more precise information on this point.

4.4. Difficulties Stemming from Weakness in Logic and Set Theory: Inclusion and Implication, the Role of Parameters, the Notion of an Algebraic Equation of a Geometric Object.

One of our hypotheses was the necessity of prerequisites concerning the prerequisites in logic and set theory for the learning of linear algebra. This hypothesis is developed in the works of Dorier (1990), and advanced in Robert and Robinet (1989) and Rogalski (1991). It was confirmed and refined by the detailed study of student performance on a linear algebra problem given to students at two different times in two different forms. We analyzed a time-limited written exam in February 1992 and, in February 1994, we recorded the work done by students working in groups of four on the same problem (but with variants) which we called "the problem of hyperplanes".

We present below both situations accompanied by their respective studies, but without drawing any conclusions. We terminate by offering some hypotheses and some suggestions.

4.4.1. The February 1992 Exam

The Question

> Let E_1, E_2 and E_3 *be the subspaces of* $\mathbb{R}^4$ *defined by the equations:*
>
> E_1 : $3x - 2y - z + t = 0$, E_2 : $x + y + 2z - t = 0$, E_3 : $5x + 3z - t = 0$.
>
> (a) *Compare* E_3 *and* $E_1 \cap E_2$.
>
> *Find all the subspaces of* $\mathbb{R}^4$ *which contain* $E_1 \cap E_2$; *explain your method and how you decided to represent the subspaces.*

The Place of the Exam in Teaching and What We Can Learn From it

This exercise was the second in a mid-year exam which required solving 4 problems (2 on numerical sequences, 1 on continuous functions, and the present exercise) and 1 giving extra credit (on the application of the notion of rank to magic squares, rarely undertaken by the students). The exam, which took place on February 21, 1992, lasted for three hours - a short time considering the problems posed. The teaching of linear algebra had already included the following: The Gaussian method for equations; geometry in $\mathbb{R}^3$ (including bundles of planes); rank and dimension theory for subspaces in $\mathbb{R}^n$ - defined by both equations and parametric representations. The teaching of explicit methods, however, only took place after the exam, but methodical procedures had already been used, both in lecture and practical courses.

The first question required having an idea of the type of arguments leading to the conclusion that one vector subspace contains another, or is equal to it, as well as choosing among the various possible strategies to do this. It was highly probable, however, considering the teaching which the students had received, that many of them would use the Gaussian method, but for what purpose? The question clearly surpassed the routine level of solving equations. This aspect is even more pronounced for the second question which required prior questioning of the type: How will I define the subspaces I am looking for? What will be their dimensions? On how many parameters will the family depend?

This was thus an exercise requiring meta-type thinking and the methodical use of knowledge to formulate questions and anticipate results. It was hoped, therefore, that a few of the students' reflections on the subject would be detected, particularly since the formulation of the second question encouraged it. This was not at all the case. The timing of this mid-year exam in the teaching process was undoubtedly the main cause: at this time the students had not yet had access to the meta level of linear algebra, nor had they sufficiently become aware of useful methods for solving problems in this area. Although the problem-solving strategies in question (a) were on average quite good - undoubtedly because the necessary choices could be made partly unconsciously - the same is not true for question (b). It was therefore an exercise that would have been better given at the end of the year and, in fact, we were not able to detect here the 'long term' effect on the students' access to the meta level. We will nevertheless report the frequency with which students used dimension, even though the central part of our analysis is on another aspect: the cause of the high failure rate among students, despite their choosing suitable procedures.

We will particularly examine question (a), since question (b) was seldom completed.

Possible Strategies and Procedures and Those Found by Students, Including Those That Were Partially False
(the dotted lines mean that the corresponding deduction was only at times implicit)

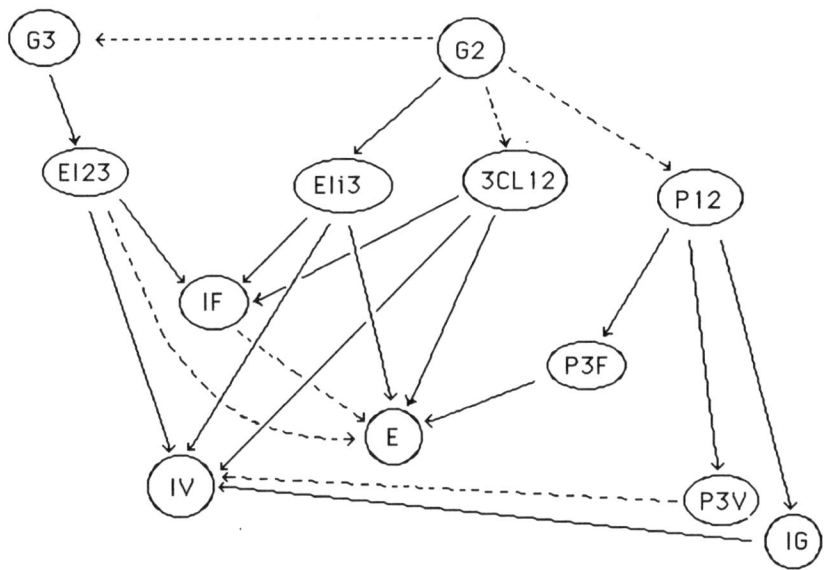

The procedures:
G3 : Apply the Gaussian elimination to the three equations
G2 : Apply the Gaussian elimination to the first two equations e_1 & e_2
EI23 : Prove that $E_1 \cap E_2 \cap E_3 = E_1 \cap E_2$
EIi3 : Prove that $E_1 \cap E_2 = E_i \cap E_3$ (i = 1 et 2)
3CL12 : Prove that the equation e_3 is a linear combination of the equations e_1 & e_2
P12 : Give a parametric representation of $E_1 \cap E_2$
IF : State the (false) inclusion: $E_1 \cap E_2 \supset E_3$
IV : State the (true) inclusion: $E_3 \supset E_1 \cap E_2$
E : State the (false) equality: $E_3 = E_1 \cap E_2$
P3F : Give an (erroneous) parametric representation of E_3.
P3V : Give a (correct) parametric representation of E_3.
IG : Check if some generators of $E_1 \cap E_2$ belong or not to E_3.

Main Types of 'Confusion' Errors Encountered
- equation/subspace confusion (induced by the wording of the problem);
- coordinate/vector confusion;
- vector/subspace confusion;
- confusion between "vectors generating the subspace" and "coefficients of equations representing the subspace";
- use of linear combinations of subspaces.

Other Types of Errors Encountered
- $y = 0$ in E_3 because its coefficient in the equation of E_3 is zero;
- the intersection of E_1 and E_2 has *one* equation;
- the 'equation' of $E_1 \cap E_2$ is a linear combination of the equations of E_1 and E_2;
- errors in the conclusion by studying the intersection or the inclusion of subspaces using generators (*if* u *and* v *are not in* F, *then* $F \cap \lin\{u,v\} = \{0\}$,...).

These types of errors were quite frequent but will not be analyzed here in detail. They were observed, essentially, in the work of students who did not know to construct a reasonable procedure to try out the question. Instead, we will examine the errors, related to logic and set theory, of students who were able to start out using the right procedure.

Overall Evaluation of 125 Exam Papers
Success:	27
Failures or unanswered problems:	92 (7 unanswered)
Both right and wrong solutions:	3
Impossible to decide (?) :	3

Strategy **1**
{ G2 and/or G3 → 3CL12 or EI23 or EIi3 → IV or IF or E }
appeared 63 times (50% of the exam papers).

Strategy **2**
{P12 → P3F or P3V → Comparison of parametric representation}
appeared 11 times (9% of the exam papers).

Strategy **3**
{ P12 → IG }
appeared 6 times (5% of the exam papers).

A reasonable strategy appeared a total of 80 times (64% of the exam papers); 'reasonable', here meaning, that which leads to a complete solution of the problem. This may seem quite high, but it is unlikely that this means truly deliberated strategic choices in all cases, particularly for strategy **1**, where it is possible that many students started out with the Gaussian method a bit haphazardly (by contract, there are equations, so...), only attempting to interpret the result afterward. This demonstrates the limitations of analyzing only written work: it would be more advantageous to know the reasoning behind certain students' choices during the exam, however this never appears on exam papers (not part of the contract in an exam).

One should also note that, despite the overrepresentation of techniques using the Gaussian method, obviously explained by teaching choices, there is a certain variety of points of view used (equations, parametric representation, generators).

An **explicit search for dimension** was found 43 times (34% of the exam papers); a fairly good result at this moment in the teaching process.

We need to say a word, however, about **question (b,)** from the point of view of the strategies and the search for dimensions. This question was never completely solved and was only attempted in 51% of the exam papers. Nevertheless, a certain number of students tackled the problem in a reasonable way with partial success or with an idea which, had it been suitably developed, would have led to success. This was the case for 34 of the students (27% of the exam papers). The quite general failure on this question makes the breakdown of the procedures fruitless, since the choice of procedure was frequently sketchy and unclear. However, only 14 students (11% of the exam papers) asked the a priori question of the dimension of the subspaces sought. The difference between questions (a) and (b) probably stems from the fact that in (a), such a search would have played a role of verification, whereas in (b), it was essential for the choice of methods of representation of the spaces sought.

In question (a), the **principal causes of failure** in putting the three strategies into practice were:
- inclusion written backwards (IF);
- equality written in the place of inclusion (E);
- difficulties comparing the subspaces represented by parametric equations (P12,P3F, P3V).

They were found in 61% of the students who had chosen a reasonable strategy.

It must be said that they involved difficulties in handling the notions of set theory and logic:
- Does adding conditions in the definition of a set mean that it becomes larger or smaller?
- What difference is there between inclusion and equality?
- What can be deduced from set equality $A \cap B = B$?
- How should an inclusion be shown (let $x \in A$, then ... so $x \in B$)?
- How do quantifiers intervene in the comparison of two subspaces represented by parametric equations? In particular, should a different name be given to the parameters?

Analyzing how the population is divided on these errors gives the following table:

	Success	24 %	
Reasonable strategies	Various errors	1 %	64 %
	Errors of logic and set theory	39 %	
No strategies, varied errors, unanswered problem, etc.			36 %

This shows that the percentage of good students; i.e., those who used a reasonable strategy but who failed due to weaknesses in logic and set theory (37 %), is nearly equal to the percentage of poor students who did not carry out any strategy at all (36 %). Since it is also very likely that nearly the entire group of these latter students were also poor performers in logic and set theory, it can be estimated that three-quarters of a section of science-oriented university students have serious problems in this area.

Among the 6 who chose strategy 3 { **P12→ IG** }, 5 answered the question correctly and 1 incorrectly, the error being of the linear algebra type and not of the logic and set theory type (the fourth among the 'other error types'). In fact, this strategy has little to do with logic or set theory: once the two generators of $E_1 \cap E_2$ have been determined, it is easy to see that they belong to E_3.

On the other hand, out of the 11 students who used strategy 2 {**P12 → P3F** or **P3V → Comparison of the parametric representations**}, 8 answered the question incorrectly (73 %). Yet this is a much more delicate strategy from the logic and set theory point of view: one must write and understand that the inclusion $E_1 \cap E_2 \subset E_3$ means that for all (s,t) there are l, m, n such that..., and this comprehension is indispensable for carrying out the calculations and interpreting the results.

Finally, out of the 63 students who used strategy 1, a little less restricting from the logic and set theory point of view, there were 41 incorrect answers (65 %).

The clear conclusion would be that *students possessing logic and set theory deficiencies will encounter serious difficulties in linear algebra, at least in analogical problems of the type studied here.* This confirms Dorier's conclusion (1990).

We were able to pinpoint more precisely the types of logic and set theory necessary for solving this type of linear algebra problem by examining the second activity based on the problem of hyperplanes - in a form that we will call "workshops".

4.4.2. The 1994 Workshop

By 'workshop' we mean small-group work as proposed by Robert and Tenaud (1988), and detailed in Robert (1990). Students worked in groups of 4, around tables of 4, forming islands in the classroom. They were given the task of solving a problem with no indication as to what methods to use, and had to write out the solution. The session lasted one and a half to two hours. Generally, teachers only intervened when requested, if there was persistent disagreement within a group, or if a stumbling block is encountered, or if the group thought it had solved the problem. (Actually, this rule was often broken since the teacher often succumbed to the desire to know how the groups were progressing).

The Questions

1) *The two subspaces E_1 and E_2 of $\mathbb{R}^3$ are given by their equations e_1 and e_2:*

e_1 : $3x - 2y - z = 0$; e_2 : $2x - y + 2z = 0$.

Find all the subspaces of $\mathbb{R}^3$ which contain $E_1 \cap E_2$.

2) *The three subspaces* E_1, E_2 *and* E_3 *of* $\mathbb{R}^4$ *are given by their equations* e_1, e_2, e_3 :

$e_1 : 2x + y - z + t = 0$; $e_2 : x - y + 2z - t = 0$; $e_3 : 4x - y + 3z - t = 0$.

(a) *Compare* E_3 *and* $E_1 \cap E_2$. *If possible find two methods.*

(b) *Find all subspaces of* $\mathbb{R}^4$ *which contain* $E_1 \cap E_2$, *indicating how you chose to represent them.*

The Objective of the Workshop
The goal was to encourage students to learn what were the important types of questions to ask in solving linear algebra problems and it was hoped that these questions would emerge through their small group discussions. For example, What are the dimensions of the subspace involved in a question? What methods are available to study the problems of inclusion and intersection between the vector subspaces of $\mathbb{R}^n$?

The Modifications from the Exam Problem in 1992
- To begin by $\mathbb{R}^3$ and, directly by the problem, to determine the subspaces containing $E_1 \cap E_2$. This caused the most problems in 1992, in $\mathbb{R}^4$, but it is easier in $\mathbb{R}^3$, especially since students knew what bundles of planes were. We hoped that a transfer toward $\mathbb{R}^4$ ($\mathbb{R}^n$?) would be favored by analogy.
- In 2) (a), 2 methods were required so as to encourage students to ask questions about the methods in 2) (b), related to the request for precision on the mode of representation chosen for a subspace of $\mathbb{R}^4$.
- Note the equations with names (e_i) different from those of the subspaces (E_i).
- Finally, the session was planned to last 2 hours, a good deal longer than students had for the exam in 1992.

Data Collection and Analysis
The oral work (discussion, writing) was recorded on audiocassette for 7 groups. The recordings of only 3 groups were clear enough to be studied. We will examine only two of the points that were analyzed: the students' use of meta and their mathematical difficulties, with the objective of better understanding those difficulties, which are related to student problems with logic and set theory.

A. The following example is <u>one group's chronological progression</u>.

Question 1	(8 min)	To determine the space $E_1 \cap E_2$, but forgetting the objective: determine the subspaces that contain $E_1 \cap E_2$.
Question 2(a)	(10 min)	To use the Gaussian method and find a parametric representation of $E_1 \cap E_2$.

	(25 min)	To try, unsuccessfully, to interpret the result as an inclusion, and with 2 minutes of help from the teacher; the wrong inclusion seems to them intuitive, but contradicts the dimensions.
Question 2(b)	(18 min)	Failure and discouragement, despite 3 minutes of help from the teacher; the teacher assisting a third time for 1 minute brings success; EUREKA!
	(2 min)	Rapid evaluation and conclusion on all points.
Writing	(12 min)	Perfect reporting of the solution, but with one ambiguity: it is not certain that the students are convinced of the reasons for the correct inclusion: $E_1 \cap E_2 \subset E_3$.

Total : 1 hour 15 min.

B. Reflection skills (the meta element)
We distinguish 4 levels in the analysis of this point:
predicting-planning-methods (before solving or before the solving steps),
explanations of changes of point of view or changes of register (during the solving process),
questioning on the objectives or the state of the solving process (during and sometimes after),
verification (when things are not working out, but also often after).
The following citations come essentially from 3 recordings.

1/ Predicting-planning-methods
Question 1:
We are in …How many?…$\mathbb{R}^3$…2 equations, so we have a vector; we have a system…, we triangularize… we find the parametric representation… that's it. "
Question 2 :
Same method for determining $E_1 \cap E_2$, we start over; we are in $\mathbb{R}^4$, we'll need two parameters
Well, we have to do the table with them, you know, the rank (to find out if a vector is a combination of other vectors).
You look for the rank, you triangularize
You have to do $E_1 \cap E_2 \cap E_3$ to compare… You have to write the system of three equations,… There will be an inclusion; we'll have to look in the other direction.
We also found:
- questioning the methods while writing the final solution;
- worries concerning coherence: "We need to be coherent";
- a clever method to construct 2 independent vectors;
- the analogy $\mathbb{R}^3/\mathbb{R}^4$, which functions as a generalization method; for example, to use the notion of bundles of hyperplanes, but with these questions: « Are we allowed to? Does it exist? »

2/ Explanations of changes of point of view or of register

We will give only 2 citations from the second question because often the point of view chosen by a student was implicit and not contested orally by the others.

Now we put vectors (after having found the parametric representation of $E_1 \cap E_2$ by solving the system of two equations, passing from the numerical coordinate register to the symbolic vector register).

The equation is a combination of the 2 others: What does that mean? I'll look with the vectors. It's easier with the vectors.

3/ Questioning on the objectives or on the state of the solution in progress

In the second exercise:

The equation is a combination of the 2 others: What does that mean?.

That means something, but we don't know what... What does that mean?

I don't know what to do with this. I no longer understand what I've found.

Why did we do that?

Also found is questioning on what 'representation' means in 2(b). Students often believe that this means 'drawing.' They find it difficult to 'represent' the intersection of two hyperplanes.

What is the intersection of two hyperplanes in $\mathbb{R}^4$? We don't understand what that means.

I would like to know what this is.

4/ Verification

Dimension considerations are often used as verification; in particular, the erroneous inclusion $E_1 \cap E_2 \supset E_3$ is very quickly rejected using dimensions Another example of verification, but this time more confused, arose in the second question:

That doesn't correspond; it should have a linear combination. The 2 others are not related (the 2 vectors which generate $E_1 \cap E_2$ and the 3 vectors which generate E_3 seem to be unrelated, while they think there is an inclusion. They try, in a confused way, to reason in terms of 'generated spaces').

Analogy also functions as verification or as clarification:

You do the same thing.

We've already done that.

Likewise, normality (concordance with the procedures already used in exercises):

This is strange.

Normally you have...

As regard this question of students resorting to meta elements, the mode of data collection used (audio recordings) obviously allows us to detect much more than perusal of the 1992 exam, for which we have only written traces. But it is very likely that the workshop form itself favors this turning toward meta elements because of the exchanges and discussions it imposes and a work pace that allows more time for the student to think.

C. The mathematical difficulties

This is clearly the main question. We will only go into the difficulties related to logic and set theory, which come out on two main points.

1/ logic and set theory and bundles of planes

(a) Difficulties with the role played by parameters describing a bundle of planes: if the equations of the bundle of planes are written lp + mq = 0, should l and m be left as 'unknowns' or should they be 'fixed'? It is probably the role and status of variables in the definition of certain families of subsets which was not well dominated, perhaps because the problem is at a higher level than that of the elements of a set; i.e., the level of the subsets. (These difficulties are not present when students work with vectors written as a 'generic' linear combination (lu + mv) of 2 given vectors, except when two sets of linear combinations should be compared).

(b) Confusion between a bundle of subspaces and the vector subspaces which are elements of this bundle:

You say that the bundle of planes is a vector subspace of the straight line D (meaning: containing the straight line D).

And the bundles of planes are the only subspaces which...

Dialogue extracts about E_3 belonging to the bundle defined by E_1 *and* E_2:

E_3 *contains the intersection.*

I'm more inclined to think that it's E_3 *which is included.*

E_3 *is included in the bundle of hyperplanes formed by* E_1 *and* E_2.

These are set theory related difficulties of the type: if F is a bundle of subspaces and E a subspace element of this bundle, should we write $E \in F$ or $E \subset F$? If D is the straight line which determines the bundle, is this $D \in F$, or perhaps $D \subset F$?

2/ logic and set theory and the notion of equations of a geometric object

(a) Difficulties in correctly expressing the relations between equations and/or systems of equations and subspaces. The error of persistently affirming the false inclusion $E_3 \subset E_1 \cap E_2$ seems to stem partly from a classic equation/subspace confusion: the existence of a linear combination $e_3 = ae_1 + be_2$ is 'naturally' translated by this erroneous inclusion[9]. This confirms Dorier's hypothesis (Dorier 1990) according to which errors in logic take on specific forms in linear algebra related to the mathematical content.

(b) The major difficulty, blocking students for a long time, stems from set theory in two forms:

- What should be concluded from the equality $E_1 \cap E_2 = E_1 \cap E_2 \cap E_3$? The students' weakness in general set theory, the theory not having functioned concretely before, does not allow students to fight against their conviction that it is the inclusion $E_3 \subset E_1 \cap E_2$ which is true. None of the students thought of drawing three Venn diagrams, for example, or of carrying out a standard set theory reasoning by saying: « let x be an element of A, transform and deduct from this that x is in B. »

- Resorting to verification by the number of restrictions imposed, i.e., the number of equations (adding restrictions also diminishes a set), is impossible for students, because the notion of 'equations of a subspace' for them is not a synonymous with 'restrictions on a point (x, y, z, t) expressing its belonging to a certain subset."

The correspondence between subspaces and equations seems to come only from the operator domain: starting with equations, we know how to calculate a basis of the associated subspace and, inversely, we know how to find equations for a subspace represented by parametric equations. The contract is to apply certain well-used algorithms on large numbers of exercises, and the strength (and therefore their didactical weakness) of linear situations is that this works even if the meaning of these algorithms is not understood. This would explain the confusion mentioned in question (a): students do not know that equations and subspaces can come from two different domains, and consequently the relations from one side must be written in the same way as on the other side (while inclusions, dual of one another, are inverted for logical reasons).

Finally, the nature of certain difficulties in logic and set theory may be specified here: the fact that an equation is a constraint that determines a subset in a set, and that a system of equations represents the intersection of all subsets determined by each equation do not seem to be acquired. This again confirms the hypothesis made in (Dorier 1990) that certain aspects of logic and set theory that are prerequisites for linear algebra are probably specificities related to concepts of this discipline.

4.4.3. Hypotheses and Suggestions

A few hypotheses on the causes of the difficulties in logic and set theory, in particular with the notion of equations of a geometrical object, are presented here. We will end with suggestions (hypothetical and therefore needing testing) on how to at least partially remedy these problems in first-year university math courses, using a particular teaching of analytical geometry.

What is the Origin of the Difficulties on Equations Representing Geometrical Objects?

This notion of equation representing a geometrical object is introduced with the notion of straight lines (when the students are between 14 and 16 years of age), and continues until they are approximately 18 years old. The characterization of an equation as a in terms of an inclusion of subsets is most likely introduced at the moment when coordinates of a point and the equation of a straight line are discussed. But nothing backs up this concept. Instead, an algorithmic practice of relations between the numerical setting of the equation and the geometrical setting is quickly substituted, in the graphic register of the straight line:

* to go from $y = ax + b$ or $ax + by + c = 0$ to the graph of a straight line, there are a certain number of tricks: use the slope and y-intercept, or $x = 0$ and $y = 0$ can be done to place 2 points, or the graphic calculator is used.

* to determine the equation of a straight line verifying geometrical conditions (going through 2 points, going through one point and being parallel to a vector), there are also tricks or formulas to use: using the slope or a determinant, for instance. In short, the equations of straight are essentially objects of algorithmic calculations. Moreover, various teaching practices aim at facilitating the acquisition of these algorithms and computer software is also often used to reinforce them.

Finally, there are hardly any inclusion problems for straight lines in a plane. Equations of straight lines and planes in space could result in such problems but few are seen in class or in textbooks.

There is Little Effective Use of the Language of Set Theory Before the Introduction of Linear Algebra.
In almost all cases, this language is not really introduced, not even partially in high school, and only incidentally in the first year of university studies. Its formalizing and generalizing features are not brought out in problems. Students rapidly move on to "true mathematics" where the work is essentially calculation (equations, derivatives, integrals, etc.), and almost never use set theory with any significance again. In particular, the translation of universal implication in terms of inclusion is rarely made explicit or, furthermore, put to pertinent use.

We believe therefore that the teaching of elementary (naïve) set theory should be seriously rethought, as well as the relation with the teaching of elementary logic (the reasoning used in mathematics). This does not mean that formal teaching is needed, but, for example, teaching supported by the use of the Legrand's 'electrical circuit' sequence (1990), closely related to the mathematical content at a given moment, can be very effective. One of the areas where these relations may be made concrete is elementary combinatorics. (It often requires working on implications of the following type: 'let x∈ A; show that x∈ B" like in the hyperplanes problem). Unfortunately, the combinatory analysis presents a symbiosis between high-level abstraction and the modeling process, making it delicate to teach.

Develop Cartesian Geometry with Particular Emphasis on the Search for Geometric Loci.
The use of analytic geometry provides mental images for a large number of concepts in linear algebra. Moreover, on certain problems it allows highlighting the advantage of linearity and abstract, and therefore general, constructions in analytical geometry.

But, of greater significance, is the fact that analytic geometry can provide a wide range of illustrations and motivations for the acquisition of the logic and set character of the relation between equations and geometrical objects, provided that this relation functions on objects and situations where overly simple algorithms, that would solve typical problems with no reference to the meaning of the equations of an object, do not exist. This means not sticking to the domain of linear objects (straight lines and planes) for which such algorithms do, of course, exist. It is therefore important to have students study curves and skewed surfaces, and to propose problems about them where inclusion, equality, and intersection come into play. From this point of view, the search for non-linear geometric loci, with the changes in setting between synthetic and analytic geometry and the change in point of view between equations and parametric representation can, in our opinion, aid in the acquisition of the logic and set status of the notion of algebraic equation of geometrical objects.

5. ANNEX - MATHEMATICAL SOLUTION OF THE HYPERPLANES PROBLEM (1992, QUESTION (a))

The Gaussian method on the equations can be presented in two different ways:

$$\boxed{G2} \rightarrow \begin{cases} -t + 2z + x + y = 0 \\ \\ z + 4x - y = 0 \end{cases} \quad \text{or} \quad \boxed{G3} \rightarrow \begin{cases} -t + 2z + x + y = 0 \\ z + 4x - y = 0 \\ 0 = 0 \end{cases}$$

since there is the following linear relation between the equations: $e_3 = e_1 + 2e_2$

$\boxed{CL12}$

- Therefore : $E_1 \cap E_2 \cap E_3 = E_1 \cap E_2$: $\boxed{EI23}$. With a variation $\boxed{G'3}$ from $\boxed{G3}$, it remains e_i and e_3 (i = 1 or 2) : this is $\boxed{EIi3}$, thus $E_1 \cap E_2 = E_i \cap E_3$. From that, one can guess a relation of inclusion or equality between $E_1 \cap E_2$ and E_3. With a minimum understanding of set theory, one can conclude by $\boxed{IV}$, if not one can conclude by $\boxed{IF}$ or $\boxed{E}$.

- From G2, it is possible to find parametric equations representing $E_1 \cap E_2$: assuming x = u and y = v , one gets :

$\boxed{P12}$: $x = u$, $y = v$, $z = -4u + v$, $t = -7u + 3v$,

In other words : $X = uU + vV$, with $U = \begin{pmatrix} 1 \\ 0 \\ -4 \\ -7 \end{pmatrix}$ and $V = \begin{pmatrix} 0 \\ 1 \\ 1 \\ 3 \end{pmatrix}$;

therefore U and V are generators of $E_1 \cap E_2$. It is then easy to check that U and V satisfy equation e_3, it is $\boxed{IG}$. Therefore: $E_1 \cap E_2 \subset E_3$.

- Equation $e_3 : 5x + 3z - t = 0$ gives a parametric representation of E_3 :

$$x = \lambda, \ y = \mu, \ z = v, t = 5\lambda + 3v \ ;$$

It is $\boxed{P3V}$, or $\boxed{P3F}$ if there is a mistake in the calculations.

To find the intersection between $E_1 \cap E_2$ and E_3, it is possible to search the condition on (u, v) in order that there exist (λ, μ, v) such that:

$u = \lambda$, $v = \mu$, $-4u + v = v$, $-7u + 3v = 5\lambda + 3v$.

For any (u, v), one can find λ and μ ($\lambda = u$, $\mu = v$!), and v in two different ways : $v = -4u + v$, and $3v = -7u + 3v - 5\lambda = -7u + 3v - 5u = -12u + 3v = 3(-4u + v)$: one gets the same value. Thus R(u, v), that is Rm $\in E_1 \cap E_2$, there exist (λ, μ, v) such that..., that is m$\in E_3$.

Therefore one gets finally $E_1 \cap E_2 \subset E_3$.

6. NOTES

[1] Unlike secondary schools, for which a national program is imposed by the Ministry of Education, the curricula of universities are decided locally, even though changes from one university to the other are usually superficial.

[2] In chapter IV, §III, we present an alternative approach to the Gregory's interpolation problem with use of 'meta lever'.

[3] For instance when asked, "What is the intersection of the two subspaces generated by U_1, U_2, and V_1, V_2", students prove that neither U_1 nor U_2 are a linear combination of V_1 and V_2, and conclude that the intersection is reduced to 0.

[4] Analysis of these surveys is presented in detail here, giving a precise idea of the difficulties students encounter on specific tasks. However, this paragraph can be skipped during an initial reading.

[5] Percentages are calculated on the number of students having answered the question.

[6] Percentages are calculated on the number of students having answered the question..

[7] Percentages are calculated on the number of students having answered the question..

[8] Percentages are calculated on the number of students having answered the question.

[9] This is comparable to the error where students think that $(e_1 - e_2)$ is the equation representing the intersection, as mentioned above.

ALINE ROBERT

PART II - CHAPTER 2
LEVEL OF CONCEPTUALIZATION
AND SECONDARY SCHOOL MATH EDUCATION

1. WHAT DO WE CALL 'LEVEL OF CONCEPTUALIZATION'?

We would like these words to refer to a stage in the field of mathematical knowledge (conceptual field) corresponding to a coherent organization of a part of this field, characterized by mathematical objects presented in a particular way, theorems on these objects, methods associated with these theorems, and problems that students can solve with the theorems at the appropriate level using these methods. Many mathematical notions can be approached at several levels of conceptualization, often partially embedded: the initial objects change and become more general, thus allowing the introduction of new, richer structures which in turn require the introduction of a new adapted formalism. Similarly, many problems can be posed and researched at several levels: often so-called theoretical exercises (i.e., general) at a given level correspond to theorems at the 'following' level. Given that there can be several different orders of presentation, there is no absolute hierarchy among these levels; they depend on the actual teaching dispensed, at least during formal instruction. It can therefore be said that levels of conceptualization are milestones identifiable throughout the teaching of the notions of a given conceptual field.

To clarify the idea behind this expression, two examples will be given of problems solved at different levels of conceptualization: one from linear algebra and the other from geometry.

1.1. Seeking Magic Squares of the Third Order

The student must find all the three-row, three-column square tables, made up of real numbers so that the sums of the elements of each row, each column, and each diagonal are equal to 0 (it is this invariant sum of the elements of the rows, columns, and diagonals which is called the 'sum' of the magic square).

Students must first be made to see that the central element of the square (denoted by 'e') is necessarily equal to zero. It is true that adding the elements in the two diagonals and the second row gives zero, but the corresponding elements can be

125

DORIER J.-L. (ed.), The Teaching of Linear Algebra in Question, 125—131.
©2000 Kluwer Academic Publishers. Printed in the Netherlands.

regrouped differently by adding the elements of the first column, the third column and three times 'e', from which the result is obtained.

Several approaches at different levels of conceptualization will now be outlined, leading to the required result (to find all the squares with zero sum), beginning with the most elementary: in other words, mobilizing the most elementary knowledge. This level is adopted, either because this is the only acquired knowledge at the moment of problem solving, or because the corresponding demonstration is economical. To bring a piano and a stool together, whatever one's physical strength, pushing the stool is obviously the best strategy !

• A first elementary attempt could be carried out empirically, without reference to any theory: students could begin with two well-chosen elements of the square and gradually determine all the other elements in order to construct a square with a null sum. For example, given the first two elements of the first row, a and b as below, the square is written:

a	b	(-a -b)
(-2a-b)	0	(2a+b)
(a+b)	-b	-a

If one begins with three elements, a contradiction will be observed, unless one comes back to only two elements.

It is therefore possible to write, from two well-chosen elements, any magic square whose sum is zero. But the problem solving stops here; students cannot go beyond nor explain this observation in any other way.

This approach originates at an elementary level and requires no use of a formalized notion of algebra or linear algebra. The problem is solved, but in a practical way, with no evidence of generalization, apart from the use of letters. This does not prevent the same method from being used with full awareness, or in any case, with justification from the properties of all magic squares.

• A second type of solution can be carried out at college level, by setting up a system of 8 equations with 8 unknowns and solving it using Gaussian elimination. The table is written as follows:

a	b	c
d	0	e
f	g	h

with:

$a + b + c = 0$

$d \quad\ + e = 0$

$f + g + h = 0$

$a + d + f = 0$

$b \quad\ + g = 0$

$c + e + h = 0$

$a \quad\ + h = 0$

$c \quad\ + f = 0$

We can immediately reduce these equations to only four unknowns: a, b, c, and d.

This produces the equations:

$a + b + c = 0$

$c + b + a = 0$

$a + d - c = 0$

$c - d - a = 0$

which in turn can be reduced to the two equations: $a + b + c = 0$ and $a + d - c = 0$. This can be solved by taking arbitrary values for a and b and writing $c = - a - b$ and $d = c - a = - 2a - b$. The same result is obtained as with the preceding method but at a richer level of elementary algebra, since knowledge of solving linear equation systems was put to use. It was even possible to keep the initial system and solve it with a general method. Note, however, that this knowledge of systems is isolated, autonomous. On the other hand, it is already conceivable that the same approach could be used with fourth order squares, for example.

• Another form of the result can be obtained at an elementary level, but in linear algebra this time. At the second-year university level, for example, it is possible to to demonstrate that the set of tables sought is a two-dimensional vector space on $\mathbb{R}$. In fact, it is the first solution, interpreted within this new setting, that shows us that we have a two-dimensional vector space. The values of a and b, found in the first method, correspond to the expression of a general magic square written as a linear combination of two particular squares, forming a basis of the space. Here a new stage of conceptualization has been adopted, which provides another way of interpreting the same ' system of equations solving' tool.

• Another demonstration, based on the eight preceding equations interpreted now as linear forms on the space of third order real square matrices, $\mathfrak{M}_{(3,3)}$, also gives the desired results, but by using notions linked to duality. It is shown that zero sum magic squares form a vector subspace of $\mathfrak{M}_{(3,3)}$ which is the intersection of the kernels of seven independent linear forms (corresponding to seven independent equations taken from the eight preceding equations). It is therefore the orthogonal of a seven dimensional subspace of the dual space of $\mathfrak{M}_{(3,3)}$. Yet these two spaces, $\mathfrak{M}_{(3,3)}$ and its dual, are nine dimensional. The subspace considered is therefore two dimensional. This is a solution at a less elementary level in linear algebra, avoiding, moreover, solving the system.

1.2. A different Example: Complex Numbers and their Use in Geometry

If students know the properties of homographic functions, i.e. functions of the complex variable z, such that: $f(z) = \dfrac{(az+b)}{(cz+d)}$, they also know that they globally conserve the set of all straight lines and circles and that they are conformal and therefore leave the angles of the curves (angles of the tangents at the intersection points of the curves) invariant.

In particular, finding the image of particular straight lines or circles is nothing more than a problem of recognizing the relations of the straight line or the circle to the characteristic points of the homographic function. Thus, if the geometric straight line passes through the point corresponding to the complex number ($-\frac{d}{c}$), its image is a straight line (which passes through the point corresponding to the complex number ($\frac{a}{c}$). If not, the image of the straight line is a circle which passes through the same point, corresponding to the complex number ($\frac{a}{c}$), etc. The problems of orthogonality in families of straight lines or circles relate to the applications of conformity.

The cross-ratio can be used to prove that four points are collinear or concyclic: if the cross-ratio of the four complex numbers is real, the corresponding points are collinear or concyclic. However, at the level of manipulations on complex numbers alone, without any more general knowledge of the properties of complex variable functions, each case requires an explicit calculation, either on the modules and arguments or on the real and imaginary parts. For example, in order to find the image of a straight line, it is necessary to actually display this image (and recognize it).

For example, to characterize the image of the real axis by the homographic function defined on $\mathbb{C}\backslash\{-1\}$ by

$$f(z) = \frac{(z - i)}{(z + 1)}$$

two choices are possible:
- The condition 'z is real' can be translated by the equality of z and its conjugate: $z = \bar{z}$, and the equation of the image of the straight line can be obtained from the equality, between $z = \frac{f(z) + i}{1 - f(z)}$ and its conjugate, which may not be an easy task.
- But if the properties of homographic functions are known, students should know that the image of the real axis, which passes through the point corresponding to the complex number (- 1), cannot be a circle and is therefore a straight line. Therefore they only need to determine the images of two points on the real axis (in fact only one point is necessary since the image necessarily passes necessarily through the point corresponding to the complex number 1).

The problem of orthogonality of families of straight lines and circles must also be treated 'by hand' in each case, through a search for the tangent angle at the points of intersection. Recognizing that four points are collinear or concyclic requires coming back to an explicit characterization in terms of angles, necessitating a calculation on arguments.

Two distinct levels of conceptualization are brought into play: knowledge of complex numbers alone, excluding the complex variable functions, or supplementary knowledge of homographs. These two levels are in fact embedded.

1.3. Outcome

Actually, many mathematical objects can intervene in solving problems at several levels, firstly, as objects in their own right, studied for their own sake and secondly, as elements of a set that contains them. The second point of view does not eliminate the first one; they overlap. A class of equivalence is only a subset of the initial set before the notion of quotient sets is known; it is later, according to the questions treated, an element of the quotient set or a subset of the initial set. For instance, functions are first studied in detail, one by one for their own interest, then they are considered as elements of various subsets of functions with specific properties. This is where different levels of conceptualization can be observed: in the elementary study of $\mathbb{R}$ and real functions, and in the study of function sets, which are only useful because of the introduction of topological notions, etc. Hence the question arises, above and beyond the identification of different levels of conceptualization, which are proposed one after the other in the teaching environment, concerning the link between these levels and teaching: How does one pass from one level to another?

2. LEVELS OF CONCEPTUALIZATION IN SECONDARY EDUCATION

We will now analyze briefly the mathematics curriculum in secondary schools in France (up to the junior and senior years of high school) using the concept of levels of conceptualization. Certain notions are taught as if students were being prepared for a presentation of these notions at a certain level of conceptualization, but without ever reaching this level during the high school years. Moreover, this is often a formal level requiring the introduction of a new specific formalism, for which students are prepared but never reach.

A few key examples will be instructive. In middle school geometry, elementary objects (point, straight line, plane) are not defined, even though an axiomatic point of view is not adopted. It is assumed that all students derive the same meaning from a common perception of these objects. Then new definitions (parallel lines, parallelograms, symmetries, etc.), Euclid's axiom and theorems are given, from which students have to construct proofs.

At the senior high school level, on the other hand, vector geometry is presented, which could replace the preceding, but which in fact overlaps with it, mixes with it. The vectors of the plane are, for example, in reference to the preceding notions. However, students are given the relation ($\overrightarrow{AB} + \overrightarrow{BC} = \overrightarrow{AC}$) and told that if an origin

O is fixed, the relation $\overrightarrow{OM} = \vec{u}$ defines a single point M, $\vec{u}$ being a given vector. This corresponds to the definition of the plane as a two-dimensional affine space on $\mathbb{R}$, however it is submerged within other information. But neither the affine spaces, nor even the vector spaces, are ever explicitly introduced in high school. Only the rules on vectors are put into operation without carrying through to complete formalization (formalization is defined here as working within a coherent theory, the theory of vector spaces, with its own formal notations, of which vectors are an element). Consequently, students remain at an elementary level in geometry, nonaxiomatic, while going 'beyond' this level with the partial introduction of objects occurring at another level of conceptualization: i.e., vector and affine geometry.

In calculus, in junior and senior science-oriented classes, the definition of the limit of a function and the convergence of series are not taught formally: teaching is based on examples of convergent or divergent series, which serve as a reference and which play a role analogous to that played by empirical objects of nonaxiomatic geometry. Then theorems are given, obviously impossible to demonstrate, but whose use makes demonstrations possible (if $f(x) \leq g(x)$ their limits are in the same order, if $f(x) \leq g(x) \leq h(x)$ and f and g have the same limit then f has a limit which is the same, convergence of bounded monotonous series, etc.). Here again, students remain at an elementary level, but can deal partially with less elementary problems in analysis.

In the final analysis, in secondary school education, it would appear that the students receive a partial, artificial, unfinished reconstruction of this type of notion: they only receive the beginning of the story, situations which can be 'properly' solved with the targeted notions, but which are treated only imperfectly, with 'hybrid' nonformalized objects requiring use of partially emerging skills, in instances where formalization can be left behind.

Before trying to comprehend the underlying intentions of these choices, the following question should be addressed: Do the notions outlined above have anything in common?

We believe the answer to be in the affirmative. The two cases, convergence of sequences and limits of function, and linear algebra as a presentation of geometry in an axiomatic fashion are mathematical concepts that involve generalizations and unifications, as outlined in the discussion on linear algebra, in the preceding chapters.

Formalization of these notions (definition in (ε, N), definition of vector and then affine spaces) is indeed a time-saver in many demonstrations, becoming 'a standard'; the same for all similar problems from this 'model'. Moreover, this formalization, which corresponds to the true introduction of the concept, not only makes working at a new level of conceptualization possible, but also allows one to attack new problems inconceivable without this formalism. But this formalism can turn out to be perfectly opaque for students, and its use can therefore be the source of a considerable loss of meaning. Indeed, we are no longer working on perception: by its very nature, the formal, unifying, generalizing level hides images associated with previous mathematical language, but this is precisely one of its economic benefits.

Given the above remarks, it is possible to venture an interpretation: the implicit hypothesis that could operate for this type of notion in secondary education would be that the student needs sufficiently numerous referents, particular, precise, perhaps dispersed, knowledge, nonformalized but available and internalized, for the need of the proper use of formalism to be seen as the most rapid, condensed, universal, economic means to solve the aforementioned problems. It could be said that this is not yet the case at the high school level, where only those students continuing with mathematics would need to appreciate the formal aspect of a solution to a problem.

This hypothesis is coupled with an epistemological position concerning mathematics: it seems to be viewed in the curricula as a place where problem solving prevails over the application of appropriate theories, whose evidence is conditioned by the problems themselves. If a theory is not evident in the problems the teacher wishes to propose, it is not presented. If the teacher considers that it is too early for students to appropriate it, even partially, it is not completely presented, but enough is worked on to familiarize students with certain corresponding problems, but without risking any loss of meaning. Once the students are believed to be ready, once they have been sufficiently exposed to a field of problems, the formalized notions are presented along with the new problems.

We do not intend to discuss the pertinence of these hypotheses at this stage, but merely present them as we see them, in the light of the rapid analysis outlined in this chapter.

3. CONCEPTUALIZATION

The underlying question in terms of teaching which ties in with the more general problem of conceptualization is one which comes from psychology: in order to construct the concept of a tree, one must have seen enough trees to be able to conceive of how a tree is different from everything else and recognize trees among other objects.

Beyond the obvious analogy, linked to the very process of conceptualization targeted, there is a great difference in level with these elementary examples borrowed from psychology: perception and action are not seen from the same perspective. Mathematical work is already symbolic; it is simply knowledge and students' familiarity with what the symbols cover that varies. Furthermore, the role of explanation is different. Mathematics students are emerged in a field of scientific knowledge - elaborated, particular knowledge - and are not learning something which concerns everyone (see Vygotski's work on scientific concepts).

The theory of learning situations or the tool/object dialectic approaches the problem of conceptualization in mathematics in its own way, and proposes an artificial genesis adapted to the notions targeted. In the last case, for example, the hypothesis is as follows: teachers help students conceptualize by having them work on problems in which the particular notion, (or concept), can be used in order to solve the problem, then by presenting the students with a decontextualized version of the concept; i.e., as a concept with certain properties.

However, it seems that the targeted notions do not lend themselves to this type of approach, justifying their particular treatment in the curricula. Indeed, a concept created to replace various efficient though varied steps, each of which partially hides the essence of the step behind the particularity of the problem, cannot, by its very nature, be associated with a particular problem. Its usefulness lies elsewhere: in its both simplifying and unifying character.

MARC ROGALSKI

PART II - CHAPTER 3
THE TEACHING EXPERIMENTED IN LILLE

The observation and analysis of the difficulties involved in teaching linear algebra, presented in the first chapter of the second part of this book, as well as the epistemological analysis, presented in the first part, were first applied in an experimental teaching of linear algebra begun at the Science and Technology University of Lille in 1984, and retested from 1989 (the term 'experimental' simply meaning that we had wide-ranging freedom in the organization of teaching and testing). Bringing these elements together has provided an improvement in our course design[1], but it has been only since 1991-92 that the experiment has become fairly stable; i.e., founded on clear bases with few changes from one year to the next. Ayats, Gaultier de Kermoal, Rogalski, and Roussignol have been the major actors in this phase of the teaching experiment.

1. THE UNDERLYING PRINCIPLES OF THE COURSE DESIGN IN LILLE

We will interpret here, through the preceding analyses on formalizing, unifying, generalizing, and simplifying elements of knowledge, the bases of the Lille course design. We have seen that for teaching such concepts, situations/problems (which exist only rarely) must often be replaced by cultural expectations, questionings, and objectives that are postponed in time. Linear algebra, for example, is not a theoretization of problem solving enabling the use of a-didactical situations (which could be tackled by students working alone, and involve validation elements in themselves); it is a cultural choice to make a theoretical detour that we hope will be fruitful, without prior certainty. Students must assimilate questionings of the following type:

What cultural objectives are involved in the construction of general linear algebra? Why make an *a priori* theoretical detour? Why formalize and unify bodies of prior knowledge in a new structure? What do these bodies of knowledge have in common and what can treating them in an abstract approach add to learning?

We hypothesize that it is some types of questioning, more than problems, that must be transmitted to students when introducing concepts to learn. More specifically, students must be led to understand and accept the formalizing and generalizing role of the theoretical detour as a response to an entire field of problems and questionings which are similar, even though they do not know this.

DORIER J.-L. (ed.), The Teaching of Linear Algebra in Question, 133—149.
©2000 *Kluwer Academic Publishers. Printed in the Netherlands.*

To sum up, we have postulated **three hypotheses** which would act as an operational starting point for developing course design for teaching linear algebra.

a) *The formalizing, generalizing, and simplifying nature of the concepts of linear algebra must be taken into account;*

b) *Learning linear algebra requires a certain number of prerequisites;*

c) *The three following ideas must be used interactively:*

* *use of the meta lever;*

* *construction of long-term course design;*

* *use of changes in point of view as a unifying stimulator, first, and as a problem-solving stimulator, after.*

Before presenting the details of the course design, these hypotheses will be elucidated.

1.1. Take Into Account the Formalizing, Unifying, Generalizing, and Simplifying Nature of the Concepts of Linear Algebra

In order to provide good learning conditions, several factors can be manipulated, for which a suitable presentation in time is essential. These conditions include the following objectives:

* construct **preliminary bodies of knowledge,** which will be unified by the theory, to teach and construct knowledge **in different settings**;

* **distinguish a set of similar questions** which students have not managed to solve within the preliminary bodies of knowledge mentioned above;

* use **changes in point of view** between these bodies of knowledge and between these questions as **reasons to unify**, even as instigators of unification;

* then bring back former knowledge in the unifying theory, use it to **reinterpret** objects and problems, and to solve new problems which were inaccessible in the former knowledge;

* make **explicit for students on the discourse level** these back-and-forth references, and the importance of changing points of view and unification;

* present **situations whose objective would be awareness of meta** on new bodies of knowledge and their role, more than the knowledge itself which has already been introduced.

Some of these points will be brought into play through taking into account our third hypothesis.

1.2. The Prerequisite Hypothesis

We should specify that 'prerequisite' does not necessarily mean 'taught beforehand' and/or elsewhere (in high school, for example). It means that certain knowledge is essential for the teaching and learning of linear algebra and therefore that we must see to it that students have it at their disposal. Certain elements of this knowledge may well be taught shortly before linear algebra (in the same year) or at the same time. But others are elements of a 'state of mind' or of competencies more than precise knowledge, and it may therefore be necessary that students have acquired them before arriving at the university (this is one of the questions of the recent changes in high schools in France).

1.2.1. Comprehension of the True Stakes in Mathematics and a Certain Practice of Elementary Logic and Language of Set Theory are of Primary Importance

Students must understand what is said, keep control of meaning for assertions that will not be immediately concrete, and be able to conduct quite delicate reasoning. It can be observed, for example, that implication and equivalence are often sources of difficulties, particularly in algebra, if they are not properly distinguished, leading students too often to conclude a demonstration of independence by « if the a_is are all zero, we have shown that $a_1u_1 +...+ a_nu_n = 0$, therefore the u_is are independent. » In addition, important problems appear with quantifiers. For example, if students are asked "What are the conditions on the parameter t such that the space $\lin\{e_1, e_2, e_3\}$ contains the space $\lin\{u(t), v(t)\}$?", it occurs that they interpret the question as: "Look for t so that $\forall$ x,y ; $\exists$ a,b,c verifying $xu(t) + yv(t) = ae_1 + be_2 + ce_3$". Clearly, dominating quantifiers is necessary in order to come to terms with this problem. We have observed in examining students' productions (presented in the first chapter) how students' weaknesses in the area of logic and set theory can lead to large-scale failure.

Among the frequent activities in linear algebra can be found : representing subspaces, proving inclusions of subspaces, and using intersections. Students must also know how to interpret the solvability or the number of solutions of equations $f(x) = y$ in terms of injectivity and surjectivity. A general notion of transformation has to be constantly used, disconnected from any numerical or graphical support: it requires being somewhat familiar with ideas and methods of elementary set theory. Using formalism is an obligation, and also intervenes in the aspect to be presented next.

1.2.2. Accepting the Algebraic and Axiomatic Process

Accepting to reason with symbols with no immediate support, considering rich mathematical objects as individual 'entities', handling simple composition laws, accepting that a formal detour can bring a later simplification and increase the fields of application, that generalizing is not a gratuitous exercise, etc., can all be considered as what we will term the algebraic and axiomatic process which to be a determining factor in linear algebra, if only for accepting the rules of the game. It is the consequence of the formalizing, unifying, generalizing, and simplifying character of the concepts of linear algebra, and we have widely developed this point in the first part of this book.

For linear algebra, as for many other fields taught in first-year university math classes, it is not at all clear whether these first two prerequisites must be acquired by the end of high-school studies or must be taught at the university level. Since they are not currently taught at the high-school level, we have integrated them into our course design.

1.2.3.   A Specific Practice of Geometry within the Sphere of Cartesian Geometry
On this point, contrary to the preceding points, there have been no precise studies. There is, however, a strong consensus among teachers, which we will adopt as a hypothesis (Robert et al., 1987), (Rogalski, 1990a). Let us mention that research is in progress on this question of the relation between linear algebra and Cartesian geometry.

Knowledge of geometry will be an important background support for language and meaning in linear algebra. It can indeed provide images of vector concepts: subspaces, linear combinations, direct sum, solutions of systems of linear equations, linear parametric representation, bases, etc. The practice of Cartesian geometry - in particular the search for geometric loci and the double representation of geometrical objects by Cartesian or parametric equations - should also be a good way to operationalize manipulations within set theory which are useful in linear algebra (see the conclusions on the perusal of students' productions presented in the first chapter). We should add that practicing graphic representations in Cartesian space to illustrate general situations of linear algebra could be useful: it undoubtedly requires particular learning parallel to linear algebra.

It should be possible, moreover, to teach this knowledge and these practices in geometry concurrently with the beginnings of linear algebra.

1.3.   Hypothesis (3) on Long-Term Course Design, Use of Meta and Changes in Point of View

By *long-term strategy* (see Robert 1992 and Rogalski 1991), we mean a type of teaching which cannot be divided up into smaller parts. The long-term is vital because mathematical preparation and changes in the didactical contract[2] have to operate over a long enough period to be efficient for students, particularly regarding evaluation. Finally, long-term is necessary in order to take into account the fact that teaching is not linear, due to changes of points of view: this implies working on a subject matter more than once.

By the *'meta lever'*, we mean the explicit use by the teacher of elements which are not formal mathematical knowledge. It concerns: mathematics as a scientific domain (organization, different uses in other fields, differentiation between general and particular fields, etc.); mathematics as a taught subject (for instance, use of methods, types of questioning in mathematics, organizing of students' reflection on concepts, etc.). Moreover, we claim that this type of information used by teachers must be followed or accompanied by activities involving its pertinent use by students. In our notion of *'meta lever'*, 'meta' means that a reflexive attitude from the student on his/her mathematical activity is expected, and 'lever' points out something which has to be used, at the right moment, in the right place, to help the student get into this reflexive attitude while achieving a mathematical task (which requires careful preparation by the teacher). The general idea is to involve students in a mathematical activity that can be solved by them and, from there, to make them analyze, in a reflexive attitude, some possibilities of generalization or unification of the methods they have developed by themselves (this point will be developed in the next chapter).

By *change of settings or points of view* (see Douady 1986), we mean that teaching is organized in such a way that course and exercises emphasize the translations of the same concept or question from one setting into another (from formal to numerical, from numerical to geometrical, etc.), or lead the students to a change of point of view on a notion (for instance, seeing the linear equation $a_{11}x_1 + a_{12}x_2 + \ldots + a_{1n}x_n = 0$ as the n-tuple $(a_{11}, a_{12}, \ldots, a_{1n})$ in order to go from the use of linear combinations of equations to the notion of rank of a set of n-tuples, etc.).

2. THE DETAILED PLAN OF THE COURSE DESIGN

First, it should be mentioned that, in addition to the choices ensuing from our three hypotheses, a more particular choice was made to organize the teaching in such a way that it was centered on the notion of rank. It has been shown both historically and epistemologically that this concept is not only central to the learning of linear algebra, but also that it is a particularly difficult concept for the students to grasp. This suggests entering linear algebra through equations; a theory for which the notion of rank is presented naturally, and is richer because it is linked to duality. It facilitates the introduction of a problem for which the notion of rank can appear as a central issue for answering the questions. This also brings about a greater insistence on the rank of a vector system than solely that of independence.

The course design comprises four steps that we will now briefly present.

2.1. First Step

Preliminary phases are developed, including an organization of the changes of settings in order to facilitate the convergence of different points of view.

At the beginning of the course, we set up the "circuit électrique" activity (see Legrand 1990), whose aim is to help students understand the rules of mathematical reasoning. We also give the basic language of set-theory, and some experience with space geometry. Then, we introduce, during the first course, the Gaussian method of elimination for solving systems of m linear equations with n unknowns, by using $\mathbb{R}^n$ and its linear structure as the set of reference in which the solutions must be found. This immediately gives the students a powerful tool for solving problems in algebra, as well as in plane or space geometry. Moreover, this introduces what will be, for us, one of the central themes in linear algebra: the solving of systems of linear equations, from which the concepts of subspaces of $\mathbb{R}^n$, of rank, of double point of view equations/parameters will be drawn. The basic questions upon which we try to induce our students to reflect upon and which will justify the introduction of the concepts of linear algebra are:

* Do I have too many, just the right number, or not enough equations ?

* How many parameters do I need to describe the set of solutions of a system of linear equations ?

We continue the teaching with Cartesian geometry in $\mathbb{R}^3$, with this double aspect equations/parameters as a central question (not only for the representation of straight lines and planes but also for other types of curves or surfaces).

2.2. Second Step

This starts by an explicit presentation to the students of all the common questions that can be asked about the preceding phase. This also implies the change of point of view which consists in seeing an equation as a n-tuple. Then we define linear independence and rank and we show the invariance of rank using the reversibility of certain linear combinations (as done in the exchange theorem). This is followed by the description, with parameters, of subspaces of $\mathbb{R}^n$, initially defined by their equations, and *vice versa*. Then we explore the notions of dimension and bases of subspaces. In this phase, some results on the systems of linear equations, which were only conjectures in the first phase, are proved. The fact that the row and column rank of an array are the same is also proved.

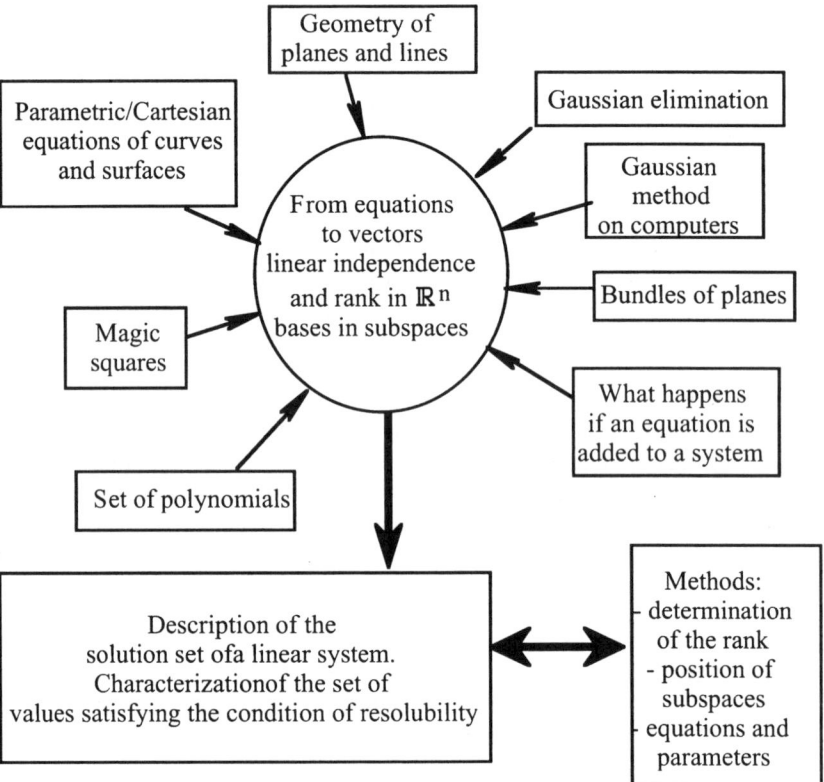

Figure 1.

The proofs of fundamental results are made with the use of abstract formulation, avoiding the use of coordinates. This is justified to the students by a systematic use of algebraic thinking (helped by some preliminary activities involving elements of group theory and illustrated by some concrete finite groups). Another justification we use is the search for generality in order to ease the transition to further developments in abstract linear algebra.

Finally, we give some elements of methodology concerning the solving of problems about subspaces in $\mathbb{R}^n$: inclusion, equality, intersection, parameters and equations of subspaces.

During this phase we also apply linear methods in other fields: polynomials, linear recurrent series, linear differential equations.

It must be noted that we remain reluctant to study the $\mathbb{R}^n$ space thoroughly, and, in particular, we avoid tackling problems whose solution would require students to understand the intrinsic nature of the notions of vectors or linear applications, independently of the representations in the canonical base of $\mathbb{R}^n$. Thus, matrices and changes in base are only introduced with the third step: general linear algebra.

2.3. Third Step

Here we teach abstract linear algebra (axiomatic theory of finite dimensional vector-spaces, linear operators) and its application as a modeling setting within mathematics. At the same time, we draw out several methods and point out some prototypical problems. These problems were solved during the previous phases, but they are now solved again in the setting of formal algebra. They have to be rich enough so that their generalization makes the use of the formal concepts almost compulsory (e.g., one can give 15 conditions for an interpolation, instead of the usual 3 or 4). The 'contract' with the students is to model the problem in linear algebra terms; e.g., the problem has to be expressed in terms of a linear equation 'T(u) = v'. In this phase, problem often arise in connection with other parts of the mathematical curriculum, outside linear algebra (e.g., differential equations).

2.4. Fourth Step

This phase is more technical and also shorter. It presents the matrices, the techniques associated with the change of bases and the inversion of square matrices. Of course, the tool 'matrix', introduced in relation to linear operator, is then used in several problems in relation to various other fields of mathematics. But matrix calculations are not an important goal for us.

2.5. Further Points

(1) It is important to see that our problem is not to set up an easy progression: $\mathbb{R}^2$ - $\mathbb{R}^3$ - $\mathbb{R}^n$ – general linear algebra, inasmuch as we believe the main difficulty to be more qualitative and global than quantitative, even if increasing didactical difficulties subsist owing to its increasingly abstract character. We have in fact insisted on the point that some problems which are treatable in $\mathbb{R}^n$ be postponed until after the introduction of the general formalism of vector spaces and linear applications. This is one of the differences with British or American practices where $\mathbb{R}^n$ is the favored model studied in depth, a choice which can be a source of later obstacles (see Harel's and Hillel's contributions in chapters 5 and 6). We believe that the problems raised by changing representation registers (especially when changing bases in $\mathbb{R}^n$) are only really solvable when the formal concepts (vectors, linear applications) are introduced and studied before or at least in parallel. The reader can refer to chapter 8 and Pavlopoulou's work for more information on these questions.

The aim of this type of teaching organization are: to single out problems, to point out phenomena that would bring these concepts closer together, to allow for changes in setting and point of view, to encourage reflection on the stakes involved, and to provide students with a methodological progression design to promote a better understanding of the concepts. From this point of view, it may seem to be an obvious paradox: in order to improve acquisitions, we develop a strategy demanding more of students with the, at least temporary, risk of decreasing their success in exams.

(2) We have chosen to teach methods in linear algebra (and in other mathematics in first year), so as to give students the possibility of tackling quite rich and difficult problems; i.e., the only ones capable of bringing out the epistemology of linear algebra knowledge and thus give it meaning. However, methods are not recipes that give answers, but **a rational means to pose pertinent questions when faced with one or several problems and to thus guide the exploration necessary to the solution**. When the general methods are seldom available to the student, we have chosen to give reference situations that can serve as models to do the same thing (particularly for modeling problems from other mathematical domains in linear algebra). For more details on teaching methods, see (Rogalski, 1990 b) and (Robert et al., 1987), and, for those concerning linear algebra, (Rogalski, 1991). See Annex 2 for an idea of what is understood by the teaching of methods.

(3) A learning software is used to visualize the Gaussian method and to give students a geometrical intuition of the linear combination of equations and the elimination of a variable. This software, developed by Carlos Sacré at the University of Lille I, has proven to be a good introduction of bundles of planes, which is then generalized to $\mathbb{R}^n$.

(4) We have students work in small groups of 4 in what we have called 'workshops' on rich and crucial problems for some aspects of linear algebra (one workshop lasts for 2 hours; 15 are scheduled for the first-year university curriculum). This type of activity, inspired by (Robert et Tenaud, 1989), better respects students' research time and encourages exchanges and discussion, and is, therefore the most advantageous time to implement the methodological procedures mentioned above. A wide variety of linear algebra themes are tackled in the workshops: for example, bases and magic squares, equation/parametric representation duality, problems of intersection and inclusion of $\mathbb{R}^n$ vector subspaces, initiation to the algebraic process on elements of group theory, use of symbolic drawings in linear algebra, introduction to the axiomatization of linear algebra (see Annex 1).

(5) We have developed situations where the activity required from students voluntarily mixed classic mathematical work and meta reflection on the stakes involved in the mathematical concepts used, their advantages or disadvantages, and the role of the theoretical detour. In other words, students are at times asked, not only to carry out mathematical techniques, but also to integrate the epistemological dimension into their work. For one of the situations, see Annex 1; for an in-depth study of another of these situations (the Gregory formula), see the chapter 4 on meta issues.

(6) Six central ideas are key points that students should acquire.

    (a) A basic technique: the **Gaussian method**
        To solve equations; for identifying the minimal number of equations for defining a subspace; to determine the rank of a family of vectors; to identify the image of a linear map; ...

    (b) A core concept : **the rank**, under various perspectives:
        rank of a family of vectors ;
        rank of a system of equations ;
        rank of a rectangular table ;
        rank of a linear map.
        Linked with this concept, four notions:
            linear independence
            generated subspace;
            vector basis;
            linear dimension.

    (c) A systematic twofold point of view : **equations and parameters**
Two ways for defining a subspace of dimension k in a n-dimension space: with $n-k$ equations, or generated by k vectors, using k parameters.

    (d) A central modeling way : **the $T(x) = y$ equation**
Many problems within the field of linear algebra, and problems concerning applications of linear algebra may be written in such a form, with $T : E \to F$ a linear map.

    (e) A permanent concern: **chosing bases adapted** to solve the problems under study.

    (f) **When unicity implies existence**: an efficient tool for proving that equations get solutions, or that decompositions of vectors do exist, unicity is sufficient under specific conditions (injectivity and surjectivity are equivalent if concerning the same finite dimension).

The objective is that ideas (a), (b), (c), and (f) become available to students, with problems and techniques allowing them to use them. 'Available' means that, if one of these ideas is useful in solving a problem, students should be capable of thinking of it themselves. For ideas (d) and (e), the objective is that they only have to be summonable; i.e., that if students are told they must use them in such a problem, they be capable of doing so, even if they are not required to be able to spontaneously use them.

3. THE POSITIVE EFFECTS OF THE LILLE EXPERIMENT

1. The notion of rank as a fundamental notion in linear algebra is generally acquired by a large majority of students and they know how to use it to determine subspace dimensions, and often to prove subspace equalities (see Annex 4 for the analysis of a written examination on this question of rank, and also the first chapter).

The techniques involved in searching for the rank are sometimes too automatized and create confusion between vector systems and equation systems in approximately half of the students. These troubles gradually disappear as the year progresses (see the analyses presented above). They seem due in part to the change in point of view that is required to grasp the unifying aspect of linear algebra. Although the issue of the transition between the practice of linear combinations of equations and the theory of vector rank seems to be understood, this transition may be at the root of the confusion between vectors and equations.

2. Changes in point of view for defining the subspaces of $\mathbb{R}^n$: roughly half of the students master equations/parametric representation in problem solving by the end of the year (see the analyses presented above). We have observed, moreover, individual differences: often, each student shows a preference for one or the other point of view.

3. Students generally succeed quite well in the study of a given linear application, in terms of the identification of its kernel and of its image, and both methods (parametric: Gaussian on the columns, and implicit: equations) are used.

4. Finally, the ability to spontaneously model a problem within the linear algebra setting seems to be acquired only by a small minority of students (see the analysis of the linear/nonlinear test in the section on meta issues). But it is clearly a difficult problem and the maturation time is perhaps insufficient since these problems are brought up quite late. Nevertheless, this ability would seem to be summonable on demand; for example, when the question is asked in the form: "model this problem as a general linear equation $T(u) = v$, specifying the E and F spaces and the T application: E→F", the success rate reaches 50% to 60% (see Annex 2).

4. ANNEXES

4.1. *Annex 1: An Example of a Workshop on the Epistemology of Axiomatic Linear Algebra*

In a workshop, small groups of four students study a problem; a group may call the teacher only if students think that they have a solution, or if they have made no progress towards a solution after a quarter of an hour.

The workshop on axioms is a new version of a 'debate in classroom' organized in the University of Grenoble 1 by Dorier (see chapter 4). Students have first to study the possible associativity of two composition rules: $x\Delta y = y^x$ in the set of positive real numbers, and $f \circ g$ in the set of numerical functions on $\mathbb{R}$. The objective is that students become aware of the fact that usual and unconscious rules on numbers are not universal; so they have to ask themselves for other rules or other mathematical objects : 'what is actually valid?'.

Then we give the students three equations to solve, in three different domains (they are linear, but this fact is not explicit): functions, vectors in $\mathbb{R}^3$, and polynomials with rational coefficients; but in the last case, there is a coefficient $\sqrt{3}$, so the equation has no solution in $\mathbb{R}[X]$. The students have to analyze the calculation rules they use, and the nature of numbers and mathematical objects on which they calculate. So they make explicit a set of rules of linear calculations, which are obviously overabundant, and are the same in the three examples.

This activity gives a motivation for the lecture which will follow, about the axiomatics of linear spaces: unifying, formalizing, generalizing and simplifying aspects of this axiomatic are valorized and better understood.

Here are the equations we give to the students:

1/ *Find if possible a real function* f *defined on* $]2,5[$ *satisfying for every x*

$$3f(x) - f(x) = f(x) + \pi[2f(x) - \frac{1}{3} f_0(x)] + f(x) + 2f_0(x) ,$$

where f_0 *is the function* sinus *on the interval* $]2,5[$.

2/ *Find if possible a vector* u *of* $\mathbb{R}^3$ *satisfying*

$$\sqrt{2} u - u = u + 3[2u - u_0] + u + 4u_0 , \qquad \textit{where } u_0 \textit{ is the vector } (1,-2,6).$$

3/ *Find if possible a polynomial* P *with rationnal coefficients, satisfying*

$$\frac{\sqrt{3}}{2} P - P = P + \frac{4}{7} [2P - P_0] + 5P_0 , \qquad \textit{where } P_0(x) = 3x^3 - \frac{6}{5} x + 4.$$

4/ *The same problem as 3/, but with* $\frac{3}{2}$ *in place of* $\frac{\sqrt{3}}{2}$.

4.2. Annex 2: Examples of the Meaning of 'Methods' and 'Methodology' in Linear Algebra

We put particular emphasis on the following : types of problems; questions one should asked oneself; different ways (often linked to different points of view) of solving problems.

These above are explicitly taught, used for solving exercises in small classes, and written in papers given to the students.

The objective of this teaching of methods is that students may be able to start off solving a problem and be able to control their solution by answering the questions posed the method. Teachers have then the possibility of giving more difficult problems to students in which concepts of linear algebra are used (and not only techniques).

Here are some examples of methods.

1)

* *Type of problem*: problem of inclusion or equality between subspaces E and F.

* *Questions to ask oneself*: What are the dimensions? How many equations for defining the subspaces? Is there a relation between equations of E and equations of F? Do we have generators of one of these subspaces?

* *Methods and points of view*: there are three points of view for the study of this type of problem:

(a) A **global point of view**, using equations (if the equations of E are part of equations of F, then $E \supset F$; if $E \supset F$ and if they have the same number of linear independent equations, then $E = F$), or dimension (if $E \supset F$ and $\dim(E) = \dim(F)$, then $E = F$).

(b) A **semi-global point of view**, using generators (if all the vectors of a generating family for E are in F, then $F \supset E$).

(c) A **punctual point of view**, using individual vectors (if each vector in E is a linear combination of vectors of F, or satisfies the equations of F, then $F \supset E$).

The choice of point of view to solve the problem depends on the answers to the preceding questions.

2)

* *Type of problem*: find kernel and image of a linear map $T: E \rightarrow F$.

* *Questions to ask oneself*: How is given map T? Is there a basis which gives for the map a simple matrix? Does the context call for finding ker(T) and/or im(T) by equations? or by generator vectors?

* *Methods and points of views*: there are two points of view, giving two methods for solving this problem:

(a) **The parametric - or vectorial - method**: 'Gaussian method on columns'. The columns of the matrix are generator vectors for im(T), and one searches a maximal free subset of them by using the Gauss method on the columns $T(e_1)$, ..., $T(e_n)$. One then easily gets im(T) and ker(T) in a parametric form.

(b) **The equations method**: 'Gaussian method on rows'. The system of equations $T(x) = 0$ and $T(x) = y$ is solved by using the Gauss' method on these equations; i.e., on the rows of the matrix. One obtains a minimal and triangular system of equations for ker(T); and the conditions of resolution on the second members form a triangular system of equations for im(T); these two systems easily give parametric forms for ker(T) and im(T).

3)
In the methodology, it is important to note the different settings where the 'objects' of linear algebra leave, and the different points of view which are useful on these objects.
We present here the text we give to the students about these notions:

Settings and points of view in linear algebra
For vectors:
* the 'pure' vectorial setting (vectors denoted by x, u,... with calculations on these symbols);
* the 'numerical' setting (with coordinates $x_1, x_2, ..., x_n$);
* the geometrical setting, in a 3-dimensional space.
For linear maps:
* the formal setting (with symbols T, u, ... and calculations on them);
* the numerical-matrix setting: (a_{ij});
* the geometrical setting for linear geometrical maps in a 3-dimensional space.
For subspaces :
* the parametric point of view;
* the equation point of view.
For matrices, for rank of vectors, for studying a linear map:
* the point of view with rows;
* the point of view with columns.
For comparing subspaces :
* the global point of view (equations, dimension);
* the semi-global point of view (with generators);
* the punctual point of view.

4.3. Annexe 3: Chronology of the Course Design (Example in 1991)

Annex 3.1 Chronology of the two first steps in the teaching of linear algebra in $\mathbb{R}^n$, equations and parameters, rank, geometry

LECTURES	EXERCISES IN SMALL CLASSES	WORKSHOPS	EXERCISES ON COMPUTER	HOMEWORK /TESTS
True and false in math.	Exercises on reasoning			
Equations and Gaussian method. Vectorial structure of $\mathbb{R}^n$, affine and linear spaces of solutions. Lines and planes in $\mathbb{R}^3$, parameters andequations, curves and surfaces, bundles of planes	Elementary set theory. Geometry of lines and planes ⌐. Exercises on the Gaussian method. Exercises on Cartesiangeometry		Visualisation of the Gaussian method with acomputer	Homework on applications of the Gaussian method to geometry
Too many equations ? Linear independence of equations, of vectors in $\mathbb{R}^n$		Links between drawings and formulas		Report on the exercises on computer
Theory of rank in $\mathbb{R}^n$, without coordinates. The two forms of descriptions: implicit and parametric, of solutions and of "conditions for a solution" for a linear system, dimension and "concrete" bases in $\mathbb{R}^n$	An activity involving polynomials ⌐ Exercises about linear independence and rank	Magic squares ⌐ Theory of groups, algebraic thinking		Test about rank, linear indépendence, dimension in $\mathbb{R}^n$, $n \leq 6$

Annex 3.2 - *Chronology of the third step: abstract linear algebra, unifying of several domains and settings of modeling internal to mathematics; introduction to methodology in linear algebra*

LECTURES	OTHER MATHEMATICS	EXERCISES IN SMALL CLASSES	WORKSHOPS	HOMEWORK TESTS
Axiomatics of linear spaces, subspaces, linear independence, rank, dimension, bases, generated subspaces	Recursive linear sequences, with variation of constant ⌐		Group theory, algebraic approach ⌐ The 'axioms' workshop ⌐	A home work on the choice of bases for polynomial spaces
Methodology: comparison of subspaces, independence of vectors (depending on the type of spaces)	Polynomials ⌐ Rational functions, decomposition ⌐	Exercises on rank Exercises in linear algebra on functions, polynomial spaces		Homework on modeling in linear algebra (functional equation or...)
Linear applications and their matrices Methodology: find the matrix of a map	Linear differential equations of first order, variation of constant ⌐	Exercises on determining the matrices of applications	Use of symbolic drawings in linear algebra	Class test: to determine the image and the kernel of a linear application in $\mathbb{R}^n$ or in polynomial spaces, or...
Kernel and image Different notions of rank Methodology: the two methods for determining the kernel and the image of an application		Exercises about kernel and image		

4.4. Annex 4 : A Test About the Notion of Rank, (1990/91)

Text

Let $X = (1,-1, 2, 0)$, $Y = (0, 3,1,-1)$, $Z = (-1,1,-1, 0)$, $T = (2,1, 5,-1)$, *and* $V_m = (1, m,1,1)$ *be five vectors in* $\mathbb{R}^4$, m *being a real number.*

1/ *Determine the rank of the following families of vectors:*
(1) $\{X,Y\}$; (2) $\{X,Y, Z\}$; (3) $\{X,Y,T\}$; (4) $\{X,Y,V_m\}$; (5) $\{X,Y, Z,V_m\}$.

2/ *Say, without any other calculations, if the following assertions are true or false (justify):*
(a) $Z \in \lin\{X, Y\}$;
(b) *the rank of the family* $\{X, Y, Z, V_{-1}, V_1\}$ *is 5;*

(c) *the dimension of* $\lin\{X, T\}$ *is 2;*
(d) $\lin\{X, Y\} = \lin\{T, V_{-4}\}$;

(e) $V_1 \in \lin\{X, Y, Z\}$.

We give only the answers of students, in the following table; for simplicity we have organized the results in three categories : total success, medium, total failure; for each question, results are in percent, for a group of 62 students.

Quest.	(1)	(2)	(3)	(4)	(5)	(a)	(b)	(c)	(d)	(e)
Success	92	86	79	55	61	89	23	71	24	74
Medium	3	6	10	26	15	5	5	16	21	11
Failure	5	8	11	19	24	6	72	13	55	15

The results are good enough, except for two questions: (b) and (d). For solving (b) with the instructions of the text (*without any other calculations*) students had to remember that the dimension was four, but the text pushed them to use the previous calculation! In   question (d), a little calculation was necessary for proving that $\lin\{X, Y\} = \lin\{2X+Y, X-Y\}$ when X and Y are independent.

We conclude that students understand the concept of rank of a family of vectors, know techniques for determining it, and are able to use the notion of rank for determining dimension of subspaces or for comparing two subspaces.

4.5. Annex 5: Capacity for Modeling by an Equation T(u) = v, (Test, May 1992)

We give here only the first question of the test, which is related to modeling.

Text:
Second problem (linear algebra)
[...] Let $\mathbb{R}_4[X]$ *be the set of polynomials of degree* ≤ 4, *with real coefficients. We will solve the following equation (e), where the polynomial* P *in* $\mathbb{R}_4[X]$ *is unknown, and the polynomial* Q *is given in* $\mathbb{R}_4[X]$:

$$(e) :   3P(0)x^4 + P''(x) + xP'(x) - 3P(x) = Q(x).$$

1°)   *Write this equation in the form* $T(P) = Q$, *with T a linear map in* $\mathbb{R}_4[X]$, *and prove the linearity of T.* [...]

(there were 7 questions: study of generators of Im(T), Ker(T), injectivity or surjectivity, resolution of the equation for a particular Q).

Context of the test

The principal part of the linear algebra course was taught before the test, except for the theory of matrices. But linear maps and modeling by the equation T(u) = v were only recently studied.

Results

This question tested the ability of students to model by a linear equation. In fact, the test was only on the summonable character of this ability, because the text said the form in which one had to model, and only the map T was to be defined (and not the spaces E and F).

But some 'functional analysis thinking' was necessary, because the map was to be found between function spaces. A cue that this level was not reached was essentially the use of the notation T(P(x)) in this question (and also in the following questions about ker(T)). But there is an ambiguity: some students use this notation and seem nevertheless to understand the notion of a map between function spaces; so, it is not easy to determine the exact boundary...

It was also necessary to justify that T took its values in $\mathbb{R}_4[X]$. Few students made this proof, but probably it is sometimes only the result of the students forgetting an 'obvious fact'.

For proving the linearity of T, it was necessary to write 'T(P+Q)(x)'; We very often found "T(P(x)+Q(x))". Was it a proof of the lack of understanding of the concept of a map between function spaces? We used the context and others questions of the problem to decide, but there remains some ambiguity.

The following table gives the results concerning these different points. Among the 122 students, 112 tackled the linear algebra problem (the test contained also a problem about differential equations), 97 tackled the first question of this problem.

Have understand modeling	57
Have proved that d°T(P) ≤ 4	17
Have proved the linearity of T	74
Attested failure about functional analysis	13
No success to define the map T	28

5. NOTES

[1] In French didactics, the term "didactical engineering" is usually preferred.
[2] This term designates the rules established, most often implicitly, between the teacher and the students with regard to what is expected in the teaching.

JEAN-LUC DORIER, ALINE ROBERT,
JACQUELINE ROBINET AND MARC ROGALSKI

PART II - CHAPTER 4
THE META LEVER

Let us first present what is understood by the expression 'meta lever', the educational justifications that are put forward for the use of this lever, and the corresponding specific work for the teacher. This will be followed by three examples in linear algebra: a lesson plan designed to make students reflect on the very notion of structure, a problem constructed so that students will become aware of the time saving that linear algebra provides in certain problem solving activities, and an evaluation of students' comprehension of linearity for those who have participated in a previously described experimental teaching situation. Finally, the difficulties encountered by researchers in their effort to evaluate the effects of this type of scenario will be discussed.

1. DEFINITIONS AND EDUCATIONAL POSITIONS

The expression 'meta lever' (or sometimes simply 'meta'[1]) designates the use, in teaching, of information or knowledge ABOUT mathematics. This can involve the operation of mathematics, its use, the learning of mathematics, and it can be its general or particular elements. In more precise terms, it means:
- information that concerns what constitutes mathematical knowledge (methods, structures, (re)organization). Methods are defined as the procedures applicable to a set of similar problems within a given field: the methods designate that which is common to problem solving and not the technique itself (the algorithm). This implies a certain classification of problems to solve and identification of available tools and techniques.
- information that concerns what constitutes a mathematical operation: for example, information on the role of interplay of settings in problem solving (see below), the role of questioning, examples and counter-examples, the role of identifying parameters in a mathematical question, the role of testing, etc.

This information can lead students to reflect, consciously or otherwise, both on their own learning activity in mathematics and on the very nature of mathematics. It is possible that such reflection helps learning (see Piaget and Beth 1961).

However, the distinction between mathematics and meta (mathematics), in the teaching of mathematics, is not considered absolute: recognizing the meta character

DORIER J.-L. (ed.), The Teaching of Linear Algebra in Question, 151—176.
©2000 Kluwer Academic Publishers. Printed in the Netherlands.

of certain knowledge depends, on the prior experiences of the students confronted with this knowledge. Thus, certain methods bring out meta for students who do not yet completely know their applications, whereas for the teacher or the more advanced student these are nearly mathematics. For them the distinction is irrelevant.

In other words, the word meta will be used if an element is contributed to the mathematics to be learned, therefore still partly non-acquired knowledge. This explains the impossibility of assimilating these elements to mathematics, or at least the usefulness of distinguishing them from 'ordinary' mathematical knowledge.

Let us give a first example, close to linear algebra, in the following exercise:

Show that the point M *verifying:* $\overrightarrow{OM} = \overrightarrow{OA} + \overrightarrow{OB} + \overrightarrow{OC}$, *where* O *is the center of the circumscribed circle* ABC, *is the orthocenter of the triangle* ABC.

This exercise is considered to be difficult for students, since if they start from the definition of the orthocenter as the point of intersection of the altitudes they will not be able to connect this knowledge to the required result. They will only be able to proceed if the teacher suggests a vectorial interpretation of the property, « M is the orthocenter. »

The following metaknowledge emerges here: mathematical notions intervene within different areas (affine geometry as opposed to point set geometry, vector geometry, analytic geometry, etc.) and solving certain problems requires that one systematically asks the question to which of these areas does the mathematical notion presently being studied belong, and then changes the area to accomodate the requirements of the problem.

Relatively systematic recourse to changing settings, especially when confronted with a deadlock, is founded on a certain organization of mathematical knowledge, specifically in terms of settings: this is what we label metaknowledge.

For example:

i) *to clarify the possible role of an epistemological reflection.*

This is preliminary work to the axiomatic introduction of the notion of vector space. As was explained in the preceding chapter, in an activity given to students, this notion is introduced to solve an entire series of problems of the same type in the same way. Then, the objective of the activities is to solve a series of exercises treating very different spaces, but all involving equation-solving (searching for particular solutions): i.e., in the set of similitudes with center O, in $\mathbb{R}^3$, in the set of polynomials with rational coefficients. At the end, students are asked to infer the common calculation rules used. This reflection precedes the statement of vector space axioms.

ii) *to inform students about the nature of the concepts that need to be introduced.*

The teacher can develop the idea that these concepts are not all of the same type, nor are they introduced for the same reason. Certain concepts are used to solve particular problems (e.g., numbers and measurements), to unify, classify, generalize a posteriori (limits, linear algebra), or are extensions of concepts that have already been introduced.

We wish to present several arguments in favor of including the concept of meta in the teaching of mathematics.

Several partial and complementary leads, from different dimensions, support the argument that such reflective elements can help students, whether it be in problem solving, in assigning meaning, or in organizing their mathematical knowledge.

• From an interactionist perspective (see Vygotski, for example), such reflection can promote greater communication between the students than the mathematics itself and may provide them with the opportunity to get more actively involved in their learning. These reflections are elements on mathematical knowledge, yet solving a problem is just that: acting on mathematical knowledge. This form of "action aid" can be particularly interesting to students if it enables them to see how certain knowledge can be reused, or recycled, in different situations, thus saving them time and energy.

• From a constructivist perspective (in the broad sense of the term) we wonder whether these interventions do not contribute to a certain imbalance, a dynamic imbalance between metaknowledge and knowledge. If the imbalance is too great, nothing happens, but if it is small enough, students are aided in taking the first step into the unknown.

• Another point to bring up here concerns the elements linked to models of construction of scientific knowledge resulting from the very formation of this knowledge. Working on access to knowledge is opening a door to working on its acquisition.

• Finally, from the point of view of "passing from old knowledge to new knowledge", certain meta elements could play the following specific role: by making explicit what is expected of the students, the metaknowledge linked to the pedagogical content would help them when a change of viewpoint or level becomes necessary. These elements could facilitate seeing relationships or making predictions, because they prepare students, by channelling their thoughts along the same wavelength, they prevent scattering of ideas. They function as an intermediary, like a bridge or a ramp between the old and the new, or even like a fertilizer (alone it serves no purpose whatsoever, but it encourages the growth of certain seeds, provided it is properly adapted, it rains enough, and the earth has been well prepared).

In fact, the elements of the meta type could be considered an explicit interface between students, a body of people in their own right, and mathematics, a body of knowledge to acquire. Students have prior representations, knowledge, and automatic reflexes, into which they have to integrate new knowledge to the old. Metaknowledge can participate in this integration, at certain precise moments, in a limited and different fashion depending on the individual, and under certain conditions.

Thus, we are working in the following general perspective: for particular cases in which the acquisition of knowledge is complex, the teacher can propose situations in which the concept of meta could act as a lever towards the efficient construction of access to the new knowledge. This is precisely the domain of mathematics education: to design and experiment with suitably well-chosen scenarios.

We admit, however, that there are certain restrictions on the way in which meta elements can be effectively used in teaching: they can only contribute to the learning process if they are properly transmitted, at a propitious moment, and within appropriate activities. This assumes that among the elements proposed to students, certain ones can be reinvested in the activities. Examples of this kind of reinvestment will be discussed later.

The characteristics of meta interventions are hard to diagnose: they are imprecise and vary from situation to situation. When they are present in teaching, they are transmitted indirectly, most often orally, and are not directly evaluated.

The optional character of this type of teaching situation should also be noted: for example, some teachers never intervene at this level, leaving students to identify (implicitly?) the methods on their own, while others intervene selectively, never distinguishing between the mathematical and meta levels in their teaching. Sometimes these interventions are very artificial, in juxtaposition to the rest of the teaching; the end result being that students have to learn more formulas without, we believe, adding anything further to their acquisition of new knowledge.

On the contrary, we use metalever to mean both explicit intervention on the teacher's part, whose specific nature is explained to students, and true 'intimate' integration of these interventions in the rest of teaching (cf. the preceding hypothesis).

There is, moreover, a certain paradox here: teachers who put into practice the metalever concept teach lessons that may seem more difficult than usual, requiring unusual reflection, but whose aim is to allow more students to succeed. The origin of this paradox may be found in the following hypothesis: it may be necessary to have students take a detour from their mathematical practices and engage in certain reflections which contribute to the enrichment of their mathematical representations in order for their practices to become more compatible with the desired outcome of the activity.

Finally, this is a long-term perspective: such teaching can only be conceived of over a certain period of time, if only to establish changes in student habits. Furthermore, this teaching includes give-and-take between teacher intervention and student action (especially in problem solving activities), and cannot be carried out rapidly or have immediate results.

Let us specify again that the metaknowledge taught to students does not aim to reproduce expert behavior, as we have seen, but rather to make the mathematics learnable to a greater number of students and to enable them to apply this mathematical knowledge, at least within a mathematics classroom situation. On the other hand, and despite diversity in the student population, we believe that it is possible to be inspired by certain elements of expert behavior, so long as we select appropriate elements that we can adapt to our hypothesis.

Hence, we have analyzed the behavior of an expert's solving process as he works on a new unresolved problem in order to find if there are elements in his behavior that are exportable to student's in a classroom situation. Indeed, the expert's activity is then of the same nature as the student's; it is during this time that a part of the meaning of knowledge is constructed, except that the expert has much more knowledge, more references, more experience than the student, and is more confident. This makes the expert go quickly, with short-cuts, verifications, almost automatic 'strategic' decision-making, even if this hides, for example, the initiation of very real methods, in terms of corresponding knowledge, of course. We have concluded from this that it would be beneficial to replace for students, artificially and partially, the expert's often completely internalized background that is used when confronted with a problem. We want to keep the richness of the problem-solving situation, but make it accessible to more students. This is our rationale for choosing to explain the mechanisms of research in mathematics, so as to encourage students to put them into practice. These mechanisms become more and more precise and diversified as the student gets older and is exposed to more complex knowledge. They are based on the mechanisms of experts, but are more unwieldy, more systematic, and must correspond to the resources of the students, and thus be adapted to their knowledge at the moment and be totally explicit.

From this perspective, part of the work of the researcher in mathematics education consists in finding, for a precise group of students, which problem-solving methods to teach, which verifications to encourage, and which automatic reflexes need to be quickly acquired.

Another aspect has been encountered: the expert reorganizing her/his knowledge, particularly to elaborate certain concepts. Here again, without 'copying' the expert, researchers have been inspired by reflection on mathematics itself, which was originally the work of the mathematician. The hypothesis that distanced analyses of this type can, in certain cases, produce facilitating effects, particularly as regards the production of the meaning of corresponding concepts, has been exported from the domain of scholarly knowledge into the domain of learned knowledge.

The work of the researcher in didactics consists of designing well-chosen mathematical activities and an appropriate contract which stimulates adequate reflection on the nature of certain concepts.

This can be stated a little differently. At the beginning of this chapter we explained that what we grouped together under the heading "meta" could in fact stem from two types of knowledge: private knowledge, corresponding to metacognition in the classic sense, when what is at stake concerns the knowledge of the individual, his own mathematical knowledge, his strategies, etc.; and public knowledge, ready to be transmitted - for example, what the teacher wishes to communicate to her/his class as elements of methods, or as epistemological references. This can be interpreted as follows: the scientist constructs private knowledge including meta elements which help him, particularly, in research situations; the teacher attempts to give students opportunities to construct and operate a part of this knowledge, thus rendering it public.

Another work for the didactics researcher is setting up corresponding scenarios, over the long term, and evaluating them.

The following sections will illustrate these ideas with three examples.

2. INTRODUCTION TO THE STRUCTURE OF VECTOR SPACE

When we set up this teaching sequence our aim was to find a way of starting the teaching of vector space theory which would be more satisfying than the usual one of stating the axioms.

2.1. Introduction

Before introducing the experimental setting, we need to present briefly the general framework of our work about linear algebra. With Harel (see chapter V of this book) or Strang (1988), we share general basic hypotheses such as:

- Linear algebra is one of the most useful theories within and outside mathematics.
- Therefore, it is indispensable in a university curriculum of mathematics.
- Students have serious difficulties in learning linear algebra, especially because it is abstract and formal.
- Although linear algebra is used in many fields, the abstractness of concepts of vector-space theory cannot be justified by only a few applications.

We have also chosen certain identical means to try to improve the teaching, such as:

- A preliminary preparatory phase in geometry (see Harel in particular).
- A substantial preliminary approach of linear concepts, in the setting of systems of linear equations, with use of Gaussian elimination (see Strang in particular).

Yet one of the main difficulties is to make the abstract and formal nature of vector-space theory less inaccessible for the students.

On this point our approach differs from that of Harel (1985) and Strang (1988).

Harel suggests a progressive increase in abstraction: from geometry to $\mathbb{R}^2$, $\mathbb{R}^3$ and then $\mathbb{R}^n$ and, finally, formal vector-space. In his teaching experiment, the first phase is fundamental as it is the basis for visual representation. The theoretical framework is the principle of multiple embodiments.

In Strang's textbook, the applications are central. The study of equations is the starting point and then the variety of applications serves to justify the abstract nature of the theory.

Our position does not contradict these two approaches. We emphasize the transition from the use of local linear tools and techniques to the formal unified theory. Our hypothesis, based on the epistemological analysis presented above, is:

- Students have to anticipate the power of generalization due to the use of vector-space theory.

In this sense, we tried to build a teaching sequence which introduces the learner to a condensed form of reflective analysis, which has been proven to be one of the fundamental stages in the genesis of unifying and generalizing concepts.

This is the context in which we set up an introductory teaching sequence to the structure of vector spaces (Dorier 1990a, 470-476/510-542 and 1991). This sequence, built on the basis of an epistemological analysis, creates an artificial context which motivates the explicitation of the vector space axioms by the students themselves.

The students were in their first year of a French science university. The sequence took place at a time when they knew very little about algebraic structures. More precisely, they knew only the axioms of a group and had learned very little theory on this structure.

2.2. Presentation of the Teaching Sequence

The artificial context used is the study of equations.

Indeed, when one says that for three elements a, b and x of a set E, and an operation T on E : « $xTa = b$ is equivalent to $x = bT(a^{-1})$ », one needs to use the three axioms of the structure of group to justify this statement :

$xTa = b$ iff $(xTa)Ta^{-1} = bT(a^{-1})$ (existence of an inverse element for each a in E)
$(xTa)Ta^{-1} = bT(a^{-1})$ iff $xT(aTa^{-1}) = bT(a^{-1})$ (associativity)

$xT(aTa^{-1}) = bT(a^{-1})$ iff $xTe = bT(a^{-1})$ (existence of a neutral element)

$xTe = bT(a^{-1})$ iff $x = bT(a^{-1})$ (property of e)

This fact highlights one of the most interesting aspects of the axiomatic approach.

Indeed, checking that a set is a group with the use of the axioms is a very easy task while, on the other hand, knowing that a certain set is a group allows one to carry out more complex operations (like solving equations) automatically, with simplified mechanisms and rules, and minimum effort of memorization.

As an introduction to the experimental sequence on vector space, students were given a five-page document which illustrated this idea. The point of view was explicitly meta, and the students were warned that the text develops quite an unusual type of argument which is not strictly mathematical; this is an explicit change in the didactic contract. The students had about two weeks to read the paper, which gave them time to read it more than once with different points of view. This being a fairly unusual activity in mathematics, it was also important that the students could have no constraint of time and could reflect at their own pace.

Even though this paper was mainly 'a talk' from the teacher, attempting to create better conditions for devolution and to make students more critical, they were also asked to solve mathematical questions connected with the meta ideas developed. Indeed, the meta level cannot exist without referring to a mathematical activity. The underlying hypothesis is that students frequently use certain local metacognitive skills to solve a problem or learn mathematics, but in a very implicit way. Moreover, the meta level cannot be initiated just by a teacher's talk; but if the talk happens when a student is solving a problem, to which the meta level is clearly related, then the conditions for its appropriation are optimum.

For instance, in the document given to the students, we wanted to explain the role of mechanisms when solving equations. So, as a preamble, we gave a short problem in which students had to solve two equations with geometrical transformations. The two equations looked, a priori, very similar. But, in one case, the equation involved only rotations, so one could operate in a group and solve the equation easily, without much competence in geometry. In the other case, there was a projection (which was not one-to-one), so that tools specific to the context had to be used. First, students were asked to solve these two problems without further indications. Then they were led to reflect on the different competences they used to solve the questions, and the reasons for these differences. In this second phase, students had less initiative, since the reflection was guided by the elements developed in the text by the teacher, but their previous mathematical activity should have optimized the impact of the talk. Thus, the devolution of the meta questioning depended on the choice of the previous or simultaneous mathematical activity on which it relied.

The end of the document concerned additive groups (i.e., Abelian groups) and their 'natural' structure of $\mathbb{Z}$-module. This last point would serve as mathematical background for the introduction of the structure of vector-space[2].

So, this first phase in the sequence is an introduction to initiate a reflection from the students, on a meta level, about the concept of structure. Actions at the meta level are still under the responsibility of the teacher since students are mainly asked to carry out mathematical tasks in order to make the meta-discourse more meaningful and relevant to their mathematical knowledge.

Two global characteristics are fundamental for the role of this part in the teaching sequence:

1. The choice of a written paper given to the students to be studied for a fortnight is motivated by the fact that it leaves more time to understand this activity which, being quite a new type of task, might require a certain period of adaptation. Students can also go back and forth, and use different levels of reading.
2. While they are presented with elements of reflection on mathematics, this 'philosophical' discourse is directly connected to mathematical questions they have to solve. In this way each throws new light on the other, in a dialectical process which gives the student a more active appreciation of the meta questioning.

It was also important that students can talk together and with a teacher about the issues raised in the text during the fortnight. In the experiment described here, students had time during tutorial sessions to talk about the paper. Their reactions and questions showed that they were able to understand the meaning of the activity and attained the essential issues we had tried to make them discuss.

The actual teaching sequence took place two weeks later, with the whole class (about 120 students), but it was not a usual lecture, it was more interactive and was inspired by the 'technique' of 'scientific debate' (Legrand 1990).

The main idea was to introduce a vector-space as the kind of additive group in which it is 'easy' to solve linear equations. The first step concerned the generalization to the rational field of the external multiplication by integers defined on groups. Again, it started with a problem involving an equation.

The teacher introduced the equation : $2 \cdot x = x_0$ in an additive group, which led to the conclusion that its solving required that some meaning be given to $\frac{1}{2} x_0$, therefore, it was decided that an analysis of the problem of equation solving in additive groups, in which $a \cdot x$ has a meaning for any rational number a, should be undertaken(x being an unknown in the group). A fairly straightforward idea was that this generalization should be compatible with the 'natural' meaning of external multiplication by integers previously introduced in additive groups.

The students were then asked to make explicit the properties they used to solve:

$$a \cdot x = b \cdot x_0 \quad (a \neq 0), \qquad by : \ x = \frac{b}{a} x_0$$

$$and \ \ a \cdot (x + x_0) = b \cdot x \quad (a \neq b) \qquad by : \ x = \frac{a}{b-a} \cdot x_0 .$$

With the properties of multiplication by integers (given in the paper on groups), we obtained a list of about ten properties:

(1) $0 \cdot x = 0$

(2) $1 \cdot x = x$

(3) $\forall n \in \mathbb{N}, \ n \cdot x = x + x + \ldots + x \ \ n \ times.$

(4) $\forall n \in \mathbb{Z}, \ n \cdot x = -(-n) \cdot x = - n \cdot (-x)$

(5) $a \cdot (b.x) = (ab) \cdot x$

(6) $a \cdot (x+y) = a \cdot x + a \cdot y$

(7) $(a+b) \cdot x = a \cdot x + b \cdot x$

(8) $(-b) \cdot (-x) = b \cdot x$

(9) $(-b) \cdot x = -(b \cdot x)$

(10) $b \cdot (-x) = -(b \cdot x)$

At the end of this first phase, the teacher defined a vector-space as an additive group in which these properties are true. He then showed that some property, for instance property (4), could be deduced from the others, which means that it is automatically satisfied if the others are. Following this remark, the next task was for the students to obtain a set of properties, as small as possible, so that if they are satisfied, all the others are also satisfied. Thus, the students had to find a set of axioms, although no special interest was given to the problem of minimality. Technically, they had to handle the questions of logic raised in this operation .

At the end of the sequence, the students obtained the definition of a vector space over $\mathbb{Q}$, which was easy to generalize to one over R. It was important to make it clear to the students that the main purpose of the sequence was not to give this mathematical result, but to focus on all the argumentation which led to it.

The context of equation-solving worked as a paradigm to illustrate the idea of simplification and generalization included in the concept of structure. Although this paradigm did not, historically, play a dominant part in the explicitation of algebraic structures, it appeared to be advantageous in teaching. Indeed, students were much more involved in an active participation than if they only had to listen to a talk from their teacher, trying to convince them of the power of simplification of the concept of algebraic structure. They had to carry out a reflective analysis in the setting of equation-solving. In addition, equation-solving being the basis of algebra, as numeration is the basis for arithmetic, it created an a priori good motive for entering the subject.

This sequence, which is based on previously acquired mathematical competences and elements of knowledge, tends to give the student knowledge with meta characteristics on algebraic structures and their 'raison d'être'. So, previous competences and elements of knowledge about equations are used as a basis to serve a purpose of a completely different nature. The sequence involves a change of point of view on the use of equations. The main task of the sequence induces the analysis of old knowledge (solving equations) from a different angle and for an unusual purpose. The work of the teacher in the sequence itself might seem quite limited. As previously mentioned, the fact that the teacher has little to do proves that the devolution of the meta questioning is effective. While most of what the students have to discuss during the sequence is strictly mathematics, the meta level is what makes the experience hold together. Indeed, the meta questioning is the only way to make the mathematical tasks meaningful. Therefore, the success of the sequence is to be evaluated through the quality of the mathematics developed by the students, even though the goal is on a meta level. Moreover, the success of the sequence is based more on the quality of the document previously given to the students, and the conditions under which it is given, as well as on the quality of the a priori analysis of what can happen with the students, than on what the teacher will say to initiate the reflective analysis. In other words, most of the work has to be done before the actual sequence. The reflective analysis will not be initiated magically, but relies on accurate previous mathematical preparation.

2.3. Results

This experimental sequence was taught twice under similar conditions. The lecture and discussion were recorded and two observers were asked to take notes about the general 'atmosphere' in the classroom. They used codes to evaluate different elements throughout the sequence, such as : How many students are actually involved in the discussion ? To what extent do the students who do not talk concentrate? How well does a new idea seem to be received ? etc.

On both occasions, the whole sequence lasted around three hours (in two sessions, on two consecutive days) and we observed very similar reactions from the students.

First, the meta level introduced was well accepted. By this I mean that students did not find that the type of activity proposed was irrelevant and, moreover, they showed great interest, getting involved quickly and actively. In the framework of the 'scientific debate', a good evaluation is provided by the number of students talking and the richness and relevance of the discussion; in other words, by the dynamic aspect of the debate. In both experiments, results were very encouraging and the observers noticed good concentration even from students at the back or on the side of the lecture hall.

Moreover, on the content, their reactions and their answers showed that they were able to follow the idea of building an axiomatic structure on the basis proposed by the sequence. They gave, without difficulty, all the properties implicitly used when solving the linear equations given.

The reduction of the set of properties brought about a very lively and rich discussion among students in which the teacher did not need to interfere, except for writing on the black-board and summarizing in a final phase of institutionalization.

This does not mean that the students reached the right point straight away. Indeed, a few false arguments were given by students: using a property implicitly without noticing, or a property already erased, etc. But there was always at least one student who corrected his or her friend with a convincing argument. The fact that the discussion was regulated by the students, without any help from the teacher, serves as an indicator of the devolution to the students.

Moreover, a fairly substantial discussion on the problem of consistency emerged. Indeed, one problem of logic, crucial in this activity, was:

Suppose property (a) has been put aside (it will not be an axiom, because it was deduced from other properties of the list). Now one knows how to deduce property (b) from the others, but only in a way using property (a). Can property (b) be put aside or should it be kept as an axiom ?

The answer is:

If (b) was used to deduce (a), then property (b) has to be kept on the list of axioms. If not, (and if the properties used to deduce (a) are still on the list of axioms), (b) can be deduced by properties remaining on the list and therefore does not need to be kept as an axiom.

This question of consistency ended up being a major issue of the discussion and was, in both cases, solved satisfactorily after a fairly long, rich debate among the students, without any substantial intervention by the teacher.

This is a very interesting result, because very often we hear that one of the main problems inherent in teaching the theory of vector-space is that students cannot use logic or the language of set theory. But this experiment shows that they are perfectly capable of debating, among themselves, on very abstract and formalized issues which involve only logic and the language of set theory. This proves that the problem is not that they cannot use formal and abstract notions, but that students need to know why they have to deal with abstraction and formalization in order to be able to do it correctly. Of course, this is too radical a statement and the results described above only allow a more moderate conclusion. In any case, it seems justified to say that through a certain type of explicit meta questioning, questions of formalization and abstraction can be solved by college level students.

This fact is important because formalization is an inevitable component of unifying and generalizing concepts. The reflective analysis initiated in the sequence allows students to anticipate, through the analysis they are led to make, the benefits of the effort of formalization required in the axiomatic approach.

2.4. General Issues

There are some characteristics of this sequence which are likely to be general conditions for similar sequences about generalizing and unifying concepts. The context used here (solving equations) is familiar to the students. The central question, which motivates the whole sequence, is unusual and its nature is essentially meta. This prompts, right from the start, the necessity for a change of point of view from the students. It is an underlying preoccupation which directs the evolution of the sequence. One of the main problems is to initiate the reflective analysis, without making completely explicit what the students have to experience themselves. As in any didactic situation, there is thus a problem of devolution: the reflective analysis is an explicit part of the knowledge, which the sequence tries to reach, so it has to become a preoccupation of the students, not imposed by didactic means, but through constraining components of the situation itself. In the case of meta knowledge, though, the question is not easy to analyze because it is dependent on the mathematical content of the activity, and especially in its pertinence for the illustration of the characteristics of the reflective analysis. It means that this situation must impose a completely new way of approaching a supposedly well-known problem, so that students have to start a new reflection on meta bases, and this has to be explicit in the terms of the didactic contract.

This has also to be visible in the material organization. One essential point is that students must have time to think about what they are asked to do and be able to discuss their findings with their peers. Indeed, it seems that, concerning a meta aspect of knowledge, discussion and confrontation with other students' views is even more necessary than usual, since there are fewer rules governing this type of knowledge, opinions are more likely to be different, at least in their external appearance. Moreover, reflective analysis suffers from difficulty of externalization, of being put into words. It is therefore important that students be given the opportunity and the necessity of explaining their position to a peer; this might be a better motivation and an easier approach than if they had to communicate with the teacher right away.

2.5. Institutionalization

In the phase of institutionalization, it has to be explicit that the goal of this sequence is not the explicitation of the axioms of vector space but the work, which leads to it. This phase, which is the conclusion, is important and is mainly under the responsibility of the teacher. He, or she, has to point out the knowledge acquired during the sequence, and has to objectify it, and match it with the institutional knowledge, of which he, or she, is the guardian. Here, the nature of the knowledge to be institutionalized is essentially meta, and it will not be a theorem or a property as is usually the case in mathematics. This is why it is even more important to concentrate on this final phase. Usually, a meta aspect in teaching mathematics is seen as secondary and very rarely belongs to official knowledge. To give the reflective analysis the role it deserves in the process of learning, it is necessary to make a special effort in the phase of institutionalization, which will consecrate it as an official means to acquire an essential epistemological aspect of any unifying and generalizing concept. It is thus important to say to the students that the axioms are the result of a search for unification and generalization, while referring to what has been done previously and anticipating what they will have to do next. In fact, after checking that the axioms are valid for various sets (n-tuples, geometrical vectors, functions, series, polynomials), one can apply general tools and methods identically to all those sets. It is thus easier to understand what one has to gain in working in the formal and abstract setting of vector space when the results can be seen to be applied straight away in many different situations. This essential point in the institutionalization is the starting point of the theory of vector space. These results can be applied straight away in many different situations.

To summarize, the institutionalization must try to show in what way the process the students have followed is a condensed version of a general process which makes mathematical activity more efficient.

2.6. Long Term Evaluation

Although the observation of the direct effect of the sequence yielded a fairly positive evaluation, such a sequence cannot be evaluated on its own. When isolated, its effect is expected to be of short duration, even if it is possible to see some positive immediate indications. Thus we encounter the problem of long term experiments (Robert 1992) and of interaction of different aspects in the teaching of a wide mathematical domain, necessarily spread out over a long time session. Initiating the students to a form of reflective analysis is useful only if followed by activities during which the ability of students to reflect on their mathematical work is exploited to optimize their learning. In other words, this type of introductory sequence is just one part of a more complex set, which inevitably has to be considered as a whole, at some stage in the didactic analysis. Tools of macro-analysis necessary for such a use might be quite different from those usually present in didactic micro-analyses.

Moreover, external effects are more important since the time span is longer. The signs of stability detected in the experimentations here are valuable results, but it would be useful to be able to evaluate the role of such a sequence in a global approach of the teaching of linear algebra.

In this connection, it was observed that students involved in this experiment made fewer mistakes when they had to check that a set is a vector (sub)space or that a function is a linear operator. We also analyzed the results to the question (Dorier 1990b):

If u and v are linear mappings of the vector space E, is the following statement true or false:

$$v \circ u = 0 \quad \text{iff} \quad \text{Im}(u) = \text{Ker}(v) \, ?$$

We compared the results of students who had followed a standard teaching with the results of our students. The difference was significant and showed that our students were less disconcerted by the formal aspect of this question than standard students.

3. AN EXAMPLE OF THE USE OF META FOR TEACHING THE GREGORY FORMULA

The unifying and generalizing aspects of linear algebra are not always ignored in teaching, but their illustration generally remains implicit and, consequently, students cannot access this dimension if the stakes are beyond their reach. In the following pages, we will give an example of this type of situation. After an evaluation of the effects, we will show how we attempted, with the same content, to explicitly introduce a meta dimension, incorporating the unifying and generalizing aspects of the linear algebra tools used.

3.1. Presentation of the Mathematical Content

Polynomial interpolation consists, knowing certain values of a function and of its derivatives, in finding the polynomial (with a degree not more than the number of these data) having the same values as those known for the function. In theory, it is possible to solve such a problem directly by posing an unknown polynomial and writing the relations of interpolation. A squared system of linear equations is thus obtained whose unknowns are the coefficients of the polynomial. Yet this method, although sufficient for isolated cases with few data, proves very costly, even prohibitive, if there are many interpolation calculations to carry out with a great deal of data - as is the case in many fields of applied mathematics. Hence there is an advantage to develop interpolation formulas to obtain a very quick result. These formulas are established on the basis of general results on the operators of finite difference

($\Delta: P \rightarrow Q$ such that $Q(X) = P(X+1)-P(X)$) and of derivations in the vector space of polynomials with degree not more than n (noted $\mathbb{R}_n[X]$). Actually, the problem is to find a basis of $\mathbb{R}_n[X]$ adapted to the type of interpolation studied. This search often uses the operators Δ and D and their iterates.

This, therefore, is a very good illustration of the unifying and generalizing aspects of linear algebra, since a more theoretical approach to the problem - using elementary tools of vector space theory - produces a formula which avoids a great number of repetitive and laborious calculations.

The formula, known as the Gregory formula, is one of the interpolation formulas. It is used in the case where $n+1$ consecutive values of the function, in increments of 1, are known. The adapted basis is thus the basis $(1, X, X(X-1), ..., X(X-1)...(X-n+1))$ (hereafter noted $(U_0, U_1, U_2, ...U_n)$), and the formula is:[3]

$$P \in \mathbb{R}_n[X] , P = P(0) + \frac{U_1}{1!} \Delta P(0) + ...+ \frac{U_n}{n!} \Delta^n P(0).$$

Elaborating such a formula includes a large heuristic component founded on prior knowledge of the operator Δ and its links with the bases of the U_i. Once the basis has been approached, the rest of the demonstration is very mechanical: verification that it is truly a basis (in fact, it is only necessary to check that the vectors are linearly independent by an argument of dimension) and calculation of the coordinates of an undetermined polynomial of $\mathbb{R}_n[X]$ in this basis.

3.2. Use of the Gregory Formula in a Classic Setting

In a first publication where we analyzed all the assignments done in a limited time frame over one year in a classic section of first year university science students (Dorier 1990a and b), we analyzed a mid-year exam subject where students had to establish the Gregory formula.

The problem (taking approximately half of a four-hour exam) was made up of two exercises. In the first one, students were asked to determine all the polynomials P (with degree not more than 3) such that $P(1)=0$, $P(2)=2$, $P(3)=10$ and $P(4)=30$. In this context, it is clear that students were expected to build a model in order to solve a 4x4 linear system. The majority of the students accomplished this task easily. In the second exercise, independent of the first one, students were asked, after three technical questions, to validate the Gregory formula (which was given). The three preliminary questions were, of course, preparatory questions, in the sense that they led to results that facilitated proof of the formula. But this aspect was not explicit in the declaration of the problem. Finally, students were asked a fifth question where they had to find all the polynomials P with degree not more than 3 such that $P(0)=0$, $P(1)=2$, $P(2)=10$ et $P(3)=30$, and to compare this with the first exercise.

Anyone informed on the subject can easily see in this problem the implicit reasoning process that aimed at illustrating the unifying and generalizing aspects of linear algebra: use of the direct method, then establishing the interpolation formula; use of the formula in an example and comparison with the direct method to show the gain brought by the more elaborated method. Nevertheless, nothing proves that the teachers had this goal in mind. Yet both the a priori analysis and the analysis of the results obtained by the students showed a dysfunction between this legitimate implicit goal and the observed student outcomes. It is true that the simplification brought by the Gregory formula is relative; it is possible only in so far as the interpolation calculations to be carried out are numerous or concern large amounts of data. One could supposedly imagine that the advantage of the formula would be felt on a purely esthetic level, but this assumes that the students fully possess the tools of linear algebra and a minimum of algebraic culture, which is not the case for most first year students. Thus, for a student, establishing the formula requires a great investment: the use of tools and concepts only partially mastered. The cost of such work seems much higher than what is gained by using the formula for a single example of interpolation with four data, compared to solving a 4x4 system. This a priori prognosis was largely supported by our observations. Moreover, another difficulty was added in the second question when students had to calculate the value of $\Delta^P U_k(0)$. The discussion of the case concerning the relative values of the two indices could only be completed successfully by a very small minority of the students. Very few students (7%) actually validated the formula, having tripped up on certain technical aspects of the preliminary questions which, though they did not directly bring into play knowledge of linear algebra, did involve associated knowledge. In this context, what happened to the fifth and last question? Even though very few validated the formula, all students had it available for use because it was given in the problem. In fact, only 45% of the students answered the question. Among these students, fewer than one-third attempted to apply the formula and only one-tenth succeeded. All the others applied the direct method and solved a 4x4 linear system. Finally, none of the students attempted to answer the question « compare with the I ».

Our aim here is not to criticize the colleagues who produced this problem, but to try to see how to construct, on the basis of this idea, an exercise which illustrates the unifying and generalizing aspects of linear algebra. It is this construction that we will now briefly sketch.

3.3. Reformulating the Problem

As we stated above, establishing the Gregory formula (therefore the use of the theory of vector spaces) does not provide an immediately recognizable simplification. In addition, if one must solve only one or two interpolations with a reasonable number of data, the direct method (solving the linear system) is undoubtedly more economical. It is true that finding the interpolation formula demands a great deal of work because it requires elaborating a relatively complex strategy and calls on quite precise knowledge of vector spaces. Nevertheless, the benefits, even if restricted to the practical problem on interpolation, are substantial, though not immediate. In fact, a more global perspective is necessary to benefit from its possible long-term advantages. One must, therefore, change one's point of view on the question of interpolation in order to shift from a vision of a local solution to the search for a global strategy, whose characteristics must be the most global possible so that they can be generalized. This is why analyzing the interpolation problem is an unavoidable step in the process of recognizing the interpolation formulas as more powerful tools than the direct method. Our goal is to analyze how students can integrate this requirement through a written exercise. It is a question of seeing how they can carry out an analysis on a mathematical problem to understand the unifying and generalizing power of the tools of vector space theory, and grasp the advantage of a theoretical detour in an approach to a problem which can otherwise be solved directly by less elaborated methods.

At the university level, we believe that students are mature enough to accept this type of approach. In addition, they have a broad enough background knowledge to be able to compare different methods and imagine different contexts so as to judge the validity of these methods. Nonetheless, it is far from easy to bring them to spontaneously ask these questions, since this type of questioning has not been part of their prior experiences. This is why this type of activity requires, first of all, that students have the time to think about it and the opportunity to exchange opinions with others, and then that the didactical contract be explicitly modified by the teacher so that students are aware of the real questions brought out by the exercise. The first requirement is easy to satisfy by giving the exercise as a homework assignment rather than in class with a preset time limit. For the second point, we believe that it is essential to differentiate the purely mathematical level of the questioning from the more meta level, within the very structure of the text of the exercise. In addition, we hypothesize that the composition of the text should be made on two interactive but clearly distinct levels, each containing information or data and questions for students. Thus, students will explicitly assume the meta dimension with this contract (as an official part of the exercise) on the same level as the mathematical aspect. More precisely, in our case, it is possible, with an oral or written introduction, to warn students that they will not only have to solve mathematical questions, but will also have to give an explicit opinion on the validity of different methods used to solve identical problems in different contexts. This clarifies, in advance, a new component of the didactical contract. Moreover, it seems necessary to explain briefly to the students, early in the text, the type of difficulty raised by interpolation. It could, however, seem a little abrupt to begin with comments on something with which the students have not yet had the opportunity to experiment. More generally, it seems preferable to give them a mathematical task before bringing out the meta aspect, which would otherwise not have a concrete foundation. We could begin, therefore, by asking students to solve an interpolation question with only three or four data. They would thus encounter the problem of interpolation actively without the fear of failure. This should facilitate student acquisition of the meta issue.

At this stage, the change in point of view on the mathematical question seems inevitable. Here, a simple and entirely efficient way to attack the problem would be to propose to the students an interpolation problem of the same type, but with much more data. Obviously students are not asked to solve the problem with a linear system. On the contrary even, the amount of data should be enough to cut short any temptation to solve this question by the direct method. Indeed, the increase in the number of data is the element that should create the need for a change in viewpoint, before the analysis. Therefore, this need is not imposed on students by didactical artefacts, but by the presentation of a mathematical question that cannot be solved by the only method they know. This is a didactical variable which, by introducing an informational leap, forces students situationally to a change in viewpoint. Thus, after presenting students with a new interpolation to carry out from a dozen data, they can be asked the size of the system to solve if the direct method were applied.

Following these two relatively short questions, students should have recognized a context in which they have skills and, almost simultaneously, have realized the limits of these skills when the context changes slightly. We hypothesize that this provides a fertile terrain for opening a meta-type reflection on this problem, especially if we encourage students in this direction. This, therefore, seems the right moment to give a few comments in the text presenting polynomial interpolation. For example, it can be said that it is used in many fields and sometimes with very large quantities of data. The need for a more effective method to solve interpolations with large numbers of data should also be rapidly explained here. We will then suggest doing this, using the tools of linear algebra.

Analysis of the preceding propositions brings out two essential hypotheses:

1. Students must be warned of the new aspects of the didactical contract: they should think about, and then express, their opinion on their mathematical activity and on the validity of certain methods.

2. The mathematical level should be used as much as possible to introduce meta-type questioning. This is fundamental so that students begin a reflexive analysis. Students cannot begin thinking on a meta level before encountering a mathematical situation that will lead them, as naturally as possible following the teacher's direction, to questioning the method they have used, the setting in which they have worked, the possibilities of generalizing from the activity, and the conditions necessary for reproducing the generalization.

The second phase consists in the actual construction of the more efficient method, in other words, the interpolation formula. It is, therefore, essentially, a purely mathematical activity. Nevertheless, its effect on the rest of the problem, and, therefore, on the meta level, is entirely essential. It does, in fact, seem important that this phase avoid putting students in a context that is too technical or academic - at least not for too long - which could bring on a passive attitude and cause them to forget the terms of the new didactical contract. This should result in an overall meta-level attention, on the effect of any mathematical activity. It is also important, to meet this objective, that the general question of the problem (on the choice of interpolation methods) not be masked by mathematical questions that are too technical. On the other hand, if students wish to compare the different methods with realistic data, they should be able to truly judge the difficulties inherent in the discovery and construction of the Gregory formula. More precisely, they must encounter real choices so that their judgment on the different methods is reasoned. In the case studied here, it would be interesting to try not giving the formula in the problem, but allow the students to discover it with the minimum of instructions. However, it is just as important to ensure success on the part of the majority of students. This teaching question is also particularly problematic to solve; it requires, in the end, the ability to precisely judge the abilities of the average student in solving such a problem with little guidance. It is also possible to have students solve certain technical points in a preliminary phase, before beginning the actual problem. This way, without guiding them too much, students can be asked to construct the Gregory basis, used as a family, verifying certain properties in relation with the operator Δ (for example, the same as those of the Taylor basis, in relation with the derivation operator, a classic case in first year university studies).

These preliminary exercises present certain advantages. First, they bring out similarities between the two formulas as simplifications through the use of the vector space theory model. Next, they distinctly separate two different types of tasks in the discovery process: modeling and technical solving. Presenting the second before the first is a didactical artefact, a means to avoid confusion in the goals to reach in each of these tasks. The preliminary part can appear a little academic and cut off from its raison d'être, since it is only justified later. However, this seems to be the best way to solve, at least partially, the contradiction outlined above.

The last phase of the problem consists in actually solving the interpolation with a dozen data, using the interpolation formula, and this should lead the students to draw their own conclusions on the validity of each method according to the context. Asking students, at this stage, to use the interpolation should pose no problems.

The second aspect of this last phase contains, for the teacher, in a sense, a form of immediate validation of the entire activity. By examining the students' responses, the teacher may be able to measure the effects of the problem on the ability of students to think about the unifying and generalizing nature of linear algebra - at least the local and instant effects. The problem statement should end with a question explicitly on the meta level in which students will have to express their opinion on the efficiency of each method according to the context. The answers to these questions may very well be extremely varied for different students. The number and the consistency of the arguments advanced by the students may be good indicators for evaluating the success of the meta issue's progress. A simple response of the type « the interpolation formula is a faster and simpler method than solving a system » is obviously only a superficial level of analysis that shows only a tacit acceptance of what is suggested in the exercise. Therefore, it reveals no true appropriation of the meta issue. On the other hand, students can expand the question to a more global perspective and consider the efficiency of the methods as tools that are useful to an entire class of problems. They could, for example, consider the difficulty of the search for the interpolation formula compared to its universality, as well as the possibility of constructing other formulas on the same model for other types of interpolation. The degree of generality contained in their arguments thus appears to be a more revealing sign of the depth of their analysis.

3.4. Results

On the basis of what we have just said, we carried out an experiment whose results were published in Dorier (1992). This experiment showed that students are capable of critical thinking on their own mathematical activities and of giving signs of appropriation of a meta issue. Thus, the answers to the last question, although they sometimes do not go beyond the level of paraphrase, often show personal reflection on the problem. On the mathematical level, students had no major problems. This can be explained in large part by the homework format. However, discussions with students revealed that many had been perplexed by the division into two parts. Indeed, the first technical part required that they invest a large amount of effort while still uncertain as to what was being asked of them. Following these criticisms, we attempted a new version, but the text of the exercise became too long, discouraging the students. This question of the technical aspect of establishing the Gregory formula is, therefore, not entirely resolved. Moreover, it was discovered that the homework assignment had not been given at the best time. Students had little previous knowledge about linear applications, thus, their comprehension of the first part was more superficial due to their lack of experience. This last remark raises, furthermore, a more general question linked to the maturation time necessary to bring a student to a point at which using a meta analysis is meaningful. Even more so than in the case of a strictly mathematical question, the positioning in time of the teaching of a meta-type activity is an important issue which, if poorly managed, can be the sole cause of the failure of the activity.

We would say that the chosen mathematical setting is very well adapted to introducing meta reflection, a claim for which we can advance two essential reasons. First of all, interpolation is a problem whose principle is easy to understand and whose usefulness can be demonstrated, or at least touched on. Next, solving a system by the Gaussian method is a tool that students master well, but only in the case of no more than 4x4 systems. The sudden increase in the number of data is, therefore, a factor of surprise and questioning which favors the acquisition by the students of the meta issue.

Although certain difficulties remain to be resolved, it seems that such an exercise reaches the goal set out; i.e., to bring students to a reflection on the epistemological nature of linear algebra. Its impact is all the stronger from the fact that this meta reflection is linked to solving a precise mathematical problem based on students' prior knowledge. However, the fundamental question of knowing how to transform this activity into teaching material, spanning several weeks, remains to be resolved, especially since this difficulty is raised again in the problem of evaluation. Indeed, the meta aspects should be evaluated foremost over the long term, in their own interactions, as well as with the strictly mathematical content of polynomial interpolation.

4. UNDERSTANDING LINEARITY AND NON-LINEARITY

In our experimental teaching in Lille, we try to encourage the general model of the linear equation: $T(x) = y$, where $T: E \rightarrow F$ is a linear mapping between two vector spaces, $y \in F$ is given, and $x \in E$ is unknown. We give a few examples here, with explanations of their solutions, specifying that students must see them as models in order to try to "do the same" when faced with similar problems. Particularly important is the relationship between the solutions to the equation and the solutions to the associated homogeneous equation; equally important is the frequent possibility of using methods such as the 'variation of the constant' in finding a particular solution. The examples given deal with numerical systems, linear differential equations, simple functional equations, linear recurrent series, and writing a vector as a linear combination of n given vectors. In addition to what will be described herein, we have recently introduced an example dealing with linear Diophantine equations, as an application of the notion of module - a generalization of the notion of vector space.

We wondered if this aspect of teaching truly brought students to an awareness, in meta terms, of the strong possibility of solving the problem if it is linear, whereas it would undoubtedly be much more difficult, if not impossible, if it were not linear. Such awareness could indeed greatly help in problem solving, in that it encourages consideration of the possible linearity of a problem to solve, and can even lead to the adoption of one of two strategies: either trying to solve a neighboring linear case, or trying to linearize the problem. We must mention here that we did not engage in any explicit meta teaching on this subject, preferring to emphasize the general and very useful aspect of the equation, given the ubiquitousness of the equation $T(x) = y$. We sought to detect spontaneous student use of meta which would have allowed evaluation of the role of linearity, based on successful use of it, but with neither true experience of the difficulties of linearity nor of direct teaching on this point.

To evaluate the degree of student awareness of this linearity/non-linearity distinction, an exercise was given in May 1993, the objective being to test if students could establish a relation between the linear character of a problem and its solvability (that is for them, with the means at their disposal).

4.1. Presentation of the exercise

The 7 following problems were given to students, who were <u>not required to solve them</u> (except one):

(a) For each of these problems, indicate whether or not you see a method that can be used to solve the problem. If you see one (or several), describe it (or them) in a maximum of 2 or 3 lines (this does not mean solve the problem).
(b) Choose one of these problems and give a complete solution .

The 7 problems:

(1) Given a continuous numerical function g on $\mathbb{R}$*, find a numerical function* f *of class* $\mathcal{C}^1$ *on* $\mathbb{R}$ *such that for every real number* x: $f'(x) = \dfrac{\cos x}{1+x^2} f(x) + g(x)$.

(2) Find the sequences of real numbers (u_n) *satisfying the relation:*

$$u_{n+1} = 3u_n^2 + 2.$$

(3) Given the polynomial Q of $\mathbb{R}_n[X]$*, find a polynomial P of* $\mathbb{R}_n[X]$ *such that for every real number* x: $P(x+1) - 2P(x) = Q(x)$.

(4) Given the real numbers a, b, c, find the real numbers x, y, z such that:

$$x^2 + y - z = a$$
$$-x + y^2 + z = b$$
$$x - y + z^2 = c.$$

(5) Given a vector V of $\mathbb{R}^3$*, find the vector X whose orthogonal projection Y on the plane with equation:* $x + 2y - 3z = 0$ *verifies* $U \otimes Y = V$*, where U is the vector of* $\mathbb{R}^3$ *with the coordinates* (-1, 1, 2) *(*$U \otimes Y$ *designates the vector product of U and Y).*

(6) *Find the sequences of real numbers* $(u_n)_{n \geq 0}$ *satisfying the relation:*

$$u_{n+1} = 2u_n - 7.$$

(7) *Solve the differential equation* $y' = 4x^2 y^3 + e^x$.

4.2. Rapid analysis of the exercise

Problems 1, 3, 5, and 6 are linear, problems 2, 4, and 7 are not, and are unsolvable or beyond the reach of the students. Among the linear problems, the first cannot be solved in a totally explicit manner, given that the primitives to be calculated cannot be expressed through usual functions. The linear problems 1, 3, and 6 were very similar to problems or exercises seen in class and were explicitly presented as linear. Only problem 5 was less common owing to its more geometric nature but the Olinde Rodrigues's formula, using the vector product to find the matrix of a three-dimensional rotation, had been seen in an exercise. Nevertheless, the solution is difficult and the problem easily classed as 'nonsolvable' for students, with or without reference to its linearity. The response, 'solvable because linear', would be a sign of great faith in the efficiency of linearity.

The problem was quite poorly formulated on 2 points:
1. problem 5 was written with a particularly complex syntax, which clearly hindered students, one of them writing explicitly: *I don't understand the problem.*
2. the introduction should have contained a clearer explanation, stating that it was normal to not know how to solve some of the problems given, so as to combat the effect of the 'didactical contract' which assumes that any problem given by the teacher should always be capable of being solved by the student. Certain student responses were clearly a result of their reluctance to write: *I don't know how to solve this problem.*

4.3. Situating the Exercise

This was the last assignment of the year and the students knew it would not be handed back before the final exam and that they would therefore not be able to benefit, in the exam, from its correction or explanations. Thus, we counted on only a small number of responses since it would have been difficult to give the assignment much earlier, as students needed a certain maturity in terms of the new knowledge of general linear algebra.

In addition, differential equations had been treated late, and a lack of detachment on the part of the students combined with the attraction of explicit algorithms for solving such problems, all resulted in students applying the procedure for solving linear differential equations to non-linear equations. This is a relatively frequent and usually transitory phenomenon that we see every year, but is usually corrected over time.

4.4. The Type of Students

Twelve assignments were turned in, representing 18 students (8 were from single student efforts, 2 from groups of 2 students, 2 from groups of 3 students), due undoubtedly to the date. These students are obviously not representative of the entire group: on the one hand, none of the students from the weakest group turned in the assignment; on the other hand, out of the 18 students having participated, 17 had been accepted into first year science-oriented university studies. These were not necessarily the best students, but they were serious in their work and continued to work up to the last moment of the year.

4.5. Analysis of the Results

The expressions and explanations of the papers were first coded in the following manner:

L = linear
NL = non-linear
Dg = degree > 1, therefore not possible to do
Mod = can be modeled linearly as follows... or cannot be modeled.

We decided not to include problems 1 and 7 on differential equations because all students remained prisoner of the idea that every differential equation is solvable by the algorithm: homogenous equation/variation of the constant, and therefore could not handle the general question of what could make such a problem solvable.

On the five remaining problems in the 12 papers, the response L was possible 36 times and the response NL 24 times; the latter at times could be replaced by the more ambiguous response Dg and the response Mod could appear even more often by counting the possible negative and positive answers. Here are the results for each of the 12 papers:

1. ∅
2. 1 Dg (problem 4)
3. ∅
4. 2 NL and 2 L
5. 2 NL and 1 L + 1 Mod
6.   1 NL and 2 L + 2 Mod
7.   2 NL + 1 Mod (negative) and 1 L + Mod
8. 2 Dg (problems 2 and 4)
9. 1 L + Mod
10. ∅
11. 1 L (problem 3)
12. 2 NL and 2 L + 2 Mod

Globally, there were eight NL responses, one of which was combined with Mod (unable to be modeled linearly), and nine L responses, seven of which were combined with Mod, as well as one response where L was sketched (coded 'L'). It is not clear whether the Dg responses can be retained as revealing a comprehension of the linear/non-linear distinction: they could just as easily relate to problems in doing the calculations.

There was no response as regards the linear or non-linear character of the geometric problem (problem 5): it was not attempted in five out of twelve papers, declared, "not understood" in one paper, and left blank in five others. Only in one paper, number 4, did the students begin to develop a method by declaring the linear character of the problem. Surprisingly, this student was the only one of the 18 students to have failed the final exam.

In general terms, we can conclude that students responded with only about one-third of the possible answers evaluating the solvability of the problem in terms of the linear/non-linear dichotomy. This occurred even though in the sample of students who handed in the assignment, serious or quite good students were overrepresented.

This undoubtedly leads to the conclusion that the possibilities of student 'meta-spontaneity' on the question of classing problems according to their solvability, using a possible or impossible linear model, are relatively reduced when they lack explicit explanation from the teacher, especially when they have not been confronted with nonsolvable problems. But the possible actions to take on this last point seem quite difficult. They stem perhaps from a more cultural approach to teaching: helping students understand that mathematics does not solve everything, that there are still open questions, that some problems are intrinsically difficult.

5. CONCLUSION: AN IMPORTANT METHODOLOGICAL PROBLEM, PERSPECTIVES.

It seems, therefore, that although the introduction of meta-type teaching is possible in linear algebra, although such scenarios are easily imagined and can even be set up without difficulty, their effects are less easily tested. In other words, their evaluation remains problematic, as much in measuring their impact as in the very methodology of this evaluation. We will come back to these evaluations and more broadly to our perspectives.

5.1. Evaluation of Meta-Type Teaching

To advance on this point, we worked in the following directions, particularly in the teaching experimented in Lille:

i) We researched the direct and indirect traces of teacher intervention in various student productions. We investigated, in particular, the types of questionings students could produce in a given mathematical situation, either alone or guided. What means of verification do they think of, alone or guided? Do they make use of the drawings or of the coherence between the number of parameters and the number of equations, between the size and the maximum number of independent vectors?

ii) We also gave students situations where the specific teaching of methods could be found. We wondered whether the very notion of setting interplay had been picked up and to what extent it was available or only summonable.

It turned out that there were few direct traces of teacher meta explanation, especially in the notes taken by students, which did not mean that students did not hear the explanation. Let us not forget the oral, optional, unevaluated character of these explanations. For the indirect traces, if there were any, they are dispersed, fragmentary, and we have not succeeded in determining their influence.

iii) There remains one avenue of exploration with respect to these ideas: the exploitation of conflicts lead to explore in this vein. It is indeed possible to make use of the necessary diversity between teachers to study whether this introduces perceptible differences. In other words, we can study whether what was attempted 'resists' to different teachers. But such a study comes down to comparing small groups of students, which is not sufficiently reliable, and to comparing different teachers and the teaching they do - also quite a delicate matter.

5.2. Perspectives on Teacher Intervention and Meta-Lever Analyses.

A predoctoral work (Praslon, 1994) led to our proposing new distinctions aimed at refining both teacher intervention and the analyses of meta borrowings (knowledge about knowledge in linear algebra): on the one hand, their function (the epistemological dimension), on the other hand, their range and form (the pedagogical dimension). We will specify below the type of commentary which we believe can be considered in this manner.

5.2.1. The Assumed Function of Teacher Comments Related to Teaching Phases: Various Scenarios and a Historical, Epistemological, and Pedagogical Study to Pursue.

Certain comments accompanying the strictly mathematical discourse have a very short-term function and it is not always only their specific relationship to the current course content which counts. This means, for example, teacher intervention to help students follow explanations of concepts, in order to help them position themselves within the line of discourse (structuring). This can be done by putting up partial titles, for example, and also includes answering student questions by soliciting the answer from the group or by encouraging general thinking: very useful for all notions, particularly when the teacher looks for exercises or engages in correction.

Whether it be illustrating a point, rephrasing, presenting arguments, or preparing the groundwork for new material, content is indeed important, but the teacher can improvise from the state of the class, from its history, and from the mathematics at play. The same is true of summarizing, synthesizing ideas, rereading texts, returning to knowledge immediately after or a long time after explanation, etc. There is no need for additional thought and planning to intervene in this way.

The same is also true of the use of certain oratory effects (emphasis, repeating, or even switching to a more informal register, for example) which help students follow. Thus, we do not believe that this type of comment can bring out sufficient specificities linked to the notions discussed here. As stated above, the teacher can, and must, improvise. Consequently, this type of comment will not be of interest to us here.

On the other hand, another function can be assigned to the meta-type comments: encouraging students to think from the point of view of a particular notion, bringing out its point of interest, its meaning, its uses, and the means of verification to be used with it or preparing questions about it. The function carried out by corresponding teacher intervention is a function of reflection on the notions to be taught. Yet these interventions seems difficult to improvise totally. They require reflection that does not flow naturally from only the mathematical knowledge presented to the students; they bring into play prior knowledge, particularly historical knowledge, even epistemological knowledge. It is not a question of presenting this knowledge to students, but of using it to elaborate comments accompanying the presentation of mathematical knowledge. Examples of this type of approach were seen in the first studies discussed above.

This is also the case in bringing out reference points or methods within a given field, or even in finding the origin of errors and commenting on them. We believe that it is the didactical analyses beforehand, accompanied, if need be, by an epistemological reflection, which can determine what there is to say. One must go beyond the simple catalog of methods, particularly indigestible, or the simple labeling of errors, often useless. For example, the more easily recognizable the student's error, the more difficult it is to formulate a theorem concerning these errors: It is by means of a genuine analysis of the mathematics involved and of the teaching received, tested and enriched by detailed interviews with the students making the errors, that the erroneous path(s) taken can be identified (see above). Justifying adequate strategies in a particular domain of mathematical knowledge by reflection on the knowledge of that domain also seems more fruitful than disembodied enumeration.

Thus the idea of working specifically on this type of comment is reinforced for the notions in linear algebra. In the case of notions such as those we are studying, we have indeed seen what is at stake in this type of intervention, since it replaces, so to speak, the usual 'good problems', proposed as an artificial origin of the notions to be taught.

5.2.2. Form and Range of Meta-Type Comments: a More Pedagogical Problem

What remains is the manner in which these interventions can be carried out: When is the best moment for their implementation? Which part should be taken on by the teacher and which part by the student? How can importance be given to knowledge which will not be evaluated?

A primary distinction was introduced following Praslon's work: Are we talking about reflections on what is taking place in the mathematics classroom or about more general reflections? Praslon has shown that general comments, even if they are perfectly adapted from the point of view of the teacher who knows the course, often go beyond students' capabilities because they do not yet have the knowledge of what is being explained and cannot perceive the connection between the particular and the general. They do not (yet) see the reason for the presentation and, therefore, risk losing the thread of the discussion. And yet the advantage of meta intervention can be, as we have seen, placing students on the right wavelength and thus introducing a certain reflection, a certain distancing from the particular case at hand.

This implies preparatory work, not only to refine the contents of these comments but, even more, to prevent the teacher from launching into generalities that are beneficial only to him. On the contrary, he must introduce a timely discourse aimed straight at its target, without losing the students' attention. The teacher must adapt his discourse to the current knowledge of the students so that it sheds light on the problem at hand for the students as well. We believe that this goes beyond simple analogy, but we propose to continue in the investigation of these questions in undertaking research in the field.

A second distinction has brought other interesting questions: the form these meta comments take, particularly their oral or written character. Writing legitimizes the idea but can become fossilized. Should we take this risk in order to have students develop a habit of meta thinking? Is it compatible with the absence of direct evaluation of this meta knowledge? These are further questions which will be investigated in our up-coming research. One aspect of our current research involves design and experiment of intervention comments, with pre-prepared scenarios, which are better adapted both to the students and to the linear algebra that we want to teach.

6. NOTES

[1] In an early stage of our work, we used the word 'metacognition', but it quickly seemed inappropriate as the mathematical dimension specific to our type of approach in the activities offered to the students was not reflected by this word. We also tried 'metamathematical', but the meaning of this term was too specific for our purpose. This is why we created this neologism.

[2] My main source of inspiration for this presentation comes from the type of approach given in the first axiomatic definition of vector space (Peano 1888).

[3] The formula given here is valid when the values are known in 0, 1, 2, ..., n; it is then easy, with a table like Pascal's triangle, to find the values of $\Delta^k P(0)$ ($0 \leq k \leq n$). Generally it suffices to perform a translation of the formula beginning with 0.

GUERSHON HAREL

PART II - CHAPTER 5
THREE PRINCIPLES OF LEARNING AND TEACHING MATHEMATICS

Particular Reference to Linear Algebra - Old and New Observations

1. BACKGROUND

Since this paper was intended primary for French readers, it begins with a short description of the movement in the US of reforming the learning and teaching of linear algebra. There has been a growing concern in the last two decades among mathematicians and mathematics educators about the quality of the undergraduate mathematics curricula. During the 80s the focus of attention was reforming the calculus curriculum; gradually, it spread to other areas, such as statistics, abstract algebra, and linear algebra.

In the summer of 1990, I had the privilege to join a group of 16 educators from mathematics departments throughout the US - called the Linear Algebra Curriculum Study Group (LACSG) - whose goal was to address concerns regarding the learning and teaching of linear algebra. This group generated a set of recommendations for the first course in linear algebra (Carlson et al., 1993).

Following the LACSG work, numerous national meetings on the learning and teaching of linear algebra were held, and several textbooks were written on the basis of these recommendations. The LACSG recommendations were based on a combination of three major sources:

The first source was research-based knowledge of how students learn, how mathematics should be taught, and what pedagogical and epistemological considerations are involved in the learning and teaching of linear algebra. For example, the LACSG recommendation for a strong emphasis on geometric interpretations was derived from research findings that the incorporation of geometric thinking in teaching the first course in linear algebra has a significant contribution to students' understanding (Harel, 1989a, 1989b, 1990). This and other pedagogical recommendations were an outgrowth of the recognition by the LACSG of the indispensability of considering the needs and interests of students as *learners*. As I will discuss later in this paper, current research findings have refined this observation.

177

DORIER J.-L. (ed.), The Teaching of Linear Algebra in Question, 177—189.
©2000 *Kluwer Academic Publishers. Printed in the Netherlands.*

The second source was the enormous individual experience that each of the members of the LACSG had in teaching linear algebra. This experience was valuable in judging the feasibility of curricular suggestions and the benefit of pedagogical approaches.

Finally, the third source was the consultants from a variety of client disciplines who spoke about the role of linear algebra in their disciplines and their views of how the curriculum could be improved. This was a precious source of information, for it helped the LACSG to learn first hand about the expectations the client disciplines have about the course and to determine its orientation accordingly.

While these recommendations have been, I believe, a big step in the right direction, they were never intended to be the final word. As David Carlson, Charles Johnson, David Lay, and Duane Porter, the founders of the Linear Algebra Curriculum Study Group, put it, the goal has been "to initiate substantial and sustained national interest in improving the undergraduate linear algebra curriculum" (Carlson et al., 1993).

In (Harel 1997), I interpreted the LACSG recommendations from my own perspective of the learning and teaching of linear algebra. The following is a summary of this perspective:

1.1. Proof

The LACSG highlights the need for an intellectually challenging course, with emphasis on proofs that enhance understanding. I have suggested that the focus on proofs should not begin in the first course in linear algebra; rather, with a careful approach and appropriate level of rigor that corresponds to students' mathematical ability, proof and justification must be the center of mathematics curricula at all grade levels. I have also made some suggestions of how we can help make proofs a tangible experience for the students.[1] For specific details on this issue, see (Harel and Sowder 1998) and (Sowder and Harel 1998)

1.2. Time Allocated to Linear Algebra

The LACSG has also recommended that at least one "second course" in matrix theory/linear algebra should be a high priority for every mathematics curriculum. This recommendation can be viewed as recognition by the LACSG that the time allocated to linear algebra in the current undergraduate mathematics curriculum is insufficient. In my opinion we should go beyond this recommendation by incorporating basic, geometrically oriented, linear algebra ideas in high school mathematics (see below). This suggestion, if implemented, would make the first course in linear algebra a natural continuation of what students have learned in high school. Accordingly, it could build on concept images (see below) of linear algebra already possessed by students. I repeat here what I have said in (Harel 1997).

The idea to introduce linear algebra in high school may seem too ambitious to some. But if we believe in the pedagogical importance of, and the need for, a continuity between high school mathematics and college mathematics, and we recognize the problem of the insufficient time allocated to linear algebra in college, then it should be clear that an introductory treatment of linear algebra in high school is a necessity. Introducing linear algebra in high school would do more than prepare students to do matrix algebra and compute determinants. It would lay the grounds for building rich and effective concept images for linear independence, spanning set, vector space, and linear transformation. Students who followed a linear algebra program in high school would become motivated and more cognitively prepared to abstract these and other ideas in their first college course in linear algebra. This is a worthy long-term investment, which requires a restructuring of the existing high school mathematics curricula. I suggest that this restructuring take place by teaching traditional high school topics, such as systems of linear equations, analytic geometry, and Euclidean space, from a linear algebra viewpoint.

The current high school mathematics is not geared toward the needs of linear algebra. This argument may not be true if examined solely from the viewpoint of *content*. High-school curricula do include the traditional topics: systems of linear equations, analytic geometry, and Euclidean space - all are part of linear algebra. But these topics are taught in high school in ways that have little to do with the basic ideas of linear algebra. High school students are not prepared for the objects, language, ideas, and ways of thinking that are unique to linear algebra. Students make little or no connection between the ideas they learn in linear algebra and the mathematics they learn in high school. In the current situation, the only connection that potentially exists between high school mathematics and linear algebra is the study of systems of linear equations. But even this connection is superficial. High school students' involvement with systems of linear equations amounts to learning a solution procedure for 2×2 and 3×3 systems. They do not deal with matrix representations of these systems, questions about existence and uniqueness of solutions, relations to analytic geometry of lines and planes in space, geometric transformations, matrix algebra and determinants, etc.

1.3. Technology

Another recommendation by the LACSG was to utilize technology in the first linear algebra course. I have suggested incorporating MATLAB (or any other similar software package) in the teaching of linear algebra. Since MATLAB's basic data element is the matrix, the use of MATLAB can help students make n-tuples and matrices concrete. This would prepare students for a matrix-oriented course, as was also recommended by the LACSG.

1.4. Core Syllabus

Finally, the LACSG recommended the following core syllabus for the first course in linear algebra (for more details, see Carlson et al., 1993):
1. Matrix addition and multiplication (including operations with partitioned matrices);
2. System of linear equations (including Gaussian elimination, elementary matrices, echelon and reduced echelon form, existence and uniqueness of solutions, matrix inverses, and LU-Factorization[2]);
3. Determinants;

4. Properties of $\mathbb{R}^n$:
   - Linear combination, linear dependence and independence,
   - Bases of $\mathbb{R}^n$,
   - Subspaces of $\mathbb{R}^n$ (including spanning set, basis, dimension, row space and column space, range of A as a mapping, null space)
5. Matrices as linear transformations;
6. Rank: rowrank = columnrank, products, connections with invertible submatrices;
7. System of linear equations revisited: solution theory, rank + nullity = number of columns;
8. Inner product (including length and orthogonality, orthogonal sets and bases, orthogonal matrices, orthogonal projection, Gram-Schmidt orthogonalization and interpretation as a QR factorization[3], and the least square solutions of inconsistent linear systems, with applications to data-fitting.
9. Eigenvalues and Eigenvectors (including the characteristic polynomial and algebraic multiplicity, eigen spaces, geometric multiplicity, similarity and diagonalization, orthogonal diagonalization, and quadratic forms).

In the remainder of this paper, I will discuss a theoretical framework on which the above perspective is based. The core of this framework is three learning/teaching principles: the Concreteness Principle, the Necessity Principle, and the Generalizibility Principle.

2. THE CONCRETENESS PRINCIPLE

In (Harel 1989a; 1989b) I have suggested that the existing approaches to teaching linear algebra do not conform to the students' pedagogical needs. In a review of elementary linear algebra textbooks, I found that there is an implicit assumption that beginning students are capable of dealing with abstract structures without extensive preparation, and can appreciate the economy of thought when particular concepts and systems are treated through abstract representation. Specifically, it is assumed that students recognize models and solve problems by translating them into isomorphic but abstract structures, and apply the principles of the abstract setting to solve problems. Through a sequence of teaching experiments with high school and beginning university students, this assumption was found to have no grounds (Harel 1989a, 1989b).

This observation led to the formulation of the Concreteness Principle (Harel, 1985, 1990, Harel and Kaput, 1991). It states:

> *For students to abstract a mathematical structure from a given model of that structure the elements of that model must be conceptual entities in the student's eyes; that is to say, the student has mental procedures that can take these objects as inputs.*

The idea of conceptual entity formation was suggested by Piaget (1977) in his distinction between form and content. This process is an instance of reflective abstraction, in which « a physical or mental action is reconstructed and reorganized on a higher plane of thought and so comes to be understood by the knower » (Beth and Piaget, 1966). Greeno (1983) gave an operative cognitive definition to the notion of "conceptual entity." A conceptual entity, according to Greeno, is a cognitive object for which the mental system has procedures that can take that object as an argument, as input. I will illustrate this notion with two examples: the first deals with the concept of vector space of functions, and the second with matrix arithmetic. Both examples appear in (Harel 1997), but it is worth repeating them here.

2.1. *Vector Space of Functions*

The reader may have observed students correctly solve problems that ask to determine whether a specific set of vectors in $\mathbb{R}^n$ is linearly independent, but have difficulties determining if a set of functions is linearly independent. Indeed, it is a common phenomenon that students set-up the equality: $ax + bx^2 + cx^3 + dx^4 = 0$ to determine if the set A=$\{x \; ; \; x^2 \; ; \; x^3 \; ; \; x^4\}$ is linearly independent, without having a clear idea of what they are doing. Their difficulty, I have found, can be explained by the fact that they cannot apply the concept of linear independence to *polynomials as functions* because the concept of function as a vector is not *concrete* to them. That is to say, these students have not formed the concept of function as a mathematical object, as an entity in a vector space. As a result, when they set up the equation $ax + bx^2 + cx^3 + dx^4 = 0$ to determine whether the set A=$\{x \; ; \; x^2 \; ; \; x^3 \; ; \; x^4\}$ is linearly independent, they do not understand that the zero on the right side of the equation is a function; the zero vector in the vector-space $P_4[x]$, not the scalar zero. Further, they cannot interpret this equation as an *identity* between two functions; rather, they view it as an *equation* in x. Once the students form the concept of function as a mathematical object, the elements of $P_4[x]$ become concrete objects, conceptual entities, which they can treat as inputs for other operations.

2.2. *Spatial Symbol Manipulation: The Case of Matrix Arithmetic*

Students in high school deal with real numbers which are treated by them as conceptual entities. Real numbers, for all purposes of high school mathematics, represent either ratio quantities, such as speed, density, price, and probability, or magnitude quantities, such as time, weight, length, and cost. Accordingly, the symbolic representations for these objects are one-dimensional. In linear algebra, on the other hand, new types of objects are added to the play: n-tuples, matrices, and functions as elements of a vector space. These, in contrast to real numbers, represent multidimensional quantities, such as probability vectors and price vectors, directed graphs, and solutions to a differential equation that models the effect of temperature change; as such, they may not be conceived as conceptual entities by the students.

According to the LACSG recommendations, the first course in linear algebra should be matrix-oriented. This poses a curricular difficulty because students would have to deal with vectors and matrices right at the beginning of the course, before they have concretized them into conceptual entities. For this, students would be required to develop, in a relatively short period of time, a spatial symbol manipulation ability they never acquired before. For example, consider the statement:

RX=0, where R is a row reduced echelon matrix with r non-zero rows in which the leading entry of row i occurs in column k(i) and X is a column vector. This system consists of r non-trivial equations in which the unknown $X_{k(i)}$ occurs with non-zero coefficient only in the ith equation.

To comprehend this statement, one needs to carry out several mental activities, among which are:
a) visualizing the matrix R and the positions of the leading entries,
b) mentally carrying out the product of R with a column of unknowns,
c) visualizing the corresponding positions of the unknowns in the system of equations RX=0, etc.

Even when each of these steps is expressed on paper, one must first imagine and carry them out mentally, otherwise they become entirely mechanical.

The inability to spatially manipulate symbols may account for students' confusion between objects; for example, the confusion between scalars and vectors. Consider the following example: JF and LV were students in an elementary linear algebra course. They had difficulties with the problem, « Prove that W, the solution set of AX=0, is a subspace. »

Instructor: What is the definition of a subspace?
JF: Has the zero vector, closed under addition and scalar multiplication.
Instructor: Why does the zero vector belong to W?
After some effort, LV indicated that he needed to verify that A0=0.
Instructor: Are the two zeros in this equality the same?
LV: The second zero is the number zero.
JF: The same zero.

Although JF and LV eventually understood that the two zeroes might not be the same, this example demonstrates that even simple matrix equalities, such as A0=0, can pose spatial symbol manipulation difficulties.

2.3. *Evidence and Instructional Treatment*

The premise of the Concreteness Principle is that students build their understanding of a concept in a context that is concrete to them. Such a context serves both as an anchor to building adequate concept images (Vinner, 1977) and a spring to further abstraction. This premise was the basis for a linear algebra program I developed for high school students (1985). The program was developed in two sets of studies:
1. A series of teaching experiments with high school students
2. One experiment/control group study with sophomore engineering students.

These studies will be discussed briefly in the remainder of this section. I will begin with the second study, which is fully discussed in (Harel 1989b).

Focusing on the understanding of the vector-space concept with university technology students, I asked the following question. Would an emphasis on the embodiment of a *geometric* system - the geometry of two and three dimensions - lead students to a better understanding of the vector-space concept than an emphasis on the embodiment of *algebraic* systems? By understanding the vector-space concept we mean the ability to determine if, and justify why, a given system is a vector space; and to solve problems whose solutions require application of the basic properties of vector space.

The students used in this study were seventy-two sophomores engineering students. They were divided randomly into two groups, A and B, consisting of thirty-six students each. All students were taught linear algebra, by one instructor, for three one-hour lectures, each week during one semester. The instructor used mainly the system $\mathbb{R}^n$ to illustrate abstract ideas he had taught, without referring to geometric interpretations of these ideas. Weekly, each group was given two separate hours of 'recitation'[4] by another instructor. Group A was presented in the two hours of 'recitation' with a variety of embodiments of the abstract ideas they were taught in the previous lecture. Group B received the same 'regular' treatment as Group A for only one-hour of 'recitation'. The other hour was devoted to a 'special' treatment of showing how vector-space ideas are represented geometrically. At the end of four weeks, the two groups were given a test on the vector-space concept, which could be solved directly by applying the vector-space definition.

I considered six variables in the students' answers: *geometric description* (GD), *algebraic description* (AD), *correct final answer* (CF), *incorrect final answer* (IF), *correct justification* (CJ), and *incorrect justification* (IJ). The following is an excerpt of the main results (for further details, see Harel, 1989b).

A comparison between the two categories, GD and AD, across the two Groups A and B, showed a geometric mode was used more often in Group B than in Group A and an algebraic mode was used more often in Group A than in Group B. A comparison between CF and IF categories showed that more correct final answers were given by students from Group B than by students from Group A. When dichotomizing the answers as belonging vs not-belonging to the CJ category, almost half of Group A but less than one-fourth of Group B gave incorrect justification.

The fact that geometric descriptions occurred more often in Group B than in Group A and algebraic descriptions occurred more often in Group A than in Group B confirms that the instructional approach and the kind of description chosen by the students were not independent. In problems where the students were required to determine if a specific model is a vector-space, many students from Group B gave a correct final answer and only a few students gave an incorrect final answer. In problems where students were required to prove general statements, only a few students from Group B gave incorrect justifications but many more from Group A gave incorrect justifications.

Despite the fact that the two groups were exposed to the same formal definition of a vector space, their performances in solving problems that could be solved directly by applying this definition were significantly different. An explanation to this is that, although the two groups received the same concept definition of vector-space, they were exposed to different experiences which resulted in the formation of different concept images. Group B handled the problems more successfully due to its superior concept image of vector space.

In conjunction with this experiment, I conducted a sequence of teaching experiments with high school students. I found that vector spaces with dimension less than or equals to three and in which the elements are visual (i.e., geometric spaces of directed line-segments) were concrete for these students and cognitively appropriate models for introducing the basic notions of linear algebra. The program began with these models. After the central concepts of dependence, independence, linear combination, basis, and dimension were thoroughly studied in the geometric spaces of line, plane, and in 3-space, the algebraic spaces $\mathbb{R}^1$, $\mathbb{R}^2$, and $\mathbb{R}^3$, along with these central concepts, were built through the idea of vector coordinates. The pedagogical value of this approach was that students could see, in concrete environments, how one mathematical system is transferred into another system that is more amenable to computational techniques. The resulting system is, of course, isomorphic to the original. Finally, these central concepts were defined a third time in $\mathbb{R}^n$ (for a parameter n) where a major concern was given to the study of n×m systems of linear equations and matrix transformations in $\mathbb{R}^2$ and $\mathbb{R}^3$.

This approach was primarily used with Israeli high school students. It proved to be effective in that students began with a concrete environment - the systems of directed-line segments, and through a process of problem-solving, came to recognize and appreciate the power of different systems - first $\mathbb{R}^1$, $\mathbb{R}^2$, and $\mathbb{R}^3$, and then $\mathbb{R}^n$. An important characteristic of this approach was that geometry was the central topic of investigation and constituted the 'raw material' for generating problems and introducing new concepts and ideas. Thus, concepts such as "span", "linear independence," "basis," "dimension," "inner product," and "linear transformations" had their origin in geometric problems. Further, even in the last phase of the program, which aimed at abstracting the central ideas of linear algebra to a "world of undefined elements," geometry was the main motive. The treatment in this phase was analogous to Dieudonné's (1964) approach. It dealt with vector spaces of dimension less than or equals to 3, but with general elements. The objective was for students to abstract the geometric models but keeping the dimensions unchanged. In this way, students could visualize and treat linear algebra phenomena by using the graphical representations they learned in Phase 1. The pedagogical importance of this phase was for students to absorb the idea that derived results in linear algebra depend solely upon the axioms of vector space, not upon the definitions of specific elements. In this part of the program, students learned, for example, that if $L_{ab}=\{a+xb \mid x$ is a real number$\}$ and $L_{cd}=\{c+xd \mid x$ is a real number$\}$ are subsets of a vector space, sharing two distinct vectors, then $L_{ab}= L_{cd}$. Students also recognized the analogy between this statement and the Euclidean postulate, « For any two different points there is exactly one line containing these points ».

2.4. Some new Observation

In a recent linear algebra teaching experiment with college students in the United States (see Harel, 1999), I have found that when geometry is introduced before the algebraic concepts have been formed, many students view the geometry as the raw material to be studied. As a result, they remain in the restricted world of geometric vectors and do not move up to the general case. To illustrate the phenomenon of how students are constrained by their physical imageries, consider the following example:

When linear algebra instructors present to the students a problem such as "Given W is a subspace of $\mathbb{R}^n$, find the projection of $\vec{c}$ onto W" along with the sketch in Figure 1, they do not mean the sketch to be literal but symbolic. It turned out that such a sketch was not conceived by many students as a REPRESENTATION of the abstract setting, but as the ACTUAL OBJECT of inquiry.

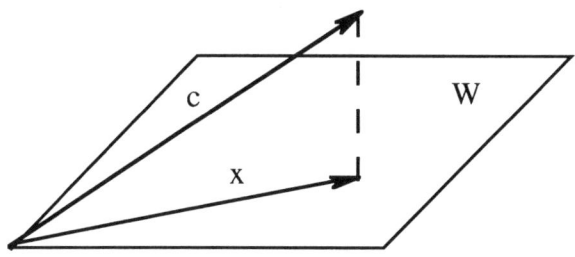

Figure 1.

My conclusions from this observation are that as we should be careful not to move students up hastily from $\mathbb{R}^n$ to more general vector spaces - the vector space of functions for example - we should be as careful how to introduce special cases of $\mathbb{R}^n$; namely, the geometry of directed segments. The student must stand on solid ground as to the world he or she is studying. In elementary linear algebra, there should be one world - $\mathbb{R}^n$ - at least during the early period of the course. As I have shown in my previous studies, geometry can be a very powerful tool in solidifying linear algebra concepts, but we need to consider carefully the way geometry is introduced and used. We, the teachers, see how the geometric situation is isomorphic to the algebraic one and so we believe that the geometric concept can be a corridor to the more abstract algebraic concept. Unfortunately, many students do not share this important insight.

3. THE NECESSITY PRINCIPLE

The second learning/teaching principle in this theoretical perspective is the Necessity Principle. It states:

> *For students to learn, they must see a need for what they are intended*
> *to be taught. By 'need' it is meant an intellectual need, as opposed to*
> *a social or economic need.*

This principle is in line with the Piagetian theory and the theory of problematique developed by French mathematics educators - see (Balacheff, 1990) and (Brousseau 1994 and 1997). It has been established that the main tool for modifying existing conceptions is true problem-solving activities where the learner applies existing conceptions to solve problems and modifies these conceptions when encountering cognitive conflicts. Such a conflict, or a disequilibrium as Piaget calls it, can lead the student to question her or his prior action and seek new ways to resolve it. It is these cognitive conflicts and the resolutions invented by the student that constitute a gradual transition toward advanced conceptions. Thus, the idea behind this principle is that instructional environments must include appropriate constraints by which students can reflectively abstract mathematical conceptions and, at the same time, keep the situation at hand realistic. The instructional activities must offer problematic situations that are realistic to and appreciated by the students. Through their activities, students must feel that what they do results in a solution of a problem (their own problem!) or in a resolution of a conflict (their own conflict!), and, if an idea (e.g., a definition of an operation, or a symbolization form of a concept) is initiated by their teacher, they must not feel that it was evoked arbitrarily.

I should point out that there is a common confusion between 'concept appreciation' and the fulfillment of the necessity principle. For example, students usually easily appreciate the Fundamental Theorem of Calculus after they have computed the Riemann Sum of several functions, but they do not have a feel for the connection between 'rate of change' (i.e. derivative) and 'accumulation function' (i.e., integral). Bringing the students to see and feel this connection in a specific context - the context of speed and distance, for example - lays the conceptual grounds for conjecturing the general relationship expressed by the Fundamental Theorem of Calculus. Conjecturing, by definition, results in a problematic situation thus implementing the necessity principle. A similar argument can be made with many ideas in linear algebra. An example of a violation of the necessity principle would be "deriving" the definition of vector-space from a *presentation* of the properties of $\mathbb{R}^n$ that correspond to the vector-space axioms. This statement does not hold for an advanced student who understands the role of the postulational approach in mathematics. It does, however, hold for a beginning student, one who has yet to witness the economy of thought in thinking in terms of vector-space axioms. For this student, properties in $\mathbb{R}^n$ are self-evident; thus they do not warrant the attention they get.

These students not only do not see the need for the concept of vector space, but also do not understand the vector space axiom statements. For example, in a one-to-one interview, students were asked to derive the property, « For any A in a vector space V, $(-1)A = -A$ », from the vector space axioms. Almost all the students who were asked this question were unable to produce the proof they had seen in class. Further, they viewed this statement as trivial, and were unable to articulate its underlying argument (i.e., the product of a vector by the scalar, -1, is the additive inverse of that vector). Similarly, these students did not understand the argument made in the axiom « For any A in a vector space V, $1A = A$ »; this condition was useless in their eyes. This, I believe, is an important observation, for there is an implicit assumption in textbooks that students understand the arguments made by the vector space axioms.

For teachers to find out what constitutes intellectual need for a particular population of students they must understand their students' ways of thinking. In (Harel and Sowder 1998), I characterized three forms of intellectual need: the need for *computation*, the need for *formalization*, and the need for *elegance*. Here I only repeat what I have said in (Harel and Sowder 1998) about the need for computation, which is the most effective among the three kinds of needs.

Generally, problems that ask to compute objects that are concrete to students or ask to determine properties of such objects are said to satisfy the need for computation. It was widely utilized in our teaching experiments in linear algebra - a subject particularly amenable to this need - to elicit specific concepts as well as to enhance students' reasoning skills. As an example, consider the pivotal cluster of concepts, 'linear combination', 'dependence', and 'independence'. We found that students entering their first course in linear algebra are familiar with system of equations and understand their importance (in solving word problems, for example). As such, they can be brought to appreciate the fact that in some cases one cannot nor does one want to solve a given system $AX = b$, and yet it *needs* to be determined whether the system has a solution and, if it does, whether or not it is unique. We pose these Existence and Uniqueness problems early in our linear algebra teaching experiments - before we mention any of these concepts - to focus students' attention on a definite goal. It is not uncommon that a few students give a correct response to the Existence problem. Namely, for $AX = b$ to have a solution, b must be expressed as $x_1A_1+x_2A_2 ...+x_nA_n$. Of course, students seldom give this clean answer but, with the teacher's help, the class as a whole understands why the suggested relation between b and the columns of A merits attention and, therefore, deserves a name - 'linear combination'. Building on this students' understanding, we help students elicit the concepts of 'dependence' and 'independence' from the solution to the Uniqueness problem. Students are now prepared to see that the relations "one of the columns of A is a linear combination of the other columns", together with its negation solve the Uniqueness problem. Once this is achieved, new concepts are born and names are designated to them: "linear dependence" and "linear independence", respectively.

4. GENERALIZIBILITY PRINCIPLE

These two principles of learning - the Concreteness Principle and the Necessity Principle - are complemented by a third principle, called the Generalizibility principle. It states:

When instruction is concerned with a 'concrete' model, that is, a model that satisfies the Concreteness Principle, the instructional activities within this model should allow and encourage the generalizibility of concepts.

This principle aims at enabling students to abstract concepts they learn in a specific model. For example, when dealing with the three-dimensional geometric model of a vector space, a basis could be defined as three non-collinear directed line segments. But such a definition is restrictive and model-dependent because it does not transfer to abstract vector spaces. In (Harel 1985), the concept of basis was explored by the students starting from the concept of minimal spanning set. Initially, such a spanning set may be less appropriate because the scalars involved might not be unique, but this in turn leads to the necessity of considering minimal spanning sets, which also need to be ordered in a specific way to give a unique scalar representation.

The application of the Generalizability Principle must be in accordance with the necessity principle. For example, some textbooks begin with the question of finding an angle between two vectors, indicating that in 2 - or 3 - dimensions the angle θ between two vectors u and v can be determined by :

$$\cos\theta = \frac{u.v}{\|u\|.\|v\|} \quad .$$

Then they point out that this formula can be carried out in any dimension, and define the cosine of the angle between two vectors in $\mathbb{R}^n$ by this formula. The problem with this approach is that the definition of cosine in $\mathbb{R}^n$ is not problem-solving driven. From the student's perspective there is no intellectual need for creating this new concept.

5. SUMMARY

I have touched upon certain aspects of each one of the LACSG recommendations. I have suggested that the focus on proofs should not begin in the first course in linear algebra, but should be emphasized throughout the mathematics curricula in all grade levels.

To compensate for the insufficient time allocated to linear algebra in the current undergraduate mathematics curriculum, I have suggested the incorporation of basic linear algebra ideas in high-school mathematics programs. This suggestion, if implemented in the spirit of the three pedagogical principles I have just discussed, would make the first course in linear algebra a natural continuation of what students have learned in high-school. Accordingly, it could build on rich concept images of linear algebra already possessed by students.

While the inclusion of geometry can aid students in building a strong understanding of linear algebra, current results suggest that the incorporation of geometry in linear algebra in the college level must be sequenced in such a way that students understand the context of investigation.

In line with the LACSG recommendations to utilize technology in linear algebra courses, I have suggested incorporating MATLAB (or any other similar software package) in the teaching of linear algebra. In particular, because MATLAB's basic data element is a matrix, programming in MATLAB can help students make n-tuples and matrices concrete, whereby implementing the Concreteness Principle, as I have discussed earlier. In addition, this would prepare students to a matrix-oriented course as was suggested in the LACSG Recommendation.

Finally, I have formulated three pedagogical principles for designing and implementing mathematics curricula: the Concreteness Principle, the Necessity Principle, and the Generalizability Principle.

6. NOTES

[1] I recognize that in France a strong emphasis on proof starts in high school geometry *and* algebra and continues in calculus at the college level. By the time they get to linear algebra, their conception of mathematical proof is reasonably strong.

[2] An mxn matrix A admits an LU factorization if it can be written in the form A=LU, where L is an mxm lower triangular matrix with 1's on the diagonal and U is upper triangular.

[3] An mxn matrix A admits a QR factorization if it can be written in the form A=QR, where Q is an mxn matrix whose columns are orthogonal and Q is an nxn upper triangular matrix with positive entries on the diagonal. QR factorization is a direct outcome of Gram-Schmidt Orthogonalization process.

[4] A 'recitation' class is run by a graduate student- a teaching assistance - whose main responsibility is to solve problems with the students.

JOEL HILLEL

PART II - CHAPTER 6
MODES OF DESCRIPTION AND THE PROBLEM OF
REPRESENTATION IN LINEAR ALGEBRA

1. INTRODUCTION

This chapter[1] examines the teaching and learning of linear algebra from the perspective of the different modes of description of vectors and operators. We identify the different modes and their associated language, as used in classroom practice, as well as discuss the mechanisms of translating from one mode to the another. We then focus more specifically on the representation of a vector and an operator relative to a basis in order to illustrate some of the difficulties faced by students of linear algebra.

The teaching of linear algebra at a university level is almost universally regarded as a frustrating experience for instructors and students alike. Many among those who teach such a course have resigned themselves to the fact that this is simply 'the nature of the beast' and that not much can be done to change things. This attitude might explain the reason why, until recently, there was a paucity of research work on the learning of linear algebra. Unlike the notions of calculus, which have been researched extensively, most of the research on learning and teaching linear algebra is relatively recent. Starting with the work of Harel (1985), other contributions have been made by, e.g., Robert and Robinet (1989), Rogalski (1990), Dorier (1990), Pavlopoulou (1993), Sierpinska (1995), Dreyfus and Hillel (1998), Sierpinska, Dreyfus and Hillel (1999).

In Canada and the US, linear algebra has traditionally been the first mathematics course that students encounter which is a full-fledged mathematical theory. It is proof-laden, and built systematically from the ground up, with all the fuss about making assumptions explicit, justifying statements by reference to definitions and already proven facts. Therefore, on one level, students' difficulties with linear algebra stem simply from their inexperience with proofs and proof-based theories. Indeed, students' proof related difficulties include: not understanding the need for proofs nor the various proof techniques; not being able to deal with the often implicit quantifiers; confusing necessary and sufficient conditions; making hasty generalizations based on very shaky and sparse evidence. What is perhaps surprising is that this phenomenon is not as local as one might have expected. So, for

DORIER J.-L.(ed.), The Teaching of Linear Algebra in Question, 191—207.

example, while the teaching of mathematics in France has been a lot more formal than its North American counterpart (or, at least, it was perceived to be), French students of linear algebra do not seem to have any easier time with proofs. Robert and Robinet (1989) asked nearly 380 students in France what they found difficult with the subject, and dealing with proofs, as well as finding that the subject contained too many definitions and new results, were two of the main responses.

The other important aspect of being a mathematical theory is its generality. Knowing linear algebra at this level demands that students start thinking about the objects and operations of algebra, not just in terms of relations between particular matrices, vectors and operators, but in terms of whole structures such as, vector spaces over fields, algebras, and classes of linear operators. Furthermore, students need to be able to appreciate that these structures can be transformed, represented in different ways, and considered as being, or not being, isomorphic. Referring to Piaget and Garcia's (1989) notion of intra-, inter- and trans-level of knowing something, we see that the level in which students need to operate is the trans-level. However, looking at students of linear algebra at our own university, the majority of whom have successfully completed an elementary linear algebra course in their pre-university studies, we can say that their understanding of mathematical objects such as matrices and linear operators can best be characterized as inter-level of thinking (Sierpinska, 1995). Similarly, as in the French study cited above, statements by students that the subject is too abstract and that the general notions such as vector spaces, endomorphisms, bases, dimension and kernel are difficult were quite prominent.

Students' proof-related difficulties are certainly not generic to linear algebra but surface in most undergraduate courses - see, for example, Selden and Selden (1987) regarding Group Theory or Leron (1993) regarding Abstract Algebra. Therefore, in this chapter, we are particularly interested in teasing out those sources of conceptual difficulties that are specific to linear algebra, which include:
- The existence of several languages or modes of description
- The problem of representations
- The applicability of the general theory.

2. MODES OF DESCRIPTION IN LINEAR ALGEBRA

A typical course will generally include several modes of description of the basic objects and operations of linear algebra. These modes of description co-exist, are sometimes interchangeable, but are certainly not equivalent. They include:
1. The *abstract mode* - using the language and concepts of the general formalized theory, including: vector spaces, subspaces, linear span, dimension, operators, kernels
2. The *algebraic mode* - using the language and concepts of the more specific theory of $\mathbb{R}^n$, including: n-tuples, matrices, rank, solutions of systems of equations, row space
3. The *geometric mode* - using the language and concept of 2- and 3-space, including: directed line segments, points, lines, planes, geometric transformations

Within each mode, vectors, vector operations and transformations have particular depictions, terminology and notation, and there are mechanisms that enable one to move from one mode to another. We first analyze these modes and the accompanying representations in more detail, restricting our analysis throughout to real, finite-dimensional vector spaces.

2.1. The Geometric Level: 1-, 2-, and 3-Dimensional Spaces

We consider any depiction of vectors by means of directed line segments, arrows or points, and of transformations in terms of geometric-spatial actions, as belonging to the geometric mode. It includes both the synthetic (coordinate-free) and analytic description, though the latter has essentially both geometric and arithmetic features.

2.1.1. Coordinate-Free Geometry
Vectors:

Non-zero vectors are directed line segments (arrows) emanating from a common point (the 'origin') hence have 'magnitude and direction' (see Figure 1), and are

often labeled as $\overrightarrow{OP}$, or $\vec{u}$, a notation which is meant to reinforce the 'vector-as-an-arrow' point of view. The zero vector has to be defined artificially as the point O with zero magnitude and no direction.

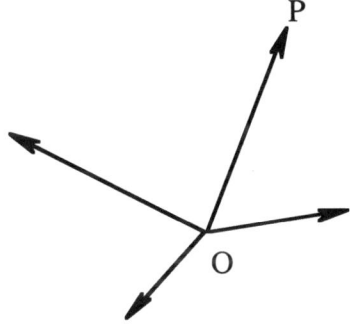

Figure 1.

Vector operations:

Addition and scalar multiplication of vectors are defined geometrically in terms of the 'parallelogram law of addition' and a stretch/shrink of a vector, in the same or the opposite direction, respectively. One can show, more or less in a heuristic way, that the operations satisfy the right properties (a complete exhaustive case-by-case proof gets very tedious).

Transformations:

Certain classes of transformations have simple and 'global' geometric descriptions; namely, rotations, projections, reflections and shears. Linearity of these transformations can be evoked by geometric arguments; e.g., rigid motions preserve parallelograms hence preserve addition.

We note that one can also start with 'free' (or 'unattached') vectors and define operations and transformations on equivalence classes of vectors 'having the same direction and magnitude'. Strictly speaking, one needs then to establish that the operations and transformations are well-defined, though, in practice, one rarely bothers since it is quite self-evident. Fixed vectors, relative to a preferred point, are then representatives of equivalence classes.

2.1.2. Coordinate Geometry
Vectors:

Vectors are arrows, lying in 2- and 3-dimensional space endowed with a coordinate system. Most often, one uses the familiar Cartesian coordinate system (Figure 2a) though there is also the option of using a more general coordinate system (i.e., non-orthogonal axes, and non-uniform units along each axis) which could serve as a precursor to the notion of basis (Figure 2b).

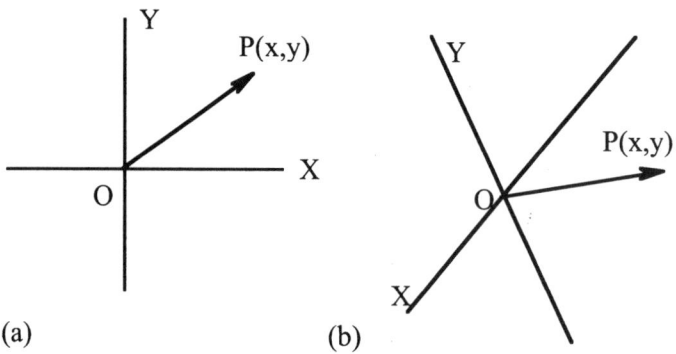

Figures 2.

Vector operations:

Addition and scalar multiplication can still be described geometrically as in the coordinate-free case. Alternatively, one can define the operations in terms of the coordinates of the endpoints and show that, geometrically, the sum of two vectors is the diagonal of the underlying parallelogram.

Transformations:

Some geometric transformations can be more specified than in the coordinate-free case by referring to the coordinate axes; e.g., 'reflection about the 45-degree line' or 'projection on the X-axis'. One can also define 'algebraic' transformations which operate on the coordinates of the end-point P, such as scaling (stretching/shrinking the x-coordinate by a factor s and the y-coordinate by a factor t).

2.1.3. *Vectors As Points*

In both the synthetic and analytic geometry cases, vectors can be stripped of the arrows and be simply points in the plane or 3-space endowed with a preferred point or coordinate system. One advantage in this mode of description is that the action of geometric transformations can be seen in relation to certain subsets of points; e.g., circles changing to ellipses, squares to rectangles, etc.

While this seems to be a minor change from the arrow notation, we will see that it presents several problems to students.

2.2. *The Algebraic Mode:* $\mathbb{R}^n$

Vectors:

Vectors are n-tuples of real numbers $[x_1, x_2, ..., x_n]$ often labeled by X, Y, Z, ...

Vector operations:

Vector addition (scalar multiplication) is the 'obvious' algebraic operation defined component-wise. The required properties are satisfied because they are essentially just the field properties of R distributed over all the components.

Transformations:

Transformations are mxn matrices, denoted by A, B, C,... Linearity is essentially the distributive law for matrix multiplication.

2.3. *The Abstract mode: Vector Space V defined by a set of axioms*

Vectors:

Vectors are the elements of a vector space, often denoted by the letters **u**, **v** (or $\overrightarrow{u}$

, $\overrightarrow{v}$, which connects them to their geometric progenitors).

Operations:

Addition (scalar multiplication) is defined as a binary (unary) operation which satisfies the axioms.

Transformations:

Linear transformations (operators) S, T, ... are defined as mappings between vectors spaces which preserve the addition and scalar multiplication.

In chapter 7 of this book, Sierpinska describes three modes of reasoning used in linear algebra. These modes of reasoning have obvious links to the three modes of description discussed above, but they are not exactly the same. It is possible, for example, for students to be working in $\mathbb{R}^n$ (the algebraic mode of description) and to be using several modes of reasoning (examples of such are found in chapter 7 of this book).

2.4. *Some Students' Difficulties with the Geometric Mode*

2.4.1. *Arrows Versus Points*

Many instructors, wanting to capitalize on students' familiarity with analytic geometry and elementary physics, use the geometric mode of description as an entry into the general theory. Most commonly, vectors are depicted as arrows lying in the Cartesian plane. Arrows are preferred to points, even though the latter is a 'leaner' mathematical form because, seemingly, there are pedagogical and psychological advantages to this particular representation. Vectors-as-arrows representation is consistent with the use of vectors in Physics; it helps reinforce the idea that a vector is a different entity than a number; it motivates the definition of vector addition and length (norm); it facilitates the interpretation of the action of some linear operators such as, say, rotations followed by a stretch. In practice, though, most instructors tend to shift back and forth between the arrow and point depiction of vectors. This shifting is done implicitly and, often, unconsciously. A single vector (or several vectors) is shown as an arrow, but it is simpler to describe and show subspaces and cosets W of $\mathbb{R}^2$ and $\mathbb{R}^3$ as lines and planes through the origin (Figure 3a), rather than the set of all vectors lying on a line or in a plane (Figure 3b).

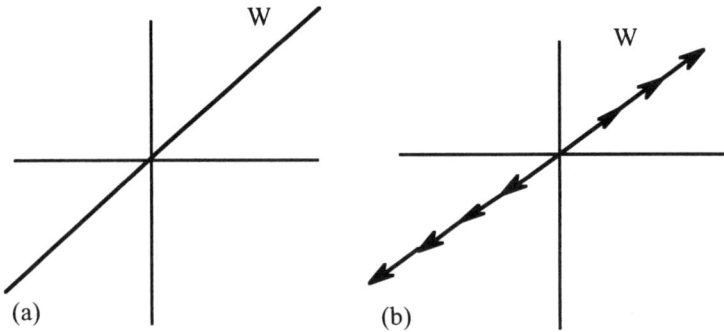

Figures 3.

While such a shift seems to be nearly trivial and can be thought of as merely a slight abuse of terminology, students, nevertheless, can be thrown aback by it. For example, students who were introduced to vectors as arrows via analytic geometry and who were asked to show vectors X=[x_1, x_2] such that $x_1+x_2=1$, drew the line $x_1+x_2=1$ which seemed, on the surface, like the correct response. However, when pressed to exhibit actual vectors, nearly all of them drew 'free' vectors lying on that line (Figure 4a), rather than 'fixed' vectors with endpoints on the line (Figure 4b). This experience showed that the relationship between arrows and points on a line (plane) is not so clear, particularly if one moves away from the origin.

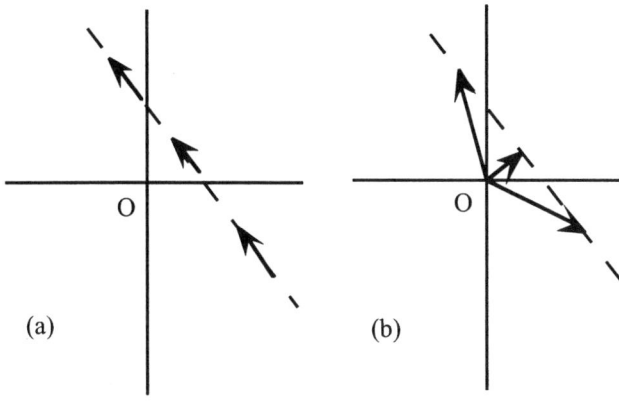

Figures 4.

2.4.2. *Standard Versus Non-Standard Basis*

The analytic geometry approach, precisely because it is familiar to students, has some other shortcomings. First of all, we found that students, even when they think of vectors in terms of arrows in the plane, are likely to continue to view the coordinate system as something determined by the axes rather than by (a pair of) basis vectors. That is to say, that, for the students, the coordinate axes are not themselves a set of vectors but an auxiliary feature, namely, calibrated lines which determine a grid. Secondly, the standard Cartesian plane, with its horizontal and vertical axes and same unit measure on both axes, is so compelling that it is hard to accept the need to choose any other axes (basis).

In our more recent research (Sierpinska, Dreyfus and Hillel, 1999) we introduced basic linear algebra concepts in a synthetic-geometric way with the aid of a dynamic geometry software (Cabri). Among our reasons for this particular approach we had hoped that the coordinate-free approach would facilitate the notion of general basis. Our efforts were thwarted somewhat by the realization that the computer screen (as well as a sheet of paper) provides, implicitly, a preferred coordinate system. We found, on several instances, that tasks on transformations given in purely geometric form were attempted to be solved by students in an analytic way, as if the standard coordinate system was in place.

2.5. *The General Use of a Geometric Language*

The geometric language is used in a kind of metaphoric way in many parts of the general theory. Thus, one speaks of projections and orthogonality in an inner-product-space; of lines, planes and hyperplanes in V; and vectors are often denoted in texts by using arrows, as in $\vec{u}$, so as to distinguish them from scalar quantities. Furthermore, abstract notions are frequently introduced using some kind of geometric depiction. For example, in (Strang, 1993), orthogonal subspaces are shown using the Figure 5:

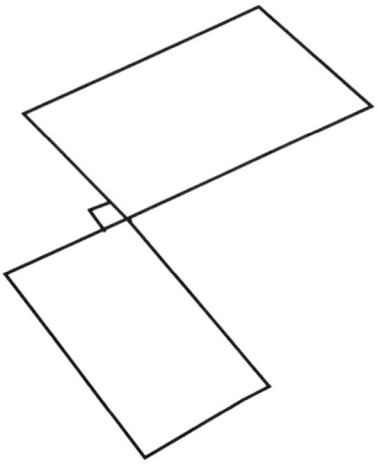

Figure 5.

In this case, the rectangles representing subspaces and the orthogonal relation between their sides are used as a heuristic geometric device to illustrate the orthogonality of the subspaces. In contrast to the above heuristic use of geometric picture, in other instances, general concepts are illustrated in the specific context of geometric vectors. For example, many texts and instructors, when introducing the notion of the projection **w*** of a vector **v** on a subspace W in an inner-product space, accompany the notion with the following (Figure 6)[2]:

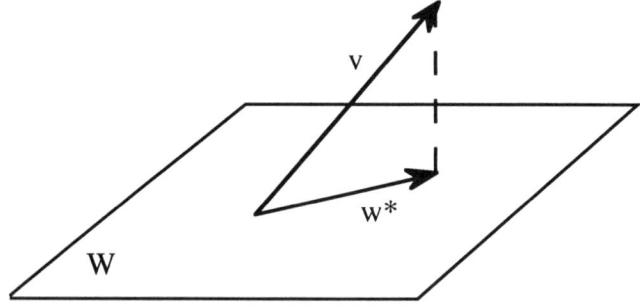

Figure 6.

Whereas this is, within some limits, a correct figure for the special case considered - $\mathbb{R}^3$ endowed with, implicitly, the standard inner-product - the reference to a geometric picture is intended as a kind of metaphor for the general case. However, students often seem to take such illustrations literally. They operate in the 'image-having level' rather than the 'property-noticing level' (Pirie and Kieren, 1994). That is, they focus on the object of the metaphor ('projection in an Inner Product Space **is** ...') rather than focus on the properties ('projection in an Inner Product Space **is like** ...'). So when we listen to students' conversation, for example, when they are working on the problem of finding the projection of sin(x) on the subspace of polynomials of degree ≤ 2 (with respect to the inner-product $\int_0^1 f(t)g(t)dt$), we find that students refer to the projection as 'closest vector **in the plane**' and that the 'plane' means for them literally the XY-plane containing the graphs of the functions (Dreyfus and Hillel, 1998).

3. THE PROBLEM OF REPRESENTATIONS

For a brief period, the tradition in North American texts followed the Bourbaki style of going from the general to the particular. The theory of vector spaces was introduced in its full generality first, followed by the more specific theory of $\mathbb{R}^n$. Since the early 1980's, most instructors and authors of introductory linear algebra texts have, by and large, abandoned this approach. The prevalent approach is to start with the geometric mode and then move back and forth between $\mathbb{R}^2$, $\mathbb{R}^3$, $\mathbb{R}^n$ and V, using the three different languages of the theory interchangeably. This practice was borne in the early stage of our research when we videotaped five experienced lecturers teaching the topic of eigenvalues and eigenvectors. These tapes illustrated how the lecturers were constantly shifting modes of description and notation; from linear operators on a finite-dimensional space to matrices; from eigenvectors of T being defined by $T(v) = \lambda v$ to arrows being stretched or shrunk by T, or to invariant lines in $\mathbb{R}^2$ and $\mathbb{R}^n$, or to solutions of a system of equations. Moreover, these shifts were usually done without a pause and without any attempt to alert the students about them in an explicit way. Nor was it always clear when the geometric description was used to illustrate a specific case (i.e., V was in fact $\mathbb{R}^2$ or $\mathbb{R}^3$), or as a heuristic for the general case.

The ability to understand how vectors and transformation in one mode are differently represented, either within the same mode, or across modes is essential for coping with linear algebra. Furthermore, because the representations are generally not unique and depend on a particular choice of basis or a coordinate system, understanding such dependence on basis (and, implicitly, understanding what remains invariant) is a major challenge for students. They not only need to be able to find, say, a matrix representation of a given linear operator in a given basis but also need to be able to think about matrix representations of linear operators as objects of inquiry in their own right, and be able to consider the general conditions under which a linear operator can have a particularly desirable representation.

3.1. Moving Between Modes

Representations allow one to go back and forth between the different modes of description as, for example, when considering an abstract 2-dimensional space geometrically, or $\mathbb{R}^n$ as an abstract space. These representations are either constructed in an explicit way, or might simply involve a subtle notational shift. We shall describe here the most common constructions, noting that Pavlopoulou (1993) has also discussed some of the difficulties in coordinating what she termed the 'registers of representation' (see chapter 8 of this book).

3.1.1. Geometric - Algebraic Modes

Analytic geometry, by its nature, links the geometric and the algebraic mode. An arrow $\overrightarrow{OP}$ is identified with the vector $X=[x_1,x_2]$ where O is the origin and P is the point $P(x_1,x_2)$.

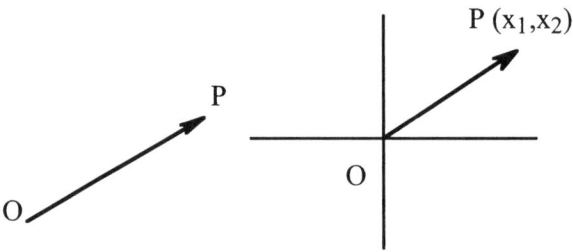

Figures 7.

This shift in mode of description necessitates showing that operations on vectors are compatible; for example, the 'parallelogram addition' of vectors corresponds to component-wise addition. Geometric transformations in the plane have to be expressible as 2x2 matrices acting on the coordinate vectors; for example, rotation by ϕ is expressible as the matrix transformation $X \rightarrow AX$ where $A=\begin{bmatrix} \cos(\phi) & \sin(\phi) \\ -\sin(\phi) & \cos(\phi) \end{bmatrix}$.

This process is not unique since any coordinate system can be used to go from the geometric mode to $\mathbb{R}^2$ and $\mathbb{R}^3$. In practice, unless there are a priori reasons to do otherwise, one starts with the usual Cartesian plane. Moving from $\mathbb{R}^2$ or $\mathbb{R}^3$ to geometric vectors calls for reversing the above steps. However, the situation is not completely symmetric when it comes to transformations. General 2x2 (3x3) matrices are not always describable in terms of meaningful geometric transformations.

3.1.2. Algebraic-Abstract Modes

Once it is established that $\mathbb{R}^n$, with component-wise addition and scalar multiplication of n-tuples, is a vector-space (tedious exercise if done in detail) then looking at it from the more abstract perspective is done via notational and linguistic shifts. $\mathbb{R}^n$ is denoted as V, and an n-tuple $[x_1, x_2, ..., x_n]$ as **v** rather than, say, X. The matrix transformation $X \rightarrow AX$ is considered as a linear transformation on V (and sometimes denoted as L_A - 'the linear transformation associated with A').

For an abstract (finite dimensional) space V, the shift to the algebraic mode is done via a choice of basis β for V. Relative to the choice of basis, a vector **v** is associated with an n-tuple of coordinates X, and a linear operator T is associated with the matrix A. Notation such as $[\mathbf{v}]\beta$) and $[T]\beta$ is used to denote the associated coordinate vector and matrix.

The basis-dependent representation is a major cause of students' difficulties in introductory linear algebra courses, and will be discussed more fully below.

3.2. The Problem of Representation in Terms of Basis

By far, the most confusing case for students is the shift from the abstract to the algebraic representation when the underlying vector space is already $\mathbb{R}^n$. In this case an n-tuple is represented as another n-tuple relative to the basis β. Thus the object and its representation are the same 'animal', if not exactly the same list of values. Similarly, a matrix transformation A; first considered as a linear operator and then as having a (possibly different) matrix representation relative to a given basis.

When students encounter vectors for the first time, mainly in the context of concrete $\mathbb{R}^n$ spaces, strings of numbers are identified with vectors. However, when it comes to representing vectors or linear operators in different bases, the identification of vector with a string of numbers becomes very much shaken. Now, an n-tuple can either be the vector or a representation of a vector relative to a basis, and the vector can be represented by different strings of numbers. For someone for whom vectors are identified with particular strings of numbers, this ambiguity may cause the whole conception of vector to fall to pieces. (Students are not easily convinced by the argument that a given n-tuple can already be thought of, by default, as a representation relative to the standard basis.)

There are notational and terminological devices to try to keep the distinction between an n-tuple X and its representation, such as writing $[X]\beta$ for the representation relative to the basis β. In some texts, a vector in $\mathbb{R}^n$ is written as $[x_1, x_2, .., x_n]$ and representations relative to a basis as $(x_1, x_2, .., x_n)$. In the former case, the x_i's are referred to as the 'components' of the vector, while in the latter, as the 'coordinates'. Some texts write a vector as a row and its coordinate vector relative to a basis as a column. However, there is so much shuffling of rows and columns in linear algebra (particularly because row and column ranks of a matrix are the same) that attempted distinction between a vector and its representation gets lost. Attempts to be notationally concise quickly become a tricky bookkeeping problem as is shown in the following excerpt (Friedberg, Insel and Spence, 1979):

Suppose now that $T : V \to W$ is a linear transformation between finite-dimensional vector spaces and that ß and ß' are ordered bases for V and that γ and γ' are ordered bases for W. Then T can be represented by matrices relative to ß and γ and relative to ß' and γ'. What is the relationship between the matrices $[T]^{\gamma}_{\beta}$ and $[T]^{\gamma'}_{\beta'}$? The answer is easily seen from the equations $[T(v)]_{\gamma} = [T]^{\gamma}_{\beta} [v]_{\beta}$ and $[T(v)]_{\gamma'} = [T]^{\gamma'}_{\beta'} [v]_{\beta'}$ given by theorem 2.16. For if Q and P are the change of coordinate matrices that change ß'-coordinates into ß-coordinates and γ'-coordinates into γ-coordinates, respectively, then from these equations it is clear that there are two methods for obtaining $[T(v)]_{\gamma}$ from $[v]_{\beta'}$, as depicted in Fig. 2.3.

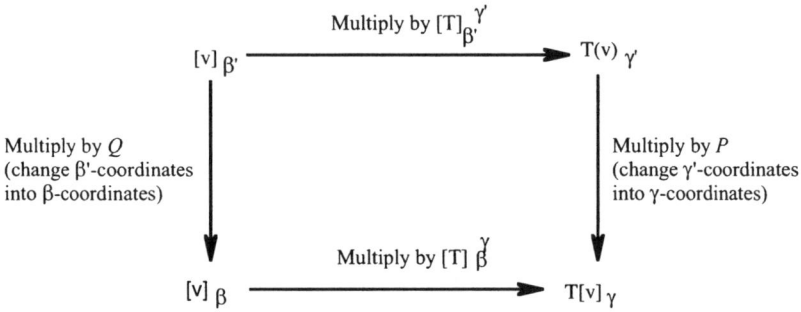

Figure 2.3.

Since $[T]^{\gamma}_{\beta} Q[v]_{\beta'} = P[T]^{\gamma'}_{\beta'} [v]_{\beta'}$ for all v of V, Theorem 2.17(b) implies that $[T]^{\gamma}_{\beta} Q$ $= P[T]^{\gamma'}_{\beta'}$. Because P is invertible (Theorem 2.26), this provides the following answer to the question posed above.

Theorem 2.27. Let T: V ----> W *be a linear transformation from a finite-dimensional vector space* V *to a finite-dimensional vector space* W, *and let* β *and* β" *be ordered bases for* V *and* γ *and* γ' *be ordered basis for* W. *Then* $[T]^{\gamma'}_{\beta'} = P^{-1}[T]^{\gamma}_{\beta}Q$ *where* Q *is the matrix which changes the* β'-coordinates *into the* β-coordinates *and* P *is the matrix which changes the* γ'-coordinates *into the* γ-coordinates.
(Friedberg et al. 1979, 98-99).

One may laud the authors of the above text for trying to be precise but it is clear that such notational gymnastics cannot hope to resolve the difficulties the students face with regard to representations.

3.3. *A Persistent Student Difficulty with Representation*

In this section we shall be somewhat more explicit about the kind of mistakes that students are making relative to the question of matrix representation of linear operators in different bases. We shall provide some evidence that reveals the existence of a teaching and learning problem. Our data come from observing three groups of students taking either the first or the second 1-semester linear algebra course at our university. Finding the coordinates of a vector and the matrix representing a linear operator relative to a basis, and the relation between two matrix representations of a linear operator are among some of the topics taught in these courses. However, as we shall show, some students can use these procedures in one way only; they are often unable to 'undo' what they have done. For example, having found a matrix representation of a linear operator T on $\mathbb{R}^3$ relative to a basis, say, $\{u,v,w\}$, they have trouble reading, from this matrix, even the value of $T(u)$.[3]

Let's look at a typical (at least in our courses) problem about matrix representation. This problem was given at the beginning of the second semester to a class of 29 students, all of whom had already completed a 1-semester course in linear algebra (and most of them in the previous semester). The problem was given as a class activity so the students could consult with their neighbours:

Let T: $\mathbb{R}^3 \rightarrow \mathbb{R}^3$ be a linear operator represented by the matrix

$A = \begin{bmatrix} 1 & 2 & 3 \\ 3 & 4 & 5 \\ 6 & 7 & 8 \end{bmatrix}$ relative to the basis $\beta=\{[1,1,1],[2,1,0],[0,5,6]\}$ for $\mathbb{R}^3$.

Find T([1,1,1]).

There were only 3 correct answers to this problem while 17 students left the question unanswered (possibly because they did not realize that [1,1,1] was already a basis vector). Three other students evaluated A $\begin{bmatrix} 1 \\ 1 \\ 1 \end{bmatrix}$ and came up with the vector

$\begin{bmatrix} 6 \\ 12 \\ 21 \end{bmatrix}$ as the answer. One student took the first column of A as an answer: 'T([1,1,1])

= [1,3,6]'; and another had the same approach, though she hesitated between choosing the first column and the first row, and wrote: 'T([1,1,1]) = [1,3,6] or [1,2,3]'. One student worked out the general formula for $[X]_\beta$, the coordinate vector in the given basis, and then obtained, by substitution, [1,0,0] as a coordinate vector for [1,1,1] and claimed that this was T([1,1,1]). Another embarked on the calculation of the general expression for the coordinate vector in the given basis but did not finish. There were few bizarre answers such as taking the dot product of vectors [1,1,1] and [1,3,6] or giving a matrix as an answer.

While the poor performance might be attributed, in part, to students having 'forgotten' their linear algebra over their Christmas holidays, the mistake of taking a column of the matrix as the image of the corresponding basis vector under the operator seems to be a persistent one; we have observed the same phenomenon with other groups of students working on similar tasks. It stems from thinking that the columns of the matrix representation of an operator are, so to speak, images of the basis vectors, not *representations* of these images relative to a basis. The difficulty is compounded by the fact that, in the case of the canonical basis, this is a correct interpretation.

With one group of students, particular attention in class and through assigned activities was devoted to the question of matrix representations of linear operators. In the first class test, administered in the 8th week of classes, and covering eigenvalue problems, diagonalization and canonical forms, the following problem related to matrix representation was given:

Let $T: \mathbb{R}^3 \to \mathbb{R}^3$ be a linear operator represented by the matrix

$$A = \begin{bmatrix} 7 & 0 & 1 \\ 0 & 2 & 0 \\ 1 & 3 & 0 \end{bmatrix}$$ relative to the basis $\{[1,3,0], [0,1,3], [0,0,1]\}$.

Find the matrix representing T in the standard basis.

Of the 29 students who wrote this test there were 15 correct answers. However, ten responses contained a familiar lapsus: these students would find the coordinates of each vector $e_1=[1,0,0]$, $e_2=[0,1,0]$, $e_3=[0,0,1]$ relative to the given basis (perhaps by first finding the general expression of the coordinate vector). But once having found that, say, $e_1=v_1- 3v_2+9v_3$, and hence has the coordinates $(1,-3,9)$, they would

go on to calculate $\begin{bmatrix} 7 & 0 & 1 \\ 0 & 2 & 0 \\ 1 & 3 & 0 \end{bmatrix}\begin{bmatrix} 1 \\ -3 \\ 9 \end{bmatrix}$ and give the resulting vector as $T(e_1)$, rather

than the coordinates of $T(e_1)$ relative to the basis. So here again they behaved as if the columns were *the* images of the basis vectors and not the *representations* of these images.

On the final exam, students were given a similar problem:

Let $T: \mathbb{Q}^4 \to \mathbb{Q}^4$ be a linear operator given by the formula:
$$T([x,y,z,t]) = [-2x-y-2z-12t, x+2y+4z+6t, -2z-3t, z+t].$$
(a) Check that T has its Rational Canonical Form in the basis $\{[0,1,0,0],[-1,2,0,0],[4,-2,1,0],[-8,4,-2,1]\}$.

Because of the particular emphasis put by the instructor on the issue of matrix representations, the overall performance on this kind of task improved. Out of 30 students writing this test, 18 gave correct answers (60%); among the rest, seven still reverted to the problem of mixing up a vector with its representation. It is tempting, at this point, to use the Piagetian language and say that the activity of representing a linear mapping by a matrix in a basis has not been *internalized as an operation*. However, the persistence of mistakes with this kind of problem points to the existence of an obstacle that is of a more conceptual nature, and not just related to a difficulty in the operationalization of a procedure.

4. APPLICABLITY OF THE GENERAL THEORY

The isomorphism between an n-dimensional V and $\mathbb{R}^n$ is a capstone result in introductory linear algebra. In a strange way, this isomorphism undermines the whole raison d'être for the abstract theory for finite-dimensional spaces, since many problems and exercises, do not require the tools of the general theory. Questions related to linear independence/dependence, basis, eigenvectors, kernel and range subspaces, or non-singularity can be solved by direct manipulation techniques (most often by reducing the problems to systems of equations and using Gauss-Jordan reduction). It is an issue that was already discussed by Dorier (1995b and 1998). In a way, students are often faced with situations not unlike those faced by mathematicians at the end of the last century. In that period, results and techniques on linear substitutions and bilinear forms were highly developed, particularly with regard to normal forms (e.g., Jordan canonical forms arising from solving linear differential equations), and without explicit use of the vector space concept. Consequently, early attempts by Grassmann, Peano and Pincherle to give an axiomatic treatment of the theory were not very influential and were not taken seriously till the 1920's.

On the other hand, problems of a more theoretical nature that are assigned to students seem, at times, to be kind of exercise in logical deduction. Instead of illuminating the meaning of new theoretical concepts, the problems, essentially, just ask students to explore some logical consequences of new definitions. For example, when the kernel of an operator is defined (Ker(T)), a typical problem is to show that Ker(T) is a subspace of $\text{Ker}(T^2)$. This fact might be useful later in the course when, say, characterizing nilpotent operators. But, for students who are trying to glean some sense of the definition of kernel of an operator, problems of this type are simply not helpful.

5. CONCLUDING COMMENTS

In the classical paper on epistemological obstacles, Brousseau (1983 and 1997) has argued that a mathematical notion is learned if it is linked with other notions and if it is utilizable, and indeed used as a solution to a set of problems that, in effect, endow the notion with a meaning. Furthermore, if a notion works well, and for a rather long time, it takes on a meaning that renders its modification, generalization or rejection more difficult. Thus, in a typical linear algebra course, we see two types of epistemological obstacles. The first stems from students' familiarity with analytic geometry and standard coordinates. Thinking about vectors and transformations in a geometric context certainly links these notions to more familiar ones. However, such geometric level of thinking, as we have seen, can become an obstacle to thinking about basis (rather than axes) and about the need for changing basis. The other obstacle comes about because specific notions related to $\mathbb{R}^n$ are learned by students. These notions do solve a variety of problems which are directly or indirectly linked to the central notion of systems of linear equations. Hence, this algebraic level of description becomes an obstacle to learning the general theory and to the acceptance of other kinds of objects such as functions, matrices, or polynomials as vectors.

This phenomenon, again, is not specific to linear algebra but happens regularly in mathematics when one increases the level of generality. For example, some students, having completed a course in abstract algebra, would still think of groups and rings only in terms of real numbers. What is more unique to linear algebra is that the process of abstraction comes back on itself; every (finite dimensional real) vector space is isomorphic to $\mathbb{R}^n$. But in the processes of returning to the familiar terrain of $\mathbb{R}^n$ something goes wrong (from the students' perspective). The status of an n-tuple as being the prototype of a vector becomes questionable - it is no longer just a vector but also a potential representation of any other vector. What we have shown is that this shift in meaning is a difficult idea for students. When working on problems involving basis, they can begin the process of changing from a vector to its coordinates but they cannot always carry the process to its conclusion; i.e., one that involves the extra step of changing back from coordinates to the vector.

The representation of vectors and linear operators involves a translation from one mode of description to another, or a translation within the same mode (i.e., change of basis, or changing from synthetic to analytic geometry). Understanding this process is a most basic prerequisite of linear algebra, particularly when it comes to problems of diagonalization and canonical forms. We have argued that such understanding must be on the 'trans-level'. It requires that the student be able not only to find a given representation of a given linear operator or a vector relative to a given basis, but also to think about these representations as objects of inquiry in their own right. It also requires thinking about the general conditions under which a vector or a linear operator can have a particularly desirable representation. However, it seems that while students are at least on the inter-level of thinking with respect to linear algebra 'objects', such as linear operators and matrices, for quite a few students, their understanding of the language of linear algebra still lags behind on the intra-level.

One way to try to help students along is by holding discussions with them on a meta-level, of the mathematical problem with which they are dealing: an open debate on the nature and status of the language in linear algebra. This idea of initiating meta-level discussions with students, in order to deepen their understanding of linear algebra, has already been suggested by Dorier et al. - see (Dorier 1995b) and chapter 4 of this book.

6. NOTES

[1] This is a modified and expanded version of the chapter « Des niveaux de description et du problème de la représentation en algèbre linéaire » which appeared in the original French text. It is based, in part, on a joint paper by Hillel & Sierpinska : On one persistent mistake in Linear Algebra, presented in PME 18, Lisbon, Portugal, 1993. The research reported in the chapter was supported by an FCAR grant (#93-ER-1535) and by SSHRC grants (#410-93-0700 and #410-96-0741).

[2] We note that, in the diagram, vectors appear as both arrows and points. The vectors v and w* are arrows, whereas the subspace W is essentially a set of points.

[3] We should remark that the difficulties we will be discussing here were not the focus of our research and were not studied systematically by setting tests or questionnaires geared towards collecting data on this subject. Most of the data that we present came from class tests or the final examination.

ANNA SIERPINSKA

PART II - CHAPTER 7
ON SOME ASPECTS OF STUDENTS' THINKING
IN LINEAR ALGEBRA

> Across the page the symbols moved in grave
> morrice, in the mummery of their letters, wearing
> quaint caps of squares and cubes. Give hands,
> traverse, bow to partner: so: imps of fancy of the
> Moors. Gone too from the world, Averroes and
> Moses Maimonides, dark men in mien and
> movement, flashing in their mocking mirrors the
> obscure soul of the world, a darkness shining in
> brightness which brightness could not comprehend.

James Joyce, *Ulysses*

1. INTRODUCTION

This chapter[1] will focus on certain aspects of students' reasoning in linear algebra that may be responsible for their perceived difficulties in the domain. On the most general plane, it will be argued that students tend to think in practical rather than theoretical ways, and several examples will illustrate how this tendency may adversely affect their reasoning in linear algebra. On the more specific plane, three modes of reasoning in linear algebra will be distinguished, corresponding to its three interacting 'languages': the 'visual geometric' language, the 'arithmetic' language of vectors and matrices as lists and tables of numbers, and the 'structural' language of vector spaces and linear transformations. Examples will illustrate students' reluctance to enter into the structural mode of thinking, and, first and foremost, their inability to move flexibly between the three modes. The examples will be taken from a series of projects conducted at Concordia University in the years 1993-1999[2].

In the years 1993-96 our research focused on the patterns of formatting of students' mathematical behaviors as they interacted individually with tutors and

DORIER J.-L. (ed.), *The Teaching of Linear Algebra in Question*, 209—246.
©2000 *Kluwer Academic Publishers. Printed in the Netherlands.*

linear algebra texts (Sierpinska, 1997). The mathematical content of the tutoring sessions was not specially designed; an existing text was used, which the student would read and the tutor would probe his or her understanding as well as assist the student in learning the material.

On the other hand, design of an entry into linear algebra was at the core of the projects in the years 1996-1999. The design was based on a geometric model of the two-dimensional vector space within the dynamic Cabri-geometry II environment. The Cabri environment appeared to make it possible for the students to achieve a more direct contact with the objects of the abstract theory without too quickly replacing these objects by computational procedures. There were three rounds of experimentation of the design. The first two were tightly controlled, with all students' work being videotaped, and they were restricted to 5-7 two hour sessions with 1-3 pairs of students. The third experiment was quite different: It was conducted within a course in linear algebra for master's students in mathematics education (there were 8 to 7 students in the course), with 13 two hour sessions, and homework assignments. The documentation comes from students' written work on the worksheets, done in the lab and at home, their diaries and the teacher's notes. All students participating in the first and the third experiment had previously taken at least a college level vectors and matrices course.

In this chapter, from the 1996-99 research, with one exception, only examples from the third experiment will be used. A rather complete account of the first experiment can be found in (Sierpinska, Dreyfus and Hillel, 1999); samples of results from the second experiment have been incorporated in (Sierpinska, Trgalová, Hillel, Dreyfus, 1999). The main problem that we attempted to solve by our design was to minimize the chances that the students develop symptoms of what we have called 'the obstacle of formalism', thus changing somewhat the notion introduced by Dorier, Robert, Robinet and Rogalski (Chapter 1 of the second part of the present book). These authors appear to reserve the term for students' erroneous reasoning in linear algebra stemming from their insufficient competence mainly in logic and elementary set theory, but also in manipulation of algebraic expressions. For us, a student would be labeled as being 'under the spell' of the obstacle of formalism if he or she behaves as if the formal symbolic representations of the linear algebra objects were the objects in themselves, yet has insufficient competence to grasp the structure of these representations, and therefore manipulates them in a manner which is incompatible with their 'grammar'; the student does not see the relationships between formally distinct representations, and thus has to deal with an often unmanageable number of objects. Students whose minds are obstrued by the 'obstacle of formalism' in our sense would make the errors of students with the 'obstacle of formalism' in Dorier et al.'s sense, but would likely produce much more nonsense writing at levels not even requiring the use of elementary set theory. From now on we shall use the term in the newly defined sense.

We had been able to avoid the development of the obstacle of formalism in students in the first two experiments, where each pair of students was assigned a tutor, but not in the third experiment, which took place in an ordinary classroom environment with one teacher taking care of the whole class. The specially designed contents of the course and the fact that all classes were taking place in a computer lab were not enough to prevent this phenomenon from occurring: The symptoms of the obstacle of formalism were apparent in three out of the eight students who started the course (one student of these three dropped the course after the eighth week).

In fact, in spite of all our efforts in improving the design, difficulties persisted. We understood that for all the innovations that we made in presenting the theory to the students, we still wanted them to understand the same theory: the structural theory of linear algebra, with, among others, its axiomatic definition of linear transformation. But the students in our experiments could not understand the theory because they appeared to want to grasp it with a 'practical' rather than a 'theoretical' mind. We started to develop the notion of 'theoretical thinking' as opposed to 'practical thinking', and the first part of this chapter is a result of our reflection on this methodological distinction.

We are aware that similar distinctions have been around in various forms for a long time in philosophy, psychology and mathematics education. In the latter domain, there are the popular distinctions of 'instrumental vs. relational understanding' (Skemp, 1978), and 'operational vs. structural thinking' (Sfard, 1987). Recently, a group of Italian mathematics educators started a project aimed at developing 'theoretical knowledge' in school-children - see, e.g. (Boero, Pedemonte and Robotti, 1997) or (Garuti, Boero and Chiappini, 1999). The theoretical framework of this research was constructed, in part, on the basis of the Vygotskian distinction between spontaneous and scientific concepts. Inspired by this research we turned to Vygotsky's work in the aim of identifying the characteristics of the practical and the theoretical modes of thinking.

The latter distinction is very general and does not capture the specificity of thinking in linear algebra. We have therefore brought in the distinction between three modes of thinking: synthetic-geometric, analytic-arithmetic and analytic-structural, whose identification is based on a historical analysis and study of the different 'languages' used in the theory of linear algebra itself.

Practical thinking as a source of students' difficulties in the learning of linear algebra will be discussed in the first part of the chapter, and examples from our research on the teaching and learning of linear algebra with Cabri will be used to illustrate our points. In the second part of the chapter, the synthetic-geometric, analytic-arithmetic and analytic-structural modes of thinking will be presented, with examples taken mainly from our research on the interactions between a student, a tutor and a text.

2. PRACTICAL, AS OPPOSED TO THEORETICAL, THINKING - A SOURCE OF DIFFICULTY IN LINEAR ALGEBRA STUDENTS

As mentioned, our understanding of the theoretical as opposed to practical thinking was inspired by the Vygotskian distinction between everyday concepts and scientific concepts (Vygotsky, 1987, Chapter 6). We shall assume that theoretical thinking is characterized by a conscious reflection on the semiotic means of representation of knowledge and by systems of concepts rather than aggregates of ideas. We further assume that, in theoretical thinking, reasoning is based on logical and semantic connections between concepts within a system; connections between concepts are made on the basis of their relations to more general concepts of which they are special cases rather than on empirical associations. The relations between concepts and objects are mediated by relations of the concepts to other concepts. In particular, definitions of concepts, comparisons between concepts and their differentiation are constructed on the basis of the relations of these concepts to more general concepts, and not, e.g., on the basis of their most common examples.

2.1. Students' Practical Thinking in a 'Linear Algebra with Cabri' Course

This section will be organized along the different ways in which the students in our research failed to think theoretically.

2.1.1. Transparency of Language

The most general difference between practical and theoretical thinking is that the latter is a specialized activity in itself, while the former is an auxiliary activity which accompanies and guides other activities (Marody, 1987, p. 82). Practical thinking expresses itself in goal-oriented, physical action. Theoretical thinking signals its existence and finds its expression through texts. Consequently, semiotic representation systems become themselves an object of reflection and analysis in theoretical thinking because they constitute the only medium through which theoretical thinking may prove its existence and convey its meanings. Practical thinking proves itself in action, and the form and structure of the sign systems it uses is transparent for it, not reflected upon; meanings of words are clarified by the contexts in which they are used.

The 'transparency of language' feature of practical thinking may manifest itself in many ways in students. Let us look at two forms of its expression: the 'inner speech' genre in writing, and the students' inability to appreciate the special sentence structure of definitions.

The 'inner speech genre' in students' writing[3]

The importance of the written speech and its distinct character with respect to the oral speech was strongly stressed by Vygotsky in his discussion of factors which contribute to the development of scientific concepts (Vygotski, 1987, Chapter 6. Development of scientific concepts).

> Even the most minimal level of development of written speech requires a high degree of abstraction.... Written speech is speech in thought, in representations.... [W]ritten speech differs from oral speech in the same way that abstract thinking differs from graphic thinking.... Written speech is more abstract than oral speech in other respects as well. It is speech without an interlocutor. (op. cit., p. 202)
> [...] Speech that lacks real sound (speech that is only represented or thought and therefore requires the symbolization of sound -- a second order symbolization) will be more difficult than oral speech to the same degree that algebra is more difficult for the child than arithmetic. Written speech is the algebra of speech. The process of learning algebra does not repeat that of arithmetic. It is a new and higher plane in the development of abstract mathematical thought that is constructed over and rises above arithmetic thinking. In the same way, the algebra of speech (i.e., written speech) introduces the child to an abstract plane of speech that is constructed over the developed system of oral speech.... [O]ral speech is regulated by the dynamics of the situation.... With written speech, on the other hand, we are forced to create the situation or -- more accurately -- to represent it in thought. The use of written speech presupposes a fundamentally different relationship to the situation, one that is freer, more independent, more voluntary. (ibid., p. 203)

Explicit and elaborate, the written speech differs from the oral and inner forms of speech by its voluntary and therefore conscious character, linking it closely with scientific thinking.

One would expect university students to have already developed the skills of written language, but all too often the written work of mathematics students carries the features of the inner rather than those of the written speech. As described by Vygotsky, the inner speech characterizes itself, from the point of view of syntax, by being maximally abbreviated, condensed, almost entirely predicative, with little or no mention of the subject or the object of the predicate (ibid., p. 204), and, from the point of view of semantics, by: 1) the predominance of the sense of a word over its meaning, of the sentence over the word, and of the context over the sentence; 2) agglutination of words, or combining several words into one to express one more complex idea all the while keeping explicit all the separate elements contained in that idea; 3) influx of sense, whereby a word, which appears repeatedly in a text, aspires into itself the generalized sense of the text as a whole (ibid., p. 277). The 'sense' of the word is understood by Vygotsky as follows:

> A word's sense is the aggregate of all the psychological facts that arise in our consciousness as a result of the word. Sense is a dynamic, fluid, and complex formation which has several zones that vary in their stability. Meaning is only one of these zones of the sense that the word acquires in the context of speech. It is the most stable, unified, and precise of these zones. In different contexts, a word's sense changes. In contrast, meaning is a comparatively fixed and stable point, one that remains constant with all the changes of the word's sense that are associated with its use in various contexts.... Isolated in the lexicon, the word has only one meaning. However, this meaning is nothing more than a potentiality that can only be realized in living speech, and in living speech meaning is only a cornerstone in the edifice of sense. (ibid., p. 276)

Let us look at an example of the 'inner speech genre' in students' writing.

Example 1.

In response to the problem: 'Verify if the translation by a constant vector a in the vector plane, defined by the formula $T(v) = v + a$, is a linear transformation', a student writes:

$T(v_1) = v_1 + a$ [1]

$T(v_2) = v_2 + a$ [2]

$a = L - w$ [3]

$a = T(w) - (k_1 v_1 + k_2 v_2)$ [4]

$\quad = T(k_1 v_1 + k_2 v_2) - (k_1 v_1 + k_2 v_2)$ [5]

$\quad = k_1 T(v_1) + k_2 T(v_2) - k_1 v_1 - k_2 v_2$ [6]

$\quad = k_1 (T(v_1) - v_1) + k_2 (T(v_2) - v_2)$ [7]

$\quad = k_1 a + k_2 a$ [8]

$a = a (k_1 + k_2)$ [9]

$\therefore a \neq a (k_1 + k_2)$, Since it depends on k_1 and k_2 [10]

It is very difficult to reconstruct the student's reasoning because the solution has no explanations. All one can do is speculate on the basis of the student's previous writing in the worksheet. It is thus quite likely that 'L' in line [3] stands for 'left-hand-side' of an equation. Maybe the student saw the pattern of the equalities in lines [1] and [2] as 'L = w + a' and deduced that 'a = L - w'. In the previous exercise in the worksheet, 'w' stood for '$k_1 v_1 + k_2 v_2$' and 'L' for the left-hand-side of the defining property of linear transformations. This would explain the passage from line [3] to line [4] and then to [5].

In line [6] the expression '$T(k_1 v_1 + k_2 v_2)$' is replaced by '$k_1\ T(v_1) + k_2\ T(v_2)$', which suggests that either the student is making the logical error of circular argument or uses the method of proof by contradiction. The passages between lines [5] and [8] could all be justified by the laws of operations on vectors, but the student does not write any justifications, so we do not know what is the basis of these manipulations for him. In line [9] he writes the scalars after the vector in what appears to be scalar multiplication, contrary to the common usage, so one has doubts whether he thinks of the letter 'a' as representing a vector and 'k_1' and 'k_2' as representing scalars. In line [10] the student contradicts line [9], with a justification, 'since it depends on k_1 and k_2', which supports the hypothesis that the student is using a proof by contradiction. However, no conclusion from this contradiction is drawn in writing, so, again, it is impossible to say if, indeed, the student consciously used this method.

For some students in our research, the equation '$T(k_1 v_1 + k_2 v_2) = k_1 T(v_1) + k_2 T(v_2)$' has acquired the sense of a magical formula, a key to all problems. It was enough to write it, and manipulate it, or to replace the left hand side by the right hand side and the problem was solved. In the solution above, the magical sense of the formula influenced also the letter 'L', first (in line [3]) used only to mean the left hand side of any equation. In fact, the expression 'left-hand-side' has the features of the 'agglutination of words' mentioned by Vygotsky.

The written text appears as a record of a certain number of actions, whose subjects and objects have been kept implicit. Thus, in grammatical terms, we could say that the text is composed of predicates only. It is not possible to say how elaborate the student's work would have been had he used written speech to express his solution. What would have been the subjects and objects of the predicates had he cared to write them down? Our experience suggests that the subject could well have been the impersonal pronoun 'it', like in 'it changes', 'it increases' (or in 'it rains'), or a vague personal pronoun 'they', and the objects - demonstrative pronouns such as 'this' or 'that' (see Sierpinska, 1992, p. 33). In the first and second experiment, where each pair of students was assisted by a tutor constantly probing their understanding, frequently the students' first response had such an abbreviated, predicative character, and it is only due to the tutor's questioning that they would be led to analyze the situation into, metaphorically speaking, 'objects doing something to other objects'.

For example, in the first experiment, there was a situation where the students were observing a Cabri representation of a transformation of vectors. First, an independent vector v was drawn on the screen. Then, with a macro construction, a dependent vector $T(v)$ was obtained. The two vectors were represented as starting from the same point, called the origin. As v was moved, $T(v)$ also moved. The students were asked to decide if the given dependence or transformation T of vectors preserves dilation; i.e., if when v is dilated by a scalar k, then $T(v)$ also gets dilated by the same scalar k.

The variable scalar k was represented on the screen by the coordinate of a thick, moveable point on a number line at the top of the screen (Figure 1a). The students dilated the vector v by the scalar k, obtained the vector kv, then obtained the vector $T(kv)$ using the given macro construction, then dilated the vector $T(v)$ by k to obtain kT(v). The students observed that kT(v) and T(kv) overlap and declared that, yes, 'the vector T(kv) is obtained by dilation by the same factor k' (Figure 1b). Then, to help the students visualize the property of conservation of dilation by transformations, the tutor joined the tips of vectors v and T(v) and kv and T(kv) by line segments and started varying the scalar k by moving the thick point back and forth on the number line (Figures 1c,d).

The students were quite impressed by the dynamic figure deployed in front of their eyes. Asked to describe what they saw, they said:

Student 1: *If k is moving, the two triangles are... not equal, but they have... uhm... they remain at the same angle... and the relationship of the sides of the triangles remains the same.*

Student 2: *They are depending on each other? When the factor changes they are changing but they are still equal in proportion ?*

Student 1 was more articulate than Student 2, but still, their descriptions cannot be understood outside the context of what they could see on the screen. Note the use of the pronoun 'they' by Student 2. The tutor then asked questions like: 'So what is changing in proportion?', 'What are the sides of the triangles representing?', 'Can you make a statement about this?'. After several attempts, the students said:

Student 2: *When the value of k changes then kv changes in proportion with T(kv)*

Student 1: *T(kv) is proportional to T(v) by k factor.*

(For a complete account of the situation, see Haddad, 1999, pp. 72-76).

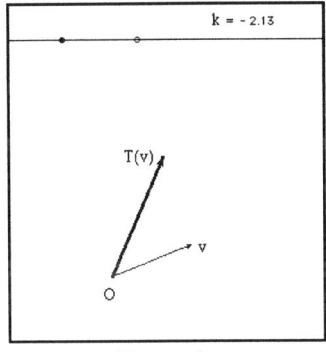

Figure 1a.

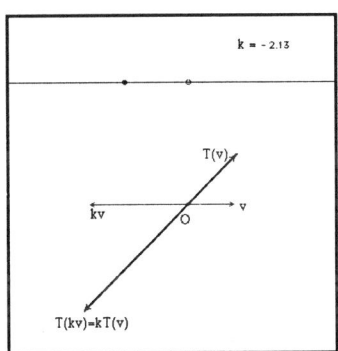

Figure 1b.

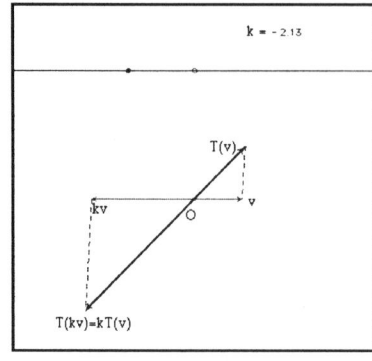 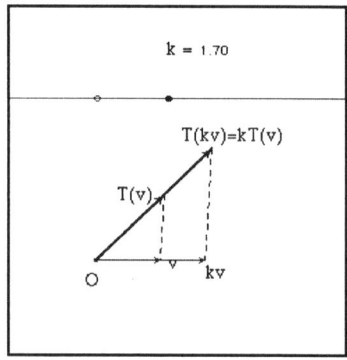

Figure 1c.   *Figure 1d.*

Figures 1. A Cabri representation of the property T(kv)=kT(v) of transformations (represented, here, without the dynamism available in the software)..

In the third experiment, in an ordinary classroom situation, such questioning and probing of individual students' understanding was not taking place with every group of students and every problem. And, therefore, students who used the 'inner speech genre' in their solutions were not led to analyze the phenomena observed in Cabri and their computations in terms of 'objects doing something to other objects'. They only saw 'things happening'. This predicative way of thinking likely fertilized the development of the obstacle of formalism in some of them.

Also other features of the 'inner speech genre', like the predominance of sense over meaning and the phenomenon of the 'influx of sense', could possibly lead to the development of the obstacle of formalism. Certainly, for some students, the sense of formulas and terms in linear algebra were heavily laden with the social and institutional context of their situation as students trying to simply 'survive the course' (an expression used by one of them). From the questions that they would pose to the teacher in the course it was apparent that they wanted to make sure that they understood the contract and knew what the teacher expected them to write in response to the questions in the worksheets. They appeared sensitive to the teacher's discourse and tried to use the terms she stressed as important in their solutions, even though, spontaneously, the students would use different words to express similar ideas. For example, the 'official' terms of 'length', 'direction', 'orientation' would be used in directly formulated exercises of comparing geometric vectors, but when spontaneously comparing vectors, they would use expressions such as 'vectors are (or are not) equal', 'parallel' and 'have same direction', respectively. Using borrowed language, these students were not able to express their own ideas, which, not mediated by verbally elaborate and explicit language, lacked the necessary support to develop into fully fledged concepts.

Definition: an unfamiliar form of statement

Related to the transparency of language for the students is their lack of sensitivity to the grammatical structure of a definition and its completeness (in the sense that the defining conditions not only just list some of the properties of the defined object but completely determine this object).

Example 2.

Asked to formulate a definition of the equality of geometric vectors using terms such as 'parallel' or 'parallelogram', students wrote statements such as:

1. « In the parallelogram, two parallel vectors are equal to each other because the length, direction and orientation are the same ».
2. « Two vectors are equal if they are parallel ».
3. « Two vectors are equal if they are opposite sides of a parallelogram ».

There was no structurally correct and complete definition among the students' solutions.

The grammatical structure of Statement 1) is not one of a definition. Statements 2) and 3) have that structure but are incomplete: 2) ignores the length and the orientation, 3) - the orientation.

In another exercise of formulating a definition, the grammatical form was already suggested in the question: « Define the operation of scalar multiplication in geometric terms: If k is a scalar and v is a vector, then the vector kv is a vector w such that _____ ..... ». Some students finished the sentence by writing: 'w = kv'. Aside from being circular, this 'definition' may also suggest an understanding of the scalar multiplication as the action of writing an expression like 'kv', and not a mathematical operation defined on vectors.

2.1.2. Lack of Sensitivity to the Systemic Character of Scientific Knowledge

For Vygotsky the most fundamental feature of scientific concepts is that they are constituted into systems. This is what makes them sensitive to contradiction.

> The key difference in the psychological nature of [scientific and everyday] concepts is a function of the presence or absence of a system. Concepts stand in a different relationship to the object when they exist outside a system than when they enter one.... Outside a system, the only possible connections between concepts are those that exist between the objects themselves, that is, empirical connections. This is the source of the dominance of the logic of action and of syncretic connections of impressions in early childhood. Within a system, relationships between concepts begin to emerge. These relationships mediate the concept's relationship to the object through its relationship to other concepts.... It could be demonstrated that all the characteristics of the child's thought identified by Piaget, characteristics such as syncretism, insensitivity to contradiction, and the tendency to place things alongside one another, stem from the extrasystemic nature of the child's concepts.... In Piaget's experiments, the child maintains that a bead sinks because it is small at one point, while he claims it sinks because it is large at another.... The contradiction [can be] noticed when the concepts expressed in the contradictory judgments are included in the structure of a single superordinate concept.... Due to the underdevelopment of the relationships of generality in the child, however, the two concepts cannot possibly be unified within the single structure of a higher concept. The result is that the child expresses two mutually exclusive judgments. From his perspective, however, these judgments relate to two different things. In the logic of the child's thought, the only relationships among concepts that are possible are those that exist among the objects themselves.... This logic of perception does not know contradiction (Vygotsky, 1987, pp. 234-235).

Students in our experiment would not contradict themselves in the same way as the above cited children in Piagetian experiments. They would not make contradictory statements close to each other in time, or statements in direct contradiction to each other (i.e., of the form: p and not p). But they did produce statements contradicting some previous experiences in more indirect ways. In linear algebra courses, even at the most elementary level, the number of new concepts grows very fast and the links between them become more and more intricate. In this situation, the detection of a contradiction requires performing a longer chain of deductions. But some of the students were unable, and perhaps sometimes even unwilling, to engage with longer reasoning. They also did not seem interested in checking their assertions, even if only in a quasi-empirical way, using the affordances provided in Cabri. Thus, with few logical and semantic links, in some students, concepts appeared organized according to their temporal appearance in the course ('this is what we did in Week 5') rather than according to 'generality relations'.

Example 3.

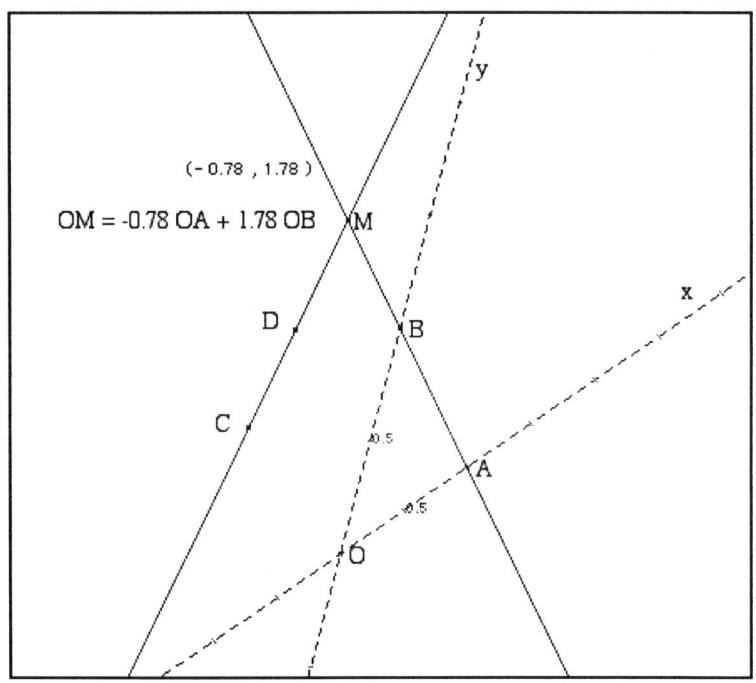

Figure 2.
A Cabri solution of the problem to express the intersection M of the lines AB and CD

in terms of the position of the vector $\overrightarrow{OM}$ with respect to the vectors $\overrightarrow{OA}$ and $\overrightarrow{OB}$.

This example is about a problem given as homework to the students after they had been working in Cabri for quite some time, and had an opportunity, among others, to define the equality of geometric vectors in various ways, decompose vectors into linear combinations of other vectors, and work with vector equations of lines. In particular, the students were familiar with the command New Axes and were given the chance to notice that a vector may have different coordinates depending on the choice of the unit vectors for the axes.

The problem was in two parts. In the first part, the students were given five points, O, A, B, C, D on the Cabri screen, so positioned that the intersection point M of the lines AB and CD was within the frame of the screen. The students were asked to describe the position of the intersection point M with respect to the point O, by writing the vector $\overrightarrow{OM}$ as a linear combination of the vectors $\overrightarrow{OA}$ and $\overrightarrow{OB}$. This part of the exercise could be done 'empirically', using the Cabri tools and just reading off the coordinates of M in New Axes built on $\overrightarrow{OA}$ and $\overrightarrow{OB}$ (see Figure 2).

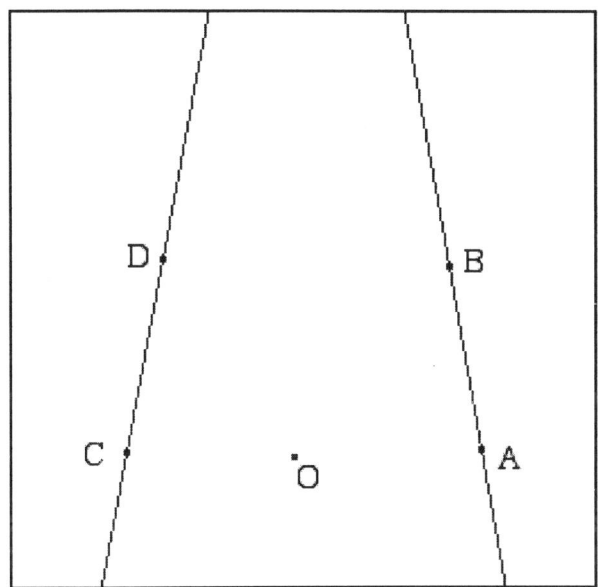

Figure 3. The Cabri screen for the problem to express the intersection point M of the lines AB and CD in terms of the position of the vector $\overrightarrow{OM}$ with respect to the vectors on $\overrightarrow{OA}$ and $\overrightarrow{OB}$, with the intersection point not visible and the condition that the screen should not be scrolled down.

In the second part of the problem, the points were so positioned that the lines AB and CD no longer intersected within the frame of the screen, but were obviously not parallel, and the question was the same (see Figure 3). The students were told that they are not allowed to scroll down the screen. This condition ruled out the possibility of solving the problem through a direct use of the Cabri tools and the students were expected to use some analytic means. Let us note that, by coincidence, the points O, A and C appeared, in the figure, as collinear. But they were not constructed as collinear and the students had the possibility of verifying this with a Cabri command.

It was expected that the students would find, with Cabri, the coordinates of C and D in the coordinate system based on the vectors $\overrightarrow{OA}$ and $\overrightarrow{OB}$, and then proceed analytically, by representing the unknown vector $\overrightarrow{OM}$ in two ways as a linear combination of the vectors $\overrightarrow{OA}$ and $\overrightarrow{OB}$, and solving a system of equations, using, for example, the uniqueness of representation of a vector in a basis:

$$\overrightarrow{OM} = s\,\overrightarrow{OA} + (1 - s)\,\overrightarrow{OB},$$

$$\overrightarrow{OM} = t\,(k_1\,\overrightarrow{OA} + k_2\,\overrightarrow{OB}) + (1 - t)\,(m_1\,\overrightarrow{OA} + m_2\,\overrightarrow{OB})$$

where (k_1, k_2) and (m_1, m_2) are the coordinates of the vectors $\overrightarrow{OC}$ and $\overrightarrow{OD}$ in the system based on the vectors $\overrightarrow{OA}$ and $\overrightarrow{OB}$. Students could either obtain concrete values for these coordinates using Cabri, or solve the more general problem with the coordinates as parameters.

This problem was typical of those used in the experimental course: Situations were so paired that a first part could be solved on the level of manipulation of objects on the screen, and the second part made the object of inquiry inaccessible to senses and thus analytical representations and the use of algebra were needed. This approach was based on the epistemological assumption that there is no direct access to objects of scientific knowledge. Scientific knowledge, by definition, was assumed to be semiotically mediated and its development regarded as triggered by the need to compare, order, and predict the behavior of objects whose identity, magnitude, appearance in time, etc. could not be evaluated with the use of the senses alone; direct observation had to be replaced by the construction of technical instruments, and of systems of representation, notation and computation, as well as of theories containing means of the validation of inferences made on this basis (Sierpinska, Trgalová, Hillel and Dreyfus, 1999).

Out of eight students in the class, three changed the positions of the points A, B, C, D, so that M would be within the screen frame, and solved the second part of the problem the same way as they did the first – using the Cabri tools. Two students attempted an analytic solution but were not successful and three others did not provide a solution.

We shall reproduce and analyze the work of one of the two students who attempted an analytic solution. We type this student's hand-written solution below:

[1]   Line $\overrightarrow{OM}$ can be defined as $k\,\overrightarrow{OA} + (1-k)\,\overrightarrow{OB}$, $k \in \mathbb{R}$

[2]   Similarly $\overrightarrow{OM} = m\,\overrightarrow{OC} + (1-m)\,\overrightarrow{OD}$, $m \in \mathbb{R}$

[3]   At the point of intersection M, it follows that

[4] $k\overrightarrow{OA} + (1\text{-}k)\,\overrightarrow{OB} = m\,\overrightarrow{OC} + (1\text{-}m)\,\overrightarrow{OD}$

[5] (Figure)

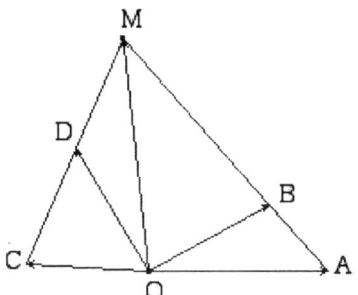

[6] Furthermore $k\,\overrightarrow{OA} + (1\text{-}k)\,\overrightarrow{OB} - m\,\overrightarrow{OC} - (1\text{-}m)\,\overrightarrow{OD} = \overrightarrow{0}$ (1)

[7] Given that vectors $\overrightarrow{OA}$, $\overrightarrow{OB}$, $\overrightarrow{OC}$, $\overrightarrow{OD}$ are all non zero, the only way (1) can be true is if k=m.

[8] Given $\overrightarrow{OM}$ is a linear combo of $\overrightarrow{OA}$, $\overrightarrow{OB}$ and $\overrightarrow{OC}$, $\overrightarrow{OD}$, it stands to reason that, at point M, k=m.
∴ we can write

[9] $k\,\overrightarrow{OA} + (1\text{-}k)\,\overrightarrow{OB} - k\,\overrightarrow{OC} - (1\text{-}k)\,\overrightarrow{OD} = \overrightarrow{0}$

[10] $\Rightarrow k\,(\overrightarrow{OA} - \overrightarrow{OC}) + (1\text{-}k)\,(\overrightarrow{OB} - \overrightarrow{OD}) = \overrightarrow{0}$ (2)

[11] Since $\overrightarrow{OB}$ and $\overrightarrow{OD}$ are non-collinear, the only way (2) will be true is for

  $(1\text{-}k)\,(\overrightarrow{OB} - \overrightarrow{OD}) = 0$ and $k\,(\overrightarrow{OA} - \overrightarrow{OC}) = 0$.

[12] But ∴ $\overrightarrow{OB}$ and $\overrightarrow{OD}$ are not collinear, the only way we can obtain this is if 1-k = 0,   that is k = 1.

[13] If k = 1 then, in (2) $(\overrightarrow{OA} - \overrightarrow{OC}) + (1\text{-}1)\,(\overrightarrow{OB} - \overrightarrow{OD}) = \overrightarrow{0}$

[14] $\overrightarrow{OA} - \overrightarrow{OC} = \overrightarrow{0} \Rightarrow \overrightarrow{OA} = \overrightarrow{OC}$

[15] In fact, we observed from Cabri that $\overrightarrow{OA} = \overrightarrow{OC}$ but $\overrightarrow{OC}$ was pointing in the oppositedirection.

[16] The intersection point, M, of two lines can be expressed as linear combinations of 2 pairs of vectors (as in the diagram) where the scalars would be equal. In addition, we noted that one vector going to each of the lines, i.e.

$\overrightarrow{OA}$ and $\overrightarrow{OC}$ must be equal, and even more so, collinear.

Except for the slip of the tongue in line [1], lines [1]-[6] are correct. Lines [7] and [8] both assert that 'k=m'. The argument in [7] is not very clear but the one in [8] could be interpreted as an implication of the statement: if $v = av_1 + bv_2$ and $v = cw_1 + dw_2$ then $a = c$ and $b = d$. This statement stands in contradiction to the observations the student must have made during previous lab activities on decomposition of vectors into linear combinations of distinct pairs of vectors and the activities with coordinate systems.

Lines [9] and [10] are obtained by substitution of k for m and factoring out k and (1-k) in (1). Line [11] would make sense had the student said, « Since

$\overrightarrow{OA}$ - $\overrightarrow{OC}$ and $\overrightarrow{OB}$ - $\overrightarrow{OD}$ are non-collinear » . In line [12] the student may have been reasoning correctly: it was possibly obvious for him that the non-collinearity of

$\overrightarrow{OB}$ and $\overrightarrow{OD}$ implies that the difference of these vectors is non-zero. But he could

have just as well applied the same reasoning to $\overrightarrow{OA}$ - $\overrightarrow{OC}$. If he didn't, it was

possibly because this contradicted the perceptual evidence of the collinearity of $\overrightarrow{OA}$

and $\overrightarrow{OC}$ in the Cabri figure, stated in line [15]. Lines [13] and [14]   are obtained through algebraically correct manipulations. In [15] it was acceptable for the student to use the equality sign between vectors 'pointing in opposite directions' even though this was not consistent with his definition of equality of vectors.

For us, this solution is an example of how a student's thinking is falling apart as a result of his sometimes making connections between ideas on the basis of theorems and definitions, and, at other times, on the basis of visual evidence provided by the figure, and arbitrary choices of premises.

2.1.3. Thinking of Mathematical Concepts in Terms of their Prototypical Examples Rather than Definitions
Theoretical and practical modes of thinking differ strongly in the manner in which they constitute the meanings of words. For the practical mind, mathematical objects are 'natural objects' not 'discursive objects': definitions and theories can only describe them, not create or construct them. Consequently, a mathematical term is interpreted, primarily, through its denotation or reference, as representing a collection of particular objects, and the connotation, or the defining property, appears as a mere common property, an interesting observation about these particular objects.

Thinking in terms of prototypical examples, rather than definitions, became an obstacle to our students' understanding, in the most striking way, at the time of their learning the notion of linear transformation. In the course, linear transformations were defined as transformations of vector spaces which conserve linear combinations; i.e., transformations T from one vector space to another such that $T(k_1v_1 + k_2v_2) = k_1T(v_1) + k_2T(v_2)$, for any scalars k_1 and k_2, and any vectors v_1 and v_2 from the domain space. The obstacle was revealed in the students' attempts to solve the so-called 'linear extension problem': given a transformation of a basis, to construct a linear transformation with those values on the basis. In the experimental course, the problem was not formulated in such general terms. It was restricted to two dimensions, and the students were not asked to 'linearly extend a transformation from a basis to the whole plane' but to assume the existence of such an extension and find some missing information about it. The details about the problem and students' solutions are in our next example.

Example 4.

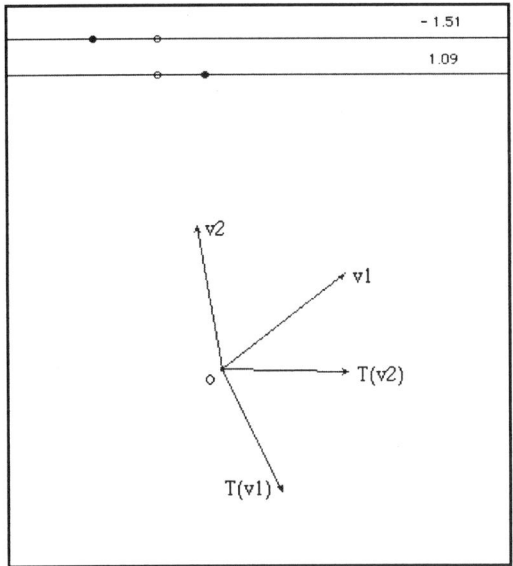

Figure 4. The figure for the 'Linear Extension Problem' printed in the students' worksheets.

The students were asked to open a new Cabri file, put two number lines in it, put the origin O and vectors labeled v_1, v_2, $T(v_1)$, $T(v_2)$ in a configuration similar to one drawn in their worksheet. The configuration was, approximately, as in Figure 4.

The two number lines allowed the students to produce two variable, independent scalars, by moving the thick points along the number lines. The numbers visible above the lines are the coordinates of the thick points. The macro menu contained functions such as 'Vector addition', 'Scalar multiplication' and 'Linear combination' which could be applied to the vectors and scalars on the screen. The problem in the students' worksheets was formulated as follows:

We assume that the vectors $T(v_1)$ and $T(v_2)$ are images of the vectors v_1 and v_2, respectively, under some <u>linear</u> transformation T.

With this information alone, can you construct

(a) $T(v_1 + v_2)$; (b) $T(2v_1)$; (c) $T(-1.5v_1 + 0.8v_2)$; (d) $T(v)$ for an arbitrary vector v.

The students were expected to use the defining property of linear transformations to obtain:

(a) $T(v_1+v_2)$ as the sum of $T(v_1)$ and $T(v_2)$, using the macro 'Vector addition' applied to $T(v_1)$ and $T(v_2)$;

(b) $T(2v_2)$ as the double of $T(v_1)$ using the macro 'Scalar multiplication' applied to $T(v_1)$ and a number 2, put on the screen using the 'Numerical edit' function;

(c) $T(-1.5v_1 + 0.8v_2)$ as the linear combination $-1.5T(v_1) + 0.8T(v_2)$, by setting the scalars on the number lines to -1.5 and 0.8 and using the macro 'Linear combination' to these scalars and the vectors $T(v_1)$ and $T(v_2)$.

In question (d), a representation of the arbitrary vector v had to be obtained first. This could be done either by

(i) drawing a free-ended arrow from the origin,

or by

(ii) constructing a linear combination $k_1v_1+k_2v_2$ with k_1, k_2 represented by the two independent variable scalars on the number lines.

In case (ii) the arbitrary vector is represented by an ordered pair of independent real variables. With the representation (i) it is necessary to find the coordinates (k_1, k_2) of the vector v in the basis v_1, v_2 (for example, by using 'New Axes' and 'Equation and Coordinates' functions). In any case, having represented v as $k_1v_1 + k_2v_2$, one can then obtain $T(v)$ as $k_1T(v_1) + k_2T(v_2)$ using the assumption of linearity of T and the 'Linear combination' function. If question (c) was solved the way it was expected, using the number lines, and option (ii) was chosen, there was practically nothing to do beyond thinking of the scalars on the number lines as variable, rather than representing the concrete scalars of question (c).

Students' responses could be divided into three main types:

1° the 'formal' solution (3 answers out of 7);

2° the 'prototype' solution (3 answers out of 7), and

3° the 'theoretical solution' (1 answer out of 7).

Only the single 'theoretical solution' was indeed a solution. The student solved the question (c) in the manner expected (by using the number lines for the scalars), and then realized that by moving the points on the number lines the scalars could be made variable, so any vector could be obtained that way, which solved (d). But this student was quite unsure whether this was correct; in fact, she felt it was 'too simple' to be true.

Three students working together produced almost identical solutions which may be interpreted as symptomatic of the 'obstacle of formalism'. These students worked exclusively on paper in this problem, starting from the equation $T(k_1v_1 + k_2v_2) = k_1T(v_1) + k_2T(v_2)$ and transforming it. They were apparently looking for conditions on the variables k_1, k_2, v_1, v_2, for which the left hand side of the equation was identical with the expressions in the questions (a) to (d) of the problem. For (a), (b) and (c) they were able to find such conditions, and their answer to the question posed in the problem was 'yes we can'. In question (d), their answer was 'No we can't as in Activity 1.2(i)'. Before we explain this last answer, we reproduce the students' solution, in type, below:

Given $T(k_1v_1+k_2v_2) = k_1T(v_1) + k_2T(v_2)$ $\forall$ k_1, $k_2 \in \mathbb{R}$ v_1 & $v_2 \in V$

(a) Yes, we can $k_1=k_2=1$

$T(k_1v_1 + k_2v_2) = k_1T(v_1) + k_2T(v_2)$

if $k_1=k_2=1$

$\therefore$ $T(v_1)+T(v_2)=T(v_1+v_2)$

(b) Yes we can if $v_1=v_2$ and $k_1=k_2=1$
$T(1v_1+1v_2)=1T(v_1)+1T(v_2)$
$\therefore T(2v_1)=2T(v_1)$

(c) Yes we can $k_1=-1.5$ and $k_2=.8$
$\therefore -1.5T(v_1) + .8T(v_2) = T(-1.5v_1 + .8v_2)$

(d) No we can't as in Activity 1.2 (i)

The 'Activity 1.2 (i)' refers to the problem of verifying if translation by a constant vector is a linear transformation, mentioned in Example 1. The activity immediately preceded the present problem. One of the three students who produced the above solution was the author of the solution discussed in Example 1. It is possible that, in this previous activity, the student was arriving at a conclusion such as: 'the answer is "no" because the equality $T(k_1v_1 + k_2v_2) = k_1T(v_1) + k_2T(v_2)$ was dependent on the values of k_1 and k_2'. In question (d) the students could not give any concrete values or conditions on the variables; so it is possible that they thought, 'the answer depends on k_1 and k_2'. And thus a connection was made with the previous activity. The answer was 'No' there, so the answer must also be 'No' here.

The 'formal' type of solution discussed above is not an example of what we understand by 'thinking about mathematical concepts in terms of prototype examples', although it also is based on certain prototypes. It is an example of a pragmatic thinking of a student who tries to survive a course by producing a discourse formally similar to one that appears to be officially approved of by the teacher. In this frame of mind, the equation $T(k_1v_1 + k_2v_2) = k_1T(v_1) + k_2T(v_2)$ takes the place of the concept of linear transformation, and the main intellectual effort goes into identifying the typical actions that can be performed on this equation and typical conclusions that can be drawn from the results of these actions (i.e., other equations). The 'prototype', in this case, refers to prototypical actions to be performed in response to prototypical exercises. It is a kind of practical thinking that, unlike the kind we wish to discuss in the present section, is not conducive to cognitive growth.

It is the solutions of type 2° that better reflect the 'thinking of mathematical concepts in terms of prototype examples'. These solutions can be seen as based on an understanding of linear transformations that can be modeled by the following statements:

Linear transformations are rotations, dilations, projections, shears, etc., and their combinations, with constant parameters; for each such transformation, for concrete vectors v_1 and v_2 and concrete scalars k_1 and k_2, $T(k_1v_1 + k_2v_2) = k_1T(v_1) + k_2T(v_2)$.

Students representing this kind of understanding easily constructed vectors $T(v_1 + v_2)$, $T(2v_1)$ and $T(-1.5v_1 + 0.8v_2)$, in questions (a), (b) and (c) of the 'Linear Extension' problem, but felt blocked in front of question (d), where the image of an arbitrary vector v was to be constructed. They would ask questions like: 'How can I find the image of any vector if I don't know what is T?'. Their efforts would then go into trying to figure out if T is not one of the known linear transformations, a rotation by a fixed angle, a reflection, or a dilation, etc., or possibly some combination of these, just by investigating the relations between the vectors they had on the screen and their images. In our second experiment in the teaching of linear algebra with Cabri, the positions of these vectors were such that no coherent conjecture could be made this way, and the students (all three pairs taking part in it) felt defeated by the problem. To illustrate this, we reproduce below a chronological sequence of excerpts from the protocol of a session with one pair of students working on question (d) in the second experiment. The excerpts are separated by the mark '(...)'.

[1] Student 1: $T(v_1)$ is to v_1

[2] Student 2: as $T(v_2)$ is to v_2.(...) I guess we have to find just the image of v which is totally independent of all the other ones, right?

[3] Student 1: Well, it's gotta be the same transformation.

[4] Student 2: But how do we, how did they get? (...) Reflection? But how do we do the reflection [on Cabri]? (Searches through the menus) Rotation? What if we, okay, isn't it like k times v

[5] Student 1: We don't know what k is. We just know it's like all that with the rotation and dilation and everything, it's transformed.

[6] Student 2: But they are not exactly the same. This one's a lot longer than this one (...) What exactly is an image? What did they?

[7] Student 1: It's the transformation. $T(v_1)$ is the image of, $T(v_1)$ is what happens to v_1 when you transform it.

[8] Student 2: And how do you, what do you do to transform it?

[9] Student 1: This is what we don't know.

[10]Student 2: Okay, we've got to figure that out. How do we, how do they get from v_1 to $T(v_1)$? They're not the same size (...)

[11] Student 1: Right, but it's the same transformation as that.

[12] Student 2: (...) Not a rotation. Just, it's symmetry, but I don't know how to do that. (...)

[13] Student 1: It's not a direct reflection (...)

[14] Student 2: We could try measuring angles, I don't know. This angle seems like the double of this angle.

[15] Student 1: If it's bisected there (...)

[16] Student 2: If there's a formula? We, we can rotate it and then dilate it, or something like that. But it still makes no sense. (...) It has to have something to do with the angles, I'm sure, to rotate some place, 'cause what else could it be? It's the length (a short silence). I don't understand why is this one, this one really seems equal to this one, doesn't it, but this one doesn't? That's what I don't understand.

[17] Student 1: But, I mean, too, this is like an obtuse angle and that's really acute, so there's no relationship.

[18] Student 2: Yeah. What does it have to do with the v? I wish we could see how they did that (...) How did they do that? Like we're given nothing, right? Aren't we supposed to have other 'données', how do you say that in English? Information. (Big sighs. Long silence)

[19] Student 1: I don't know.

[20] Student 2: *Me neither. (...) There's nothing (...) v_1 and v_2, (...)*
$T(v_1)$ and $T(v_2)$, (...) they have nothing to do with each other.

A theme that popped up in lines [2] and then [18], without being pursued further, was: How can an arbitrary vector v be related to the other vectors (and, in particular, v_1 and v_2)? It would have been more fruitful to look into this particular relation rather than the one the students engaged into; namely, the relation between the pairs v_1 , $T(v_1)$ and v_2 , $T(v_2)$. Student 2 wanted to know what was the macro construction used by the designers of the problem to create $T(v_1)$ from v_1 and $T(v_2)$ from v_2 (lines [4], [6], [10], [18]). Student 1 appeared to be slightly more theoretically minded. At least, the notion of transformation was the main reference point. She was seeing beyond the concrete representation in Cabri of the transformation of the basis v_1 and v_2 to understand that it was supposed to generate the same transformation (lines [3], [11]). She appeared to think of rotations and dilations as examples of the general notion of transformation. Thus, her concepts seemed to be linked by relations to generality rather than by perceptual and situational associations and contingencies (line [5]). In line [7], she explained the notion of image in general terms, not through concrete examples. Yet, she was unable to solve the problem; possibly because she had not thought of going back to the definition of linear transformation as preserving linear combinations and of representing the arbitrary vector v as a linear combination of the vectors v_1 and v_2. Unable to cope with the problem theoretically, she finally succumbed to joining the other student's game of chasing different commonly known transformations and trying to fit them between the elements of the pairs v_1, $T(v_1)$ and v_2, $T(v_2)$ (line [15]). But the configuration of the pairs was such that no simple relation could be seen, even in approximation.

In the third experiment, one student, who followed a line of thought similar to that of Student 2 above, was led to believe that he had thus successfully solved the problem. He had reproduced the configuration from Figure 4 in Cabri in such a way that the angles between vectors v_1+v_2 and $T(v_1+v_2)$ and between vectors v_1 and $T(v_1)$ were, by coincidence, approximately equal to -100°. Having measured the ratio of lengths of $v_1 + v_2$ and $T(v_1 + v_2)$, 5.1/5.8 ≈ 0.87, the student then claimed that $T(v_1 + v_2)$ was obtained from $v_1 + v_2$ by a 'scalar multiplication by 0.87 and then by R(O,-100°)'. He then also measured the ratio of the lengths of $T(2v_1)$ and $2v_1$, 7.6/7 ≈ 1.08, and wrote that 'by multiply[ing] $2v_1$ by scalar multiplication 1.08 and R(O,-100°) we get the image $2T(v_1)$'. His answer to part (c) was: 'Yes, by chang[ing] the number line[s] such that k_1 = -1.5 & k_2 = 0.8 and same like (a)'. In part (d) he wrote: 'Yes, multiply v by k and then R(O,-100°) we get T(v) such that T(kv) = kT(v)'. His drawing contained a vector v, not starting from the origin, a vector labeled 'kv' of length approximately 0.72 times the length of v, drawn over v, and a vector labeled 'kT(v)', apparently obtained from kv by rotation around O by about -100° (see Figure 5). It may thus appear that, for this student, T was a combination of a dilation by a variable scalar and a rotation by -100°.

It may appear rather amazing that the student was not bothered by the fact that the factor of dilation was not only variable but undetermined in his transformation. However, this phenomenon could be explained by a notion of vector that appeared very clearly in a student in the second experiment (see a detailed description in Sierpinska, Trgalová et al., 1999, authored by Trgalová). For this student, a vector did not have a fixed length; it could be made longer or shorter by 'changing its scalar quantity' (using scalar multiplication). The symbol 'kv' denoted not a set of vectors, but one single vector with varying length. This notion of 'elastic arrow' influenced the student's understanding of other basic notions. In particular, a linear combination of vectors was understood as a sum of two vectors with variable lengths. This student also expressed her belief that a combination of a rotation by an angle α and a dilation by factor k is identical with a combination of a rotation by the same angle and a dilation by a different factor, because the image vectors 'differ only in magnitude'. If the author of the solution to the Linear Extension problem presented in Figure 5 also thought of vectors as 'elastic arrows', then his combination of a dilation by an arbitrary scalar k and a rotation by -100° around the origin was a well defined transformation. Let us note, moreover, that, for this student, the term 'arbitrary vector' meant a vector with arbitrary direction, orientation, length *and* point of application.

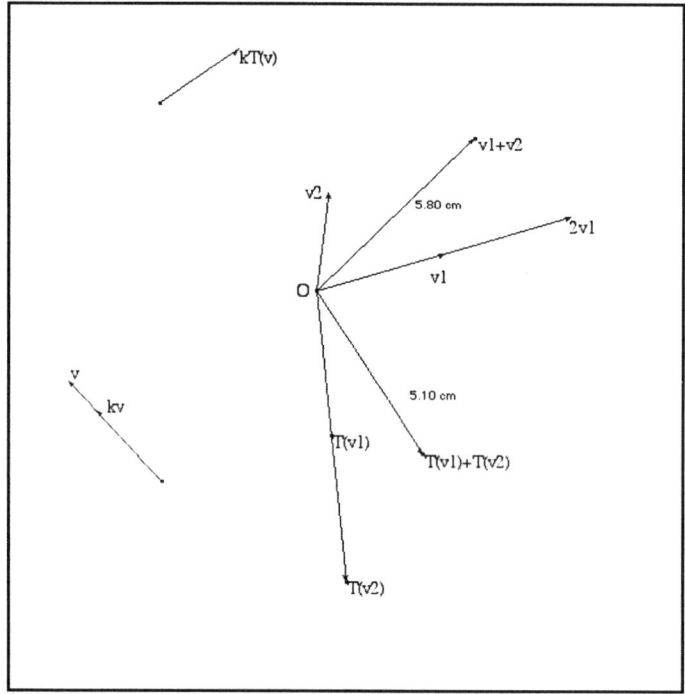

Figure 5. The drawing accompanying a student's solution of the Linear Extension problem in the third experiment.

In the Linear Extension problem there is a big gap for the students between finding images of concrete linear combinations and finding the image of any vector. There seem to be two sources of this gap. One is the need to construct a representation of the object 'any vector'; it is not given in the problem. Another is the need to have a notion of axiomatic definitions. But the students perceive the defining condition of linear transformations as only a property of a transformation; a transformation must be given by some other means, like a macro or a formula (i.e., they need a descriptive definition). They do not think that saying 'T is a linear transformation with such and such values on the basis' does anything to define the transformation. This does not tell them 'what T is'.

It is true that the 'prototype example' type of solution of the Linear Extension problem is a symptom of some progress in understanding the notion of linear transformation, as opposed to the 'obstacle of formalism' type of solution. But it cannot be considered satisfactory if the students are expected to further learn even such elementary notions as that of the matrix representation of a linear transformation. After all, understanding that a matrix, whose columns are coordinates of the values of a linear transformation in a basis, determines this transformation uniquely with respect to that basis, is equivalent to understanding the Linear Extension Problem.

Of course, students are able to learn the definition of the matrix representation and apply it correctly in directly formulated exercises immediately following its introduction. But this information does not stay in their long term memory because it cannot be reconstructed rationally on the basis of its logical and conceptual links with other pieces of students' knowledge about linear transformations. We shall give an example of this 'forgetfulness' in the next section.

2.1.4. Reasoning Based on the Logic of Action, and Generalization from Visual Perception

The practical mind is geared to dealing with situations rather than concepts. The practical student grasps a mathematical concept together with the whole didactic, social and psychological context in which she or he has been introduced to it. The student does not discriminate between the problem of knowing what a new mathematical term means and the problem of knowing what the teacher expects him or her to do and write in response to typical questions. The constraints of the situation determining the meaning of its mathematical content and the constraints pertaining to the didactic contract are treated as equally important.

Also, the visualizations provided by the teacher are taken holistically by the students without their trying to discriminate between the mathematically relevant features and those contingent on the technical support used to create them.

Example 5.

One week after the introduction of the definition of the matrix of a linear transformation in a basis, and two weeks after the Linear Extension problem was first attempted by the students and then discussed with the teacher, the students were given the following problem:

Find the matrix in basis $<v_1,v_2>$ of a reflection R in the line through v_1.

A figure accompanied the problem (see Figure 6).

The students were not expected to work in Cabri. It was enough to have the experience of Cabri to understand that the numbers (-1.50, -1.00) are the coordinates of $R(v_2)$ in the axes through v_1 and v_2 and thus read the matrix directly off the figure: In the reflection through the direction of v_1, $R(v_1) = v_1$; moreover, according to the figure, $R(v_2) = -1.5v_1 -v_2$. Hence, the matrix of R in basis $<v_1, v_2>$ is $\begin{bmatrix} 1 & -1.5 \\ 0 & -1 \end{bmatrix}$.

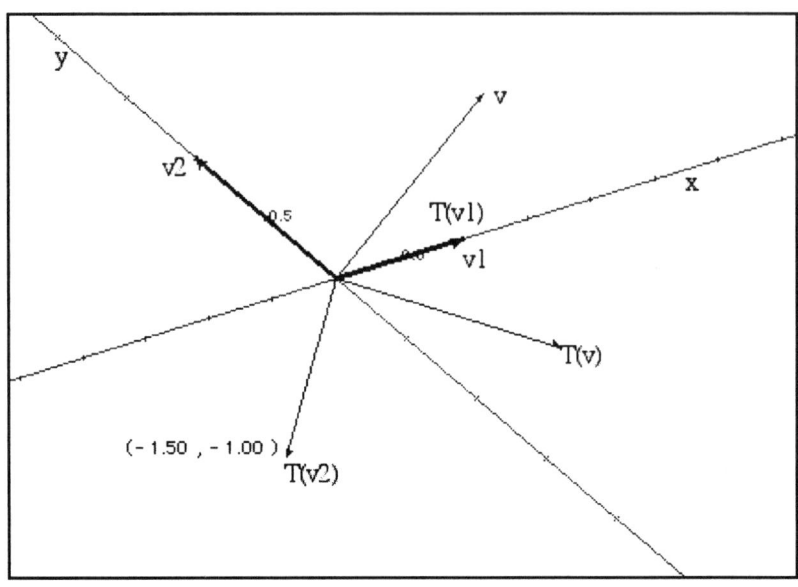

Figure 6. Figure accompanying the 'Matrix of a reflection' problem

Out of the 7 students who were in class that day (the 'theoretical solution' student from the Linear Extension problem was not there), only one student gave the expected answer. Two students answered by the 2x2 identity matrix, possibly copying with a mistake a matrix of reflection in an orthonormal set of axes. Four other students produced answers based on a rather complicated idea. The idea was, not to study the given figure, but to go to Cabri, put four vectors labeled v_1, v_2, w_1, w_2, draw axes through v_1 and v_2, find coordinates of vectors w_1, w_2 in these axes, and write a matrix with these coordinates in columns. The students labeled this matrix 'A'. Some of them wrote concrete entries in the matrix and some of them only put $\begin{bmatrix} a & b \\ c & d \end{bmatrix}$. Then they reflected w_1 and w_2 in the line through v_1, and obtained the coordinates of the image vectors w_1', w_2'. They wrote: « After reflection through v_1, the reflected matrix (i.e., reflected vectors w_1, w_2) $A' = \begin{bmatrix} a© & b© \\ -c & -d \end{bmatrix}$ ». There were no concrete entries for A' in most students' work. The students were, possibly, generalizing from a configuration like in Figure 7.

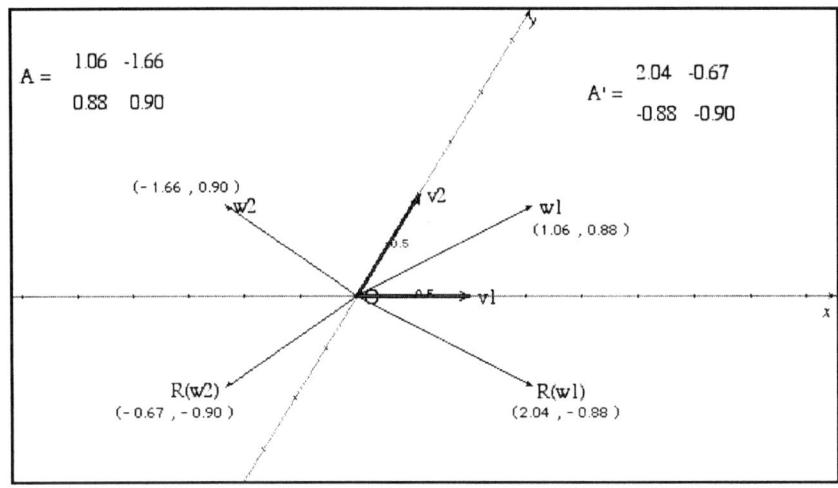

Figure 7. A possible Cabri screen created by the students in solving the problem of the matrix of a reflection.

It is possible that, through some quite complicated thinking process, for these students, the notion of the matrix of a transformation became reduced to just 'the coordinates of the vectors w_1 and w_2'. This hypothesis could be justified by the context in which the students were acquiring their experience with finding matrices of linear transformations defined on a basis. The context was that of exercises in the Cabri environment. In these exercises, there would be five vectors on the screen, (v_1, v_2, w_1, w_2, v) and it would be assumed that $w_1 = T(v_1)$ and $w_2 = T(v_2)$ under some linear transformation T. The students would then find $T(v)$ using a macro-construction named 'Lin-basis'. They studied or modified the transformation and often would have to find the matrix of the transformation in v_1 and v_2. For example, they would explore how the matrix of the transformation depended on the configuration of the vectors v_1, v_2, w_1, w_2. To find the matrix of the transformation in basis v_1, v_2 it was enough to write down the coordinates of the vectors w_1, w_2 in the axes drawn through v_1 and v_2. Maybe the students were not aware that by varying the positions of the vectors w_1 and w_2 they were changing the transformation (although this question was explicitly discussed in class), and thus were led to believe that any pair of vectors w_1 and w_2 defines a unique transformation (which is absurd, but only logically absurd, not absurd from the point of view of someone who is just playing with arrows on the screen and producing all sorts of 2x2 matrices).

This process could perhaps be described as one of the substitution of the context of exercises and of the logic of action for the theory.

2.2. General Reflection on the Distinction Between Theoretical and Practical Thinking

The distinction between theoretical and practical thinking is a methodological distinction, not a distinction between actual processes of thinking, which seem to always have both a practical and a theoretical component. Scientific knowledge is theoretical but scientists do not always think in the theoretical mode. Most of the time scientists think in 'practical' ways, where the practice has to be understood, not as everyday life practice, but their practice as scientists working in a familiar paradigm. They switch to theoretical thinking in situations which challenge their common mathematical sense; e.g., when confronted with a contradiction, or a particularly confusing question, or when they have to defend their theory against criticism.

Moreover, practical thinking can be regarded as a necessity from an epistemological point of view if we assume that theoretical thinking is about something other than itself. It is the practical thinking that leads to the identification of problems. These can then be framed and formulated within a theory and possibly solved or understood on the basis of its technical and analytical tools.

But practical thinking may lead to confusion and mistakes, not only in our students, but also in scientists, as in the well known example of research psychologists who tended to think about probability as a measure of 'confidence' and 'placed greater "confidence" in experimental results with a large sample size than with a small sample size when the probability value of rejecting the null hypothesis was held constant' (Johnson-Laird and Wason, 1977, p. 263). Thus, practical thinking functions as an epistemological obstacle in the classical sense of Bachelard (1983); i.e., as a way of thinking which is useful or even necessary for the development of scientific knowledge, but has a limited domain of validity.

3. SYNTHETIC-GEOMETRIC, ANALYTIC-ARITHMETIC AND ANALYTIC-STRUCTURAL MODES OF THINKING IN LINEAR ALGEBRA

Three modes of thinking and reasoning co-exist in linear algebra which we call synthetic geometric, analytic-arithmetic and analytic-structural. While these modes of thinking appeared in the history of mathematics in a sequential manner, it did not happen that one of them eliminated the other two. The development of algebra owes a lot to their constant interaction; from the analytic geometry of Descartes to the geometric intuitions underlying the theory of Banach spaces. Indeed, the story of linear algebra can be told so as to illustrate this continuous activity of change of perspective, done in the hope of gaining a better and deeper understanding of the domain. For example, in his work on quaternions, Hamilton would frequently switch from an algebraic approach to a geometric approach and vice versa. His discovery of quaternions occurred as he was examining the geometric properties of the multiplication of complex numbers (see Dorier, Part One, this volume). Also, the intuitive roots of Grassmann's formalism are geometric. But the most interesting fact is that linear algebra can be seen as the result of an overcoming of two obstacles or two opposed dogmatic positions: one refusing the entry of numbers into geometry, and the other that of 'geometric intuition' into the pure domain of arithmetic. It is with Leibniz and his Geometry of Situations that mathematics started to overcome the apparent contradiction of the two perspectives and endeavored to take full advantage of the complementarity of the geometric and arithmetic thinking (Otte, 1990).

Thus, rather than regarding the above mentioned modes of thinking in linear algebra as steps or stages in the development of algebraic thought, we consider it preferable to see them as modes of thinking that are equally useful, each in its own context, and for specific purposes, and especially when they are in interaction.

3.1. A Characterization

In our naming of the three modes of thinking - synthetic-geometric, analytic-arithmetic and analytic-structural – the main difference between the 'synthetic' and the 'analytic' modes of thinking about mathematical objects is that in the synthetic mode the objects are, in a sense, given directly to the mind which then tries to describe them, while, in the analytic mode they are given indirectly: In fact, they are only constructed by the definition of the properties of their elements. For example, in the synthetic mode, a straight line is seen as a pre-given object of a certain shape lying somewhere in the space. One can speak of the properties of the straight line, but these properties will only describe the line, they will not define it. In the analytic mode, the straight line is defined as a certain specific relationship between the coordinates of points or vectors in a space of a given dimension. Thus, the synthetic mode belongs to the practical way of thinking, and the analytic – to the theoretical way of thinking.

The development of linear algebra started as a process of thinking analytically about the geometric space. Taking a rather broad perspective[4], we could distinguish, in this development, two large steps related to two processes. One was the arithmetization of space, as it took place in the passage from the synthetic geometry to the analytic geometry in $\mathbb{R}^n$. The other was the de-arithmetization of space or its structuralization, whereby vectors lost the coordinates that anchored them to the domain of numbers and became abstract elements whose behavior is defined by a system of properties or axioms.

Structuralized, a 'space' becomes an algebraic system, a set closed with respect to certain operations. This abstraction allows one to conceive of spaces that are infinite dimensional and of spaces over arbitrary fields. The familiar objects of analytic thinking in linear algebra - vectors and matrices – lose their numerical substance. They are no more 'boxes with numbers', but units whose internal structure is not of much interest in the reasoning. An nxm matrix is now, first of all, an element of the algebra $\mathfrak{M}_{n,m}(K)$ of nxm matrices over the field K. By the same token, the theory of determinants and the techniques for their calculation lose the prominent position they hold in analytic-arithmetic thinking. It is far more important, for example, that the determinant of a linear operator on $\mathbb{R}^n$ is a 'geometric property' (Walker, 1978, p. 35): It does not depend on the choice of basis.

Before we get into a more detailed description of the three modes of thought, let us have some examples of them.

If somebody is thinking about the possible solutions of a system of three linear equations in three variables by visualizing the possible relative positions of three planes in the space, he or she is in the synthetic-geometric mode (or mood). If one thinks about the same problem in terms of the possible results of a row reduction of a 3x4 matrix, one is in the analytic-arithmetic mode. Thinking in terms of singular and non-singular matrices would be a symptom of the analytic-structural mode.

Another example: Given two matrices A and B. The problem is to check if B is the inverse of A. One student calculates the inverse of A using the well-known formula with the determinant and co-factors, and compares the result with B. Another student multiplies A by B. To deal with the inverse of a matrix, the first student uses a technique for computing the inverse. The second student uses the defining property of the inverse. We shall say that the first argument was analytic-arithmetic, while the second was analytic-structural.

One can also compare the proofs of the Jordan Canonical Form theorem: one calculating, for a given $n \times n$ matrix M, a non-singular matrix P such that the matrix $P^{-1}MP$ has the Jordan canonical form (Turowicz, 1973, p. 99), and another using the theory of invariant subspaces, the decomposition of a vector space into a direct sum of invariant subspaces, etc. (Lipschutz, 1992, p. 373). Again, the first would be called analytic-arithmetic, the second, analytic-structural.

Much of the analytic-arithmetic reasoning goes along the line: Show that two processes lead to the same result. On the other hand, a paradigmatic example of the analytic-structural thinking is: To show that a is b, prove that a satisfies the characteristic properties of b. While the analytic-arithmetic thinking aims at simplifying calculations and making them accurate, the structural thinking aims at extending our knowledge about concepts. If we know that a is b, we get to know more about a and b.

In analytic-arithmetic thinking an object is defined by a formula that allows one to compute it; in analytic-structural thinking an object is best defined by a set of properties. For example, formulas and techniques which allow one to calculate inverses of non-singular matrices, important results of analytic-arithmetic thinking, become almost footnotes in the structural mode. The attention shifts from the process of calculating the inverse matrix to the property of having an inverse.

The notion of linear transformation used to be traditionally defined as substitution of variables (e.g. variables y are expressed as linear combinations of variables x , see Daintith & Nelson, 1989, p. 201), which we consider as reflecting an analytic-arithmetic mode of thinking about transformations. In this approach, it is understood, or taken for granted, that x and y are numbers. However, in the modern undergraduate texts, the definition is most of the time structural: the linear transformation becomes a mapping from one vector space to another, satisfying a certain condition. This condition does not give a formula for the calculation of the image because it does not get into the nature of the vectors – which do not have to be n-tuples of numbers – but can be elements of any vector space. For example, they can be matrices or functions. In this approach, vectors have meaning only as elements of larger structural wholes.

Each of the three modes of thinking in linear algebra uses a specific system of representations. The synthetic-geometric mode of thinking uses the language of geometric figures - planes and lines, intersections – as well as their conventional graphical representations. In the analytic-arithmetic mode, geometric figures are understood as sets of 'n-tuples' of numbers satisfying certain conditions that are written, for example, in the form of systems of equations or inequalities. In this mode, the numerical components of geometric objects, like points or vectors, are important. Thus, for example, a general system of equations would be written with all the coefficients written out explicitly, e.g.

$$a_{11} x_1 + \ldots + a_{1n} x_n = b_1$$
$$\ldots\ldots\ldots\ldots\ldots\ldots\ldots\ldots$$
$$a_{m1} x_1 + \ldots + a_{mn} x_n = b_m$$

Analytic-structural thinking goes beyond this type of analysis and synthesizes the algebraic elements of the analytic representations into structural wholes. Thus, a system like the one above would be written in a matrix form as $Ax=b$, or in a vectorial form $x_1 A_1 + \ldots + x_n A_n = b$.

With respect to systems of equations, there is yet another difference between analytic-arithmetic and analytic-structural modes of thinking. What is important from an analytic-arithmetic point of view is to find methods of solving systems of equations. In the structural mode of thinking, the questions asked would be related, for example, to conditions on the matrix A and the vector b for the existence and uniqueness of a solution. The properties of the matrix would be more important than the nature of its numerical components.

It seems that while, historically, analytic-arithmetic thinking in linear algebra produced a system of rules, or a certain Art, the endeavor of structural thinking was more ambitious: to achieve a system of truths or the status of a 'science, like geometry' (Hamilton, 1837/1967)[5].

We are making here a distinction between analytic-arithmetic and analytic-structural thinking in linear algebra but, in fact, the dialectic of the analytic-arithmetic and the structural could be a broader phenomenon in the history of mathematical ideas. A similarity can be seen in the way in which Spalt (1992) sees the development of the Calculus in the eighteenth and nineteenth centuries.

> Au 18e siècle, le siècle d'Euler, l'analyse développe essentiellement des méthodes de calcul avec l'infini. Les séries..., les différentielles et les intégrales sont des objets qui, chacun à sa façon, formulent un infini, le saisissent sous l'aspect d'une formule. La formule algébrique, avec ses transformations multiples par le calcul est placée au centre. [Dans] la nouvelle forme que prend l'Analyse pendant la première moitié du 19 siècle... les objets qui se trouvent au centre de l'analyse ne sont plus présentés comme formule, mais comme concept, c'est-à-dire par des propriétés. (Spalt, 1992, p. 85).[6]

In principle, synthetic-geometric arguments do not belong to linear algebra proper. But they are used as heuristic tools, and in teaching, for the sake of visualization, shortness, or grasping 'the essence' of a fact.

As mentioned above, linear algebra grew, partly, from analytic thinking about geometry (Dorier, Part One, this volume). But analytic thinking about geometry both extends the range of considered relations between figures and changes the meaning of certain intuitive geometric concepts. Synthetic geometry is interested only in 'geometric properties' of figures: It studies relationships between only those properties of figures that would be invariant under the change of coordinates, if a system of coordinates was introduced. For example, the distance between two points would be studied, but the difference between the coordinates of a vector would not be an interesting property. Linear algebra studies properties that do not have to be geometric. However, on the structural level, there is a return to the concern about 'geometricity': Which properties do not depend on the choice of basis in the vector space? It will be stressed, for example, that linear independence of vectors, the determinant, the trace, the characteristic polynomial and eigenvalues of a linear operator are 'geometric properties' (Walker, ibid.).

There is some kinship, therefore, between the synthetic-geometric and the structural ways of thinking in linear algebra. One important common feature is the *independence from a coordinate system*. Another is, *based on properties, not calculations*. For example, in the structural mode, the notion of eigenvalue cannot be reduced anymore to that of a root of a polynomial. It must be thought of as a scalar related to invariant one-dimensional subspaces of a linear operator. It is an object of reflection and a concept; not an outcome of a calculation.

Both the synthetic-geometric and the structural modes of thinking are visual, although in very different ways. The latter is more metaphoric and/or diagrammatic than the former. Think of the usual representation of a linear transformation in the form of two round shapes linked by an arrow, as opposed to the more literal representation of a straight line or a plane in a system of coordinates being rotated or transformed in some other 'concrete' way.

3.2. The Tension Between the Synthetic, the Analytic-Arithmetic and the Structural Modes of Thinking in Students of Linear Algebra

Each of the three modes of reasoning and thinking in linear algebra leads to different meanings of the notions involved, because each of them is related to a different theoretical perspective. These meanings are not equally accessible to beginning students. Modern academic linear algebra textbooks often offer structural arguments to justify certain basic statements. While usually short and elegant, they represent a level of theoretical sophistication that leaves a beginning student with a feeling that either nothing has been proved, or that the proved fact is of little significance (cf. Sierpinska, 1995, 1996). Structural thinking has a complex epistemology that accounts for students' difficulties in algebra at all levels (Sfard, 1994). But also the analytic-arithmetic thinking is no easy hurdle to overcome, and a synthetic-geometric argument can be more often a challenge than a visual aid. Spontaneously, students use other modes of reasoning, some of which are intermediate between the three historic ones. Even if they have access, to a certain degree, to each of these modes of thinking, they do not see why they would have to use one or the other in a pure and consistent way[7], and they prefer to make up some intermediary forms which appear to them as more convenient and making more sense. They have also trouble switching and translating from one mode to another. We shall see an illustration of these phenomena in the examples below[8].

Example 6: The pitfalls of synthetic geometric thinking.

It is a problem for the students to see exactly how the graphical representational system relates to an analytic representational system.

Looking at a diagram representing three planes intersecting in one line (see Figure 8), one of our students, labeled Sandy 2, said that it represented a system of equations with a unique solution. There was a unique line common to all planes. This student was generalizing from two dimensions to three dimensions in a synthetic-geometric, not an analytic, way. The solution of a system of equations was, for him, an intersection of hyperplanes in a Cartesian space; i.e., straight lines in two dimensions and planes in three dimensions (see lines 387, 389 in the episode transcribed below). Very visually-minded, the student had to struggle to make his thinking more analytic: to think of a line as a set of points, and of points as tuples of numbers, and to understand that counting the number of solutions means counting points or such tuples of numbers. He had yet to see that, in linear algebra, the generalization from a two-dimensional space to a three-dimensional space does not go geometrically from points to lines and from lines to planes, but arithmetically, from points with two coordinates to points with three coordinates.

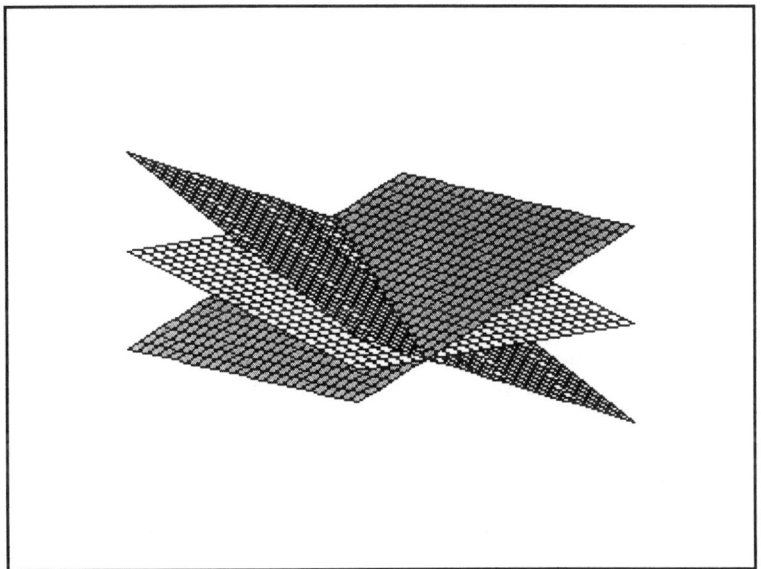

Figure 8. A representation of three planes intersecting in one line

May 31, Sandy 2, Tutor T2

375. Sandy 2: [Reads Exercise 3, p. 34 from Text 2 (Lay, 1994)] "What types of linear systems and what types of solution sets are illustrated by the sets of planes shown in the accompanying figures?"

[Looks at figure (a) described as "Three planes intersecting in a line" similar to figure 8.] In here I see one solution.

376. Tutor 2: How do you mean? ... You are saying one solution. What do you mean exactly by one solution?

381. Sandy 2: I should say one solution set. One value for x, one value for y, one value for z.

382. T2: A single value for those numbers?

383. Sandy 2: In here?

384. T2: Yeah. Look at the place where they intersect.

385. Sandy 2: It's one... It's one particular point.

386. T2: Is it a point?

387. Sandy 2: No. Well, it's a wide point, if you like.

388. T2: Okay.

389. Sandy 2: It's like a point. It's like where three lines would intersect.

390. T2: Three lines could intersect in a point, yeah.

391. Sandy 2: Yeah, but a plane, it's...

392. T2: But if you took the line, like you said, a plane is like a "wide line"... A plane extends out in all directions.

395. Sandy 2: But there is still this one value of x, one value of y, and one value of z, because the way that it expands isn't it just the x-y-z values that make it three dimensional?

396. T2: Yeah, yeah. But now, I mean, look at what it says, "Three planes intersecting in a line"

397. Sandy 2: Oh, in a line! So then there's... So in that case there's infinitely many solutions... And it's a line.

For Sandy 2, there is no inconsistency between his statement that unique solution means 'one value for x, one value for y, one value for z' (line 381), and his claim that the diagram represents a system with a unique solution. Sandy 2 sees the line as 'one particular point' (385). It is only the puzzled expression on the face of the tutor that makes him revise his opinion a little and suggest that 'it's a wide point'. The tutor does not make any direct correction following this declaration that would make the student see his mistake. In fact, the event that triggered the understanding of a line as representing an infinite solution set was when the tutor drew the student's attention to the title of the figure (396) and stressed the word 'line'. The student became aware that a 'line' should be understood as an infinite set of points, and not as something like 'one single place where all the three planes intersect'.

3.3.   Intermediate Forms of Argumentation Used by Students

It is natural to expect that the structural thinking poses the greatest challenge to students. In this mode, the computational techniques, methods of solving systems of equations, etc. - achievements of the analytic-arithmetic thought - have obtained, through structuralization, elegant and concise 'synthetic' definitions and proofs. But these proofs have not been what originally justified these techniques and methods which were accepted as valid because they worked, leading to correct results. These definitions and proofs are a result of 'second thoughts', ironically, for the sake of communication, in a drive towards simplicity, unity, generality, esthetic value, and a better organization of the whole theory. Thus, for the sake of communication (between mathematicians), some of the meaning was lost (for the students), which can account for some of the students' difficulties in making sense of the theory.

But it would be unjustified to claim that students (even only those who participated in our research) prefer an analytic-arithmetic argument over a structural one, or a synthetic-geometric over an analytic one. Of course, in an analytic-arithmetic proof, it is usually clear 'where to start'. But it is often quite difficult to go on and finish (calculations are sometimes so tedious!), so some students escape into another mode of reasoning. For example, in our observations, the student labeled Sandy 3 spontaneously used a synthetic-geometric argument to justify the fact that a convex linear combination of convex linear combinations of a set of vectors in $\mathbb{R}^n$ is also a convex linear combination of this set of vectors and fought very hard against developing an analytic-arithmetic proof to which he was pushed by the tutor. He finally accomplished it, but kept saying that he found this proof superfluous. He was already convinced and satisfied by his reasoning where he spoke of 'closed geometrical figures' and argued that 'a convex linear combination of the things from inside of the figure will stay inside this figure' (Sierpinska, 1996).

At another moment, in solving an application problem related to population movement, the same student constructed ('arithmetically') a linear transformation as a transformation of one set of variables into another set of variables and not ('structurally') as a transformation of one vector space into another. The problem was about a population movement between a city and its suburb: In a year, 5% of the suburban population would move to live in the city and 3% of the city dwellers would move to live in the suburb. The question was: If the population of the city and the suburb in 1990 is given by the vector $x_o = \begin{bmatrix} r_o \\ s_o \end{bmatrix}$, and the population vectors

for the subsequent years are x_1, x_2, etc., 'give a mathematical description of the relation between these vectors (Lay, 1994, section 2.7)'. The expected solution was to write the 'migration matrix' M, and the formula $x_{k+1} = Mx_k$. But the student did not read the proposed solution and decided to solve the problem with his own means. He did not use the matrix to represent the relations. He only wrote a system of equations:

$$r_1 = T(r_o) = 95\% \ r_o + 3\% \ s_o$$
$$s_1 = T(s_o) = 97\% \ s_o + 5\% \ r_o$$

and explored the development of the situation from there, taking concrete values for the initial state and using numerical computations.

The student saw no reason for writing the transformation T as a transformation of two-dimensional vectors defined by a matrix. For him, 'this was the same'. But, in fact, it was not 'the same': the two solutions engage quite different conceptions of linear transformation.

Arguments such as these are rather easily classified as belonging to one mode of thinking or another. But most of the time it is quite difficult to qualify students' arguments: They seem to lie somewhere in between the established categories. We shall see some examples of this phenomenon below.

Example 7: *Justifying a certain technique for finding the inverse of a matrix.*
There exists a technique for finding the inverse of a matrix which consists in row reducing the augmented matrix $[A|I_n]$, where I_n denotes the nxn unit matrix, until one gets the matrix $[I_n|B]$; then $B = A^{-1}$ (Lay, 1994, p.113). Sandy 3 had the following explanation of why this technique works:

> February 15, Sandy 3
> 83 - 85. Sandy 3: If the matrix is non-singular, then . . . by reducing it to the row reduced echelon form . . you will certainly get I_n, right?
> 86. Tutor 3: Uhuh.
> 87. Sandy 3: And this other matrix, since it is exactly the reduced echelon form of the other, it acts in an inverse way, right?   This is why you obtain a matrix which is inverse to the other.

This argument is rather vague but let us consider it 'as is'. It is not what we would call purely analytic-arithmetic because it does not get into the formulas for computing the row echelon form. But it is not structural either because it is not based on definitional properties. It is based on a reflection on the results of the operations performed on the inverse matrix, namely the matrix   B. Thus it has certain analytic-arithmetic properties: it refers to the technique of reduction of matrices. It also has certain structural characteristics: The student was not performing calculations, he was reflecting on the properties of the possible effects of a calculation.
A purely structural argument could look as follows: Let A be an nxn matrix for which there exists a matrix   B   of the same dimensions such that $AB = I_n$ (call it property (*)). Let $B_1, B_2, . . ., B_n$ be the columns of B. Then AB has columns $AB_1, AB_2, . . . , AB_n$. By the property (*), these columns should be equal, respectively, $e_1, e_2, . . . , e_n$, where e_i is the i'th unit vector. Finding the matrix B thus amounts to simultaneous solution of equations $Ax = e_1, . . ., Ax = e_n$, which can be done by row reducing the matrix $[A|I_n]$.

Example 8. Proving that $(A^{-1})^{-1} = A$.
Structural proofs of facts such as $(AB)^{-1} = B^{-1}A^{-1}$, the commutativity of the transpose and inverse operations on matrices, or the fact that $(A^{-1})^{-1} = A$ have become a part of the iron repertoire of linear algebra courses. Students, in general, do not understand these proofs and treat them as 'tricks' that they have to memorize. In one of our classes, when presented with the usual structural proof of $(A^{-1})^{-1} = A$, several students could not believe the proof was done: 'That's it? You mean this is a proof?'. Asked for an alternative proof, one of the students proposed the following argument based on the technique of calculating A^{-1}, mentioned in Example 7: To calculate A^{-1} you row reduce the matrix $[A|I_n]$ until you get $[I_n|B]$. Then B is the

inverse of A. If you now undo what you have done, row reducing $[I_n|B]$ until you get $[A|I_n]$, A will be B^{-1} or $(A^{-1})^{-1}$.

This looks like an analytic-arithmetic argument. It starts from a technique of computation, not from properties. But it is already one foot in the structural mode because it reflects on the technique and makes use of its reversible character.

We could observe students escape into synthetic-geometric thinking when they were supposed to give an analytic-arithmetic proof, or into the latter when a structural proof was expected. Quite often, analytic-arithmetic proofs are very tedious, involving multiple indices and a great deal of calculations. The need for economy of thought is a good motivation, in such cases, to use a more structural argument.

Example 9. The invariance of the solution of a system of equations under the so-called "elementary operations" on the system.

Normally rather early in their linear algebra courses, the students learn that operations on a system of linear equations such as (1) replacement of an equation by a sum of itself and a scalar multiple of another in the system; (2) multiplication of an equation by a non-zero scalar; (3) interchange of two equations on the system, do not affect its solution set. In our observations of tutoring sessions, on five students, only one (labeled Endy 1) asked the question 'Why?' with regard to this claim. It is worth noting that this student had had serious difficulties on the level of the notions of variable and solution of an equation. Two other students who, like Endy 1, worked with Lay's textbook (Lay, 1994), read the explanation without much motivation and did not seem to understand much of it. What turned out to be the biggest obstacle for them was the notion of 'row equivalence of matrices' (two matrices are equivalent in this sense if one can be obtained from the other through operations on rows analogous to the operations 1, 2, 3 above on equations). This notion has a strongly structural character. It involves the classical structuring operations such as taking equivalence classes and creating new structures by picking representatives of the classes. Row equivalence is an equivalence relation in the set of all matrices of given dimensions. As such it breaks down the set into classes. Seen as augmented matrices of systems of equations, two matrices belonging to the same class have their corresponding systems of equations define the same solution set.

The argument used to justify the claim mentioned above on the invariance of solution was based, in the text, on the fact that the row equivalence relation is symmetric: The elementary row operations can be reversed. The students could not see the idea of the argument and were blocked by the question: 'Of course, one can come back to the previous matrix, but why would one want to?'.

In offering his explanation, the author of the text asks of the reader to make abstraction of the context of solving equations and think of matrices and row operations as objects and operations in their own right. He also expects the students to consider the relation of row equivalence between matrices, a relation which is symmetric not only in the sense that 'if A is row equivalent to B then B is row equivalent to A', but also in the sense that it does not convey the idea of a direction or purpose as, for example, the operation of row reduction does. In row reduction, a matrix B is obtained from a matrix A, but this matrix B is not just any matrix row equivalent to A. It is expected to be simpler than A, 'sparse', with lots of zeros.

The concept of row equivalence as given by the definition does not and cannot, at this point, answer any practical questions for the students. Therefore, it cannot be of any use in the justification of the elimination method. It is no wonder, then, that in our observation, some students either could not remember what the term meant or associated it mechanically with 'having the same solution set', and not with the existence of a sequence of arbitrary row operations.

A different text was used in the sessions on linear equations with the student Sandy 3 (Bazanska, Karwacka, Nykowska, 1978). This text did not contain an explanation of the fact that the solution set of a system of equations is invariant under elementary operations, but the student did not even feel the need for it. He saw the claim as self-evident.

> 221. Sandy 3: [reads] 'The following operations transform the system into an equivalent one, i.e. having the same solution set: 1° multiplying the two sides of an equation by a non-zero scalar, etc.' Isn't this all the same thing? I don't know, you've got an equation, you transfer everything on one side... We call it 'transposing' [in school jargon] but we are really subtracting it from one side and subtracting it from the other. [Reads on] '2° changing the order of equations'. Of course, what would it change, indeed, if this is written here and this is written there... All this is so elementary...

In the text that Sandy 3 was using, the problem of solving linear systems was attacked via the concepts of dependence and rank. Even the terminology was suggestive: A consistent system can be 'dependent' (infinitely many solution) or 'independent' (a unique solution) and, more importantly, what is considered as eliminated in the process of solving the system are not variables from equations, as in the text by D.C. Lay, but redundant equations, those that are linear combinations of other equations (and are therefore dependent on others). With this approach, it is the question of deciding whether the system is dependent or independent that becomes important, more important maybe than the very process of finding the solutions.

The discussion, between Sandy 3 and his tutor, that followed the reading of the statement, was focused on dependent systems, parameters, general solution, particular solutions, basic solutions, etc. The question of invariance of the solution set under elementary operations was not thoroughly discussed at that time.

Looking now at Sandy 3's comments above, one may wonder, if he even read well what the elementary operations on a system of equations are and whether he noticed the last operation; i.e., replacing one equation 'by the sum of itself and a multiple of another equation'. He spoke of operations on single equations, like subtracting the same thing from both sides of the equation.

In order to clarify what his understanding could have been, Sandy 3 was interviewed eleven months after the session of February 20.

Asked why the elementary operations (he was reminded of what they are) do not change the solution set of the system, his first answer was 'because this does not change the effect, I mean the x's will not change their values', which is nothing but a restatement of the fact to prove. He had to be pushed a lot to produce some more substantial argument. For Sandy 3, there was nothing to prove. The tutor then resorted to the question, 'why can't we square both sides of an equation?'. The answer was: 'Because then you change your x's, from an x^2 you can get x and also -x'. When asked: 'And why can't you multiply by zero?', he answered: 'Because then you get 0=0 and you can't tell what it was before'.

The student was still reluctant to go into any analytic detail, but the tutor asked him to give it a try and to perform at least the 'replacement' operation on an arbitrary system of equations and explain why the new system must have the same solution as the old one. The student wrote:

$$A_1x_1 + B_1x_2 +... + Z_1x_n = b_1$$

....

$$A_mx_1 + B_mx_2 +... + Z_mx_n = b_m \quad (*)$$

and then

$$aA_1x_1 + aB_1x_2 +... + aZ_1x_n = ab_1 \qquad (**)$$

....

$$aA_1x_1+aB_1x_2+...+aZ_1x_n+A_mx_1+B_mx_2+...+Z_mx_n = ab_1+b_m \qquad (***)$$

The way he explained how he obtained the last equation in the second system throws light on why, in his comment eleven months ago, he spoke about 'subtracting the same thing from both sides of the equation'. Namely, the last equation is obtained by adding, to both sides of the equation (*), the term $aA_1x_1 + aB_1x_2 +... + aZ_1x_n$. By (**), this term is equal to ab_1, so we can replace it by this expression on the right hand side of the new equation, obtaining (***). The way the student then explained why any x's that satisfy the second system satisfy also the first, was by showing how one can get to the first system from the second. He wrote:

$$A = B$$
$$A + W = B + P$$

saying that this represents the second system. Now, to obtain the first one, 'we do the same thing as before, only we now subtract A from both sides of the second equation'. Using the fact that A=B, he wrote:

$$-A + A + W = -B + B + P$$
$$W = P$$

For him, this was the end of the proof.

This reasoning is certainly a good example of one which is somewhere between the analytic-arithmetic and the analytic-structural. This argument is not purely structural, because it does not operate on the system as a whole. But it is not arithmetic either because it does not enter into the 'arithmetic detail' of the system. It operates on equations as 'structures' of the type $A = B$, seeing the equations of the system in a more holistic way.

3.4. General Remarks on the Students' Ways of Thinking in Linear Algebra

The predominance of the intermediary forms of thinking in students can be viewed as another symptom of their tendency to think in 'practical' rather than 'theoretical' ways. There is no concern for the concepts and techniques to belong to one consistent system: one uses and takes whatever appears to be easier at the given time and situation.

But all our observations show that students can be quite creative in their thinking, if only they are not paralyzed by the 'obstacle of formalism'. It is up to us, didacticians and teachers, to help transform this potential for creativity into a real intellectual development. But we may have to become more realistic in our choice of what to teach to different groups of students.

4. CONCLUSIONS

In our research on the teaching of linear algebra we found that no matter how we tried to approach the content, students' difficulties seemed to persist. The reason could be that we introduced all sorts of changes but never gave up teaching the structural theory of linear algebra. It is not enough to just make the structural content more concrete through working in low dimensions and using visualizations. In fact, visualizations themselves are problematic; they may lead to irrelevant interpretations which make the understanding more, not less difficult (Sierpinska, Dreyfus and Hillel, 1999). It is possible that much more radical changes of the content of teaching have to be introduced.

Hamilton had once remarked on the study of algebra as arising most fruitfully in conjunction of three perspectives: technical computational, technical notational, and theoretical:

> The Study of Algebra may be pursued in three very different schools, the Practical, the Philological, or the Theoretical, according as Algebra itself is accounted as an Instrument, or a Language, or a Contemplation; according as ease of operation, or symmetry of expression, or clearness of thought (the agere, the fari, the sapere) is eminently prized and sought for. The Practical person seeks a rule which he may use, the Philological person seeks the formula which he may write, the Theoretical person seeks a Theorem on which he may meditate. It is not here asserted that every or any Algebraist belongs exclusively to any one of these three schools, so as to be only Practical, or only Philological, or only theoretical. Language and Thought react and Theory and Practice help each other (Hamilton, 1837/1967).

The common opinion is that most students adhere to the 'practical school'. Our research does not provide any statistical evidence but certainly supports this opinion. Moreover, teaching experience shows that 'The one thing [linear algebra] students seem to be competent in and comfortable with is making computations once a problem has been put into an appropriate form. They find reading mathematics difficult and tend to limit their reading to examples and exercises... They are inexperienced and uncomfortable with axiomatic definitions (or with definitions in terms of properties satisfied), with concepts, and with proofs' (Carlson, 1994).

It is probably high time to take this evidence and experience seriously, and start building on it rather than ignoring it or trying to fight it. For the large numbers of practically minded students in our universities, a 'practical' rather than 'theoretical' and especially 'structural' approach is advisable.

On the one hand, the 'practical' approach can lead to working on the applications of linear algebra, and why not a 'linear algebra through applications', where concepts are not introduced explicitly by definitions but implicitly as tools in thinking about problems in various contexts (Fletcher, 1972).

On the other hand, building on what the students seem to understand best computation it could be possible to make the analytic-arithmetic thinking evolve in the direction of 'numerical thinking'. The numerical thinking makes a rigorous distinction between the definition of a magnitude and a way of computing this magnitude, and it is concerned with the error of computation. Seen from a numerical point of view, certain axioms of the vector space take on a different meaning, which may make more sense for the students. For example, the property, usually considered as 'obvious' of distributivity $(a+b)v=av+bv$, leads to a discussion of the economy of two algorithms. If v is an m-tuple, the computation indicated by the left-hand side of the equality requires one addition and m multiplications, while the right-hand side computation involves 2m multiplications and m additions (Kielbasinski and Schwetlik, 1992, p. 15).

But the inadequacy of the curriculum is only a part of the problems of the ineffectiveness of our teaching of linear algebra. The prevailing authoritarian teaching style at the university leaves the students with the impression that theory is not their business at all. There is an atmosphere of a clear-cut division of labor, whereby the student is given no control over the validity of the statements on which he or she bases the reasonings and calculations. This atmosphere is greatly enhanced by the assignment of 'exercises' (even 'proof exercises') where the student is not given a chance to choose a method of solution, or to decide whether to prove or not to prove. Knowledge acquired this way is very likely a 'school survival' knowledge, not scientific knowledge of any kind.

The question that remains is whether, in the long run, we still want our students to be able and willing to engage in theoretical thinking. Is theoretical thinking worthy of our educational efforts? Maybe, there is just no scientific knowledge without theoretical thinking, but a blind application of techniques, or, at most, a technology without reflection on its relevance and possible consequences. In this case, if education has anything to do with scientific knowledge, then, yes, something has to be done. But, it seems that something has to be done a lot earlier in the educational process. Theoretical thinking needs special nurturing; nature and everyday socialization do not suffice. There are many occasions for that in school, not only in mathematics and science but also in the learning of languages and their grammars. Vygotsky thought that the learning of the written language (which is different from learning the technical skill of writing letters and words) is of the greatest importance for the mental development of the child (Vygotsky, 1997, Vol. 4, p. 131). There is already a body of educational experience and research on the development of theoretical thinking in children (e.g. Pallascio, 1992; Garuti et al., 1999) from which one can draw in planning for the 'education of tomorrow's university students'.

5. NOTES

[1] The chapter incorporates parts of and extends the matter included in "A propos de trois modes de raisonnement en algèbre linéaire", by A. Sierpinska, A. Defence, T. Khatcherian and L. Saldanha, published in the French version of the present book. In its present version, the chapter refers to the research on the teaching and learning of linear algebra conducted at the Concordia University by A. Sierpinska and J. Hillel with the collaboration, at various stages, of visiting scholars T. Dreyfus and J. Trgalová, and graduate students, in alphabetical order: A. Defence, M. Haddad, T. Khatcherian, R. Masters, L. Saldanha.

[2] These projects have been subsidized by the following research grants: FCAR, n° 93-ER-1535, SSHRC, n° 410-93-0700, and n° 410-96-0741.

[3] In literature, classical examples of the 'inner speech genre' are found in Joyce's 'Ulysses'.

[4] This is a rather drastic simplification of the history of linear algebra, and we are aware of it. However, this simplification helps us to make certain distinctions that are helpful is understanding what is involved in constructing the concepts of linear algebra by students. For a detailed account of the complex routes of evolution of linear algebraic ideas, the reader is referred to Dorier (1995, and Part One, this volume).

[5] Looking at linear algebra as a certain mathematical practice, and using Chevallard's notion of praxeology (1999), we could say that the analytic-arithmetic thinking led to the development of a 'technology' (a theory of techniques of solving systems of linear equations) while the analytic-structural thinking achieved the 'theory' level of the practice, i.e. a theory of the technology.

[6] Here is an English translation: 'In the eighteenth century, the century of Euler, analysis essentially develops methods of calculations with infinity. Series, ... , derivatives and integrals are objects which, each in its own fashion, formulate infinity, grasp it with a formula. The algebraic formula, together with its multiple transformations through the calculus, holds a central position. [In] the new form undertaken by Analysis during the first half of the nineteenth century ... the objects which take centre stage are no longer presented as formulas but as concepts, that is, through properties'.

[7] This is a symptom of the students' practical thinking.

[8] Examples in this part of the article are based mainly on our 1993-96 research on interactions between a tutor, a student and a text. In all examples but one the students were college students (a level which, in Québec, is between secondary school and university: Collège d'Enseignement Général et Professionnel). In Example 8, exceptionally, an account is given of a reaction of a class of university level linear algebra course.

MICHÈLE ARTIGUE, GHISLAINE CHARTIER
AND JEAN-LUC DORIER

PART II - CHAPTER 8
PRESENTATION OF OTHER RESEARCH WORKS

In this final chapter we will present the major points of five research projects on linear algebra which, for various reasons, could not be included in the preceding chapters (the reader can obtain further information by consulting the bibliographical references).

1. COORDINATION OF SEMIOTIC REPRESENTATION REGISTERS[1]

The work described here was the work of Kallia Pavlopoulou's doctoral dissertation directed by François Pluvinage, entitled *'Propédeutique de l'algèbre linéaire : la coordination des registres de représentation sémiotique'* (*Propaedeutics of linear algebra: coordination of registers of semiotic representation*), presented in June 1994 at Louis Pasteur University in Strasbourg. The reader can also consult Pavlopoulou (1993).

This work draws on Raymond Duval's research (1993 and 1995) about registers of semiotic representation and cognitive processes. According to Duval, the word representation is both important and marginal in mathematics. He defines « semiotic representations as productions made by use of signs belonging to a system of representation which has its own constraints of meaning and functioning » (Duval 1993., 39). Semiotic representations are absolutely necessary to mathematical activity, because its objects cannot be directly perceived and must, therefore, be represented. The author maintains that « semiotic representations are not only a means to externalize mental representations in order to communicate, but they are also essential for the cognitive activity of thinking » (ibid., 39). In fact, they play a role in developing mental representations, in accomplishing different cognitive functions (objectification, calculation, etc.), as well as in producing knowledge. He borrows an idea from Granger (1979) which stipulates that the development of science is linked to the development of more and more specific semiotic systems, independent of natural language. He thus differentiates *semiosis*, comprehension or production of a mental representation, from *noesis*, conceptual comprehension of an object, while maintaining their inseparability. « There is no *noesis* without semiosis, whereas mathematics is taught as though *semiosis* was an insignificant operation compared to *noesis* » (ibid., 40). In cognitive activities linked to

247

DORIER J.-L. (ed.), *The Teaching of Linear Algebra in Question*, 247—264.
©2000 *Kluwer Academic Publishers. Printed in the Netherlands.*

semiosis, he distinguishes three types of activity: the formation of a representation which can be identified as belonging to a given register; the processing and transformation of a representation within the register where it was created; and finally *conversion*, i.e., the transformation of a semiotic representation from one register to another. He underlines the importance of the third activity by describing it as a necessary passage for coordinating registers attached to one same concept. Nevertheless, although the first two activities seem to be taken into account in mathematics teaching, the third is usually ignored. It acts as an underlying assumption that is never made explicit as long as the rules of operation within each register are acquired by the learner. Duval shows, however, that many student difficulties arise from their inability to carry out conversions of registers of semiotic representation. He also maintains that the possibility of conversion is one of the essential conditions of conceptualization and, therefore, of *noesis*. « (Integrative) comprehension of a conceptual content rests on the coordination of at least two representation registers, and this coordination is revealed by the rapidity and spontaneity of the cognitive activity of conversion » (ibid., 51). His interests lie in the conditions of learning involving *semiosis*.

Pvalopoulou's work emerges as an application and verification of Duval's theory in the context of teaching linear algebra. In this field, the author distinguishes between three registers of semiotic representation: the graphical register (arrows), the table register (columns of coordinates), and the symbolic writing register (axiomatic theory of vector spaces). We can see here a connection to Hillel's work as presented in the preceding chapter. Moreover, Pavlopoulou's registers are determined by very strict rules of formation. We give an example below of a 'situation' in the three registers, which illustrates what is meant by this claim:

Graphical register

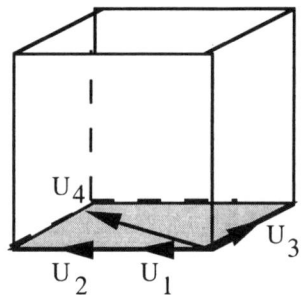

Table register:

1	k	0	m
0	0	1	n
0	0	0	0

$$U_1 \in \mathbb{R}^3, \quad U_2 \in \mathbb{R}^3, \quad U_3 \in \mathbb{R}^3, \quad U_4 \in \mathbb{R}^3$$

Symbolic
writing
register:

$$U_1 = 1\,U_1 + 0\,U_3$$

$$U_2 = k\,U_1 + 0\,U_3$$

$$U_3 = 0\,U_1 + 1\,U_3$$

$$U_4 = m\,U_1 + n\,U_3$$

This example shows that each register obeys precise codes, which are indispensable keys to understand each register. In the following examples specific references will be made exclusively to these three registers. For each evaluation or experiment the rules that governed each register were taught to the students.

Pavlopoulou begins by presenting a brief panorama of certain earlier works in which she points out the lack of questioning in the domain of registers - she then draws on teaching observations in Strasbourg to show a few examples of student difficulties; seemingly due to an inability to convert from one register to another or to an implicit confusion between an object and its representation. Here, contrary to what was seen with Hillel, it is the confusion between a vector and its geometric representation, and not its writing in n-tuples, which is identified as the major source of problems. She then meticulously examines the characteristics of each register, as well as the link between the symbolic register and the natural language register, for the object 'vector'. She also outlines what is involved in the activities of converting from one register to another. This detailed analysis reveals the difficulties belonging to the beginning register and the final register. She shows, among other things, that for two given registers the conversion difficulty is not at all the same in one direction as in the other. This detailed analysis gives a precise idea of what is at stake. However, the tasks studied are never proposed within a problem-solving context; they are presented with explicit instructions for conversion, with nothing else at stake. It would, nevertheless, be interesting to check the consistency of this type of analysis, carried out in a simplified context, when applied to a more real, and therefore complex, teaching situation. In particular, in teaching, situations are rare where the vector concept is the only one at play, yet all the examples and tests presented in this work are limited to this single concept, and implicitly to the notions of linear independence/dependence and basis. In this way, it would be interesting to examine certain epistemological aspects of the history of linear algebra in the light of registers of semiotic representation. For example, in the development of functional analysis, seen in the first part of this book, the analytical aspect was favored at the beginning. Later the introduction of geometrical language provided a formalism that was better adapted to sets of functions for solving functional equations by taking into account the topological aspects. In this phase of the history of linear algebra it is clear that the coordination of different representation registers is important but it is also clear that this coordination interacts with problem solving. The relationship between different registers depends on the types of problems presented. These problems are an essential driving force of the conceptual progress which cannot be reduced to the simple interaction between different registers.

Next, Pavlopoulou presents an analysis of four studies carried out over a two-year period with first year science students at the University of Strasbourg. This analysis very clearly points out students' 'spontaneous' level of competence in conversion activities (she proposed problems in one register and asked for translations into another imposed register). The level was globally very weak. Moreover, it would seem that the presence of the symbolic register is an important factor in student failure, especially when they must convert this register. This brings out the asymmetry of the conversion activity: 7% success in converting from the table register toward the symbolic register, 72% for the opposite conversion. The difference is also very significant with the graphical register. One question, however, remains to be answered. That the symbolic register appears as the most difficult to process is hardly a surprise; but that it is more difficult to convert into this register than from it, is more surprising. However, this is true whatever the beginning register (table or graphical), so one wonders whether the difficulty is only linked to the question of conversion or to the symbolic register itself. In addition, this register is a constituent element of the modern theory of vector spaces and it is therefore difficult to separate the problems linked to the use of the symbolic register (even in conversion activities) from the problems linked to conceptualization in linear algebra. Thus, a delicate question that the analysis raises, without really resolving it, is knowing which is the cause or the effect, between the conceptual comprehension of the vector as an element of a formal vector space and the ability to convert towards the symbolic register. Can these two aspects of cognitive activity, indisputably linked, be reduced, moreover, to a simple relation of cause and effect? It is the very role of *semiosis* in *noesis* that is at stake here. This question is all the more important that in the more complex tasks, several conversions may be necessary, but these tasks also bring into play other cognitive activities. Hillel's work sheds light upon this point.

After identifying a few typical errors in each of the conversions, Pavlopoulou analyzes several linear algebra textbooks, examining how they treat the three registers of semiotic representation retained and how they treat the different conversions. It was very clear that authors generally privileged one register over the others, often the symbolic register, for reasons of economy. Pavlopoulou shows particularly that conversions were never made explicit in the textbooks and never received specific treatment.

Following these analyses, Pavlopoulou presents a teaching experiment that she carried out with first year science university students in Strasbourg. The analysis of this experiment compared pretest and posttest results of an experimental population (2 groups totaling 45 students) with a control population (3 groups totaling 99 students). The control groups received traditional teaching, whereas the experimental groups received a specific 8-hour teaching module between the pretest and the posttest.

The pretest essentially concerned explicit conversion activities (students were given a problem in one register and were asked to translate it into another). It also included two linear algebra exercises and one open question: « I think a vector is… (complete the sentence). »

Both populations were recognized as equivalent for the pretest.

The experimental teaching module dealt essentially with activities of conversion between registers. Out of twelve exercises, the first seven covered tasks similar to those in the pretest. Exercises 8 to 10 dealt with conversions from natural language into symbolic writing and vice versa. The last two involved simple linear algebra problems where certain student difficulties seemed to be closely linked to problems of conversion between registers. This last aspect, however, does not appear to have been used a great deal.

The posttest essentially involved activities that were based on the recognition of registers or an explicit conversion between registers. It also included seven more general questions, most of which were not 'classical' linear algebra course questions.

For the 'pure conversion' questions, the demonstration of the effectiveness of the experimental teaching is very convincing. Although the control groups improved very little compared to the pretest (they still had great difficulties in dissociating the vector object from its geometric representation), it is clear that the experimental groups reached a success curve very close to total success. In particular, factor analysis on all the pretests and posttests for all groups shows that the first factor axis is not the usual success-failure axis (it will show up in the second) but rather an axis setting apart the pretest and the posttest, thereby making the gap in skills between experimental and control students stand out, before and after the experimental teaching module. It would seem, then, that the module proposed was adapted to help students acquire skills in activities of explicit conversion between registers. The results of the experiment also show a significant difference in success on the other posttest questions (in favor of the experimental groups), although less marked than the results described above. One must nevertheless remain cautious about the conclusions drawn. It must be remembered that experimental teaching induces de facto a greater investment on the part of the teacher which spills over to the students. In other words, students in any experimental teaching situation whatsoever are always more receptive and benefit positively from the success in the areas linked to this teaching. Moreover, and especially, the results of the seven questions evaluated in the posttest seem insufficient for drawing overly general conclusions, even if the analysis globally corroborates the hypothesis that teaching linear algebra centered around the coordination of registers of semiotic representation improves skills, and even in tasks not explicitly linked to activities of recognition or conversion of registers of semiotic representation.

Pavlopoulou's work provides a solid grasp on the question of registers of semiotic representation within linear algebra. The demonstration that this activity is not taken into account in teaching and the failure of students to succeed without preparation is totally convincing. The analysis of different registers and of the nature of each type of conversion, as well as their associated difficulties, is exposed in great detail. In addition, proposing classroom work on the coordination of the different registers remains persuasive, as much for its relevance in relation to the preceding analysis as for the results obtained on the improvement of student skills in the activities of explicit conversions. Furthermore, this substantial research opens the way to several future research questions. The two most important are:

- Reflection on the links between *semiosis* and *noesis*, within global teaching of linear algebra. The work accomplished thus far shows the importance of register conversions, but it must be pursued on a broader scale, including other dimensions of analysis, to better determine the role that conversions play in the comprehension of a concept, including when it intervenes in solving a problem. It would also be interesting to continue Pavlopoulou's work to other concepts of linear algebra in order to go beyond the vector.

- The question of conversions from the symbolic register to the natural language register and vice versa has been treated, but the need for a true experimentation in this direction remains. Moreover, Duval stresses in his work that « it is not possible to neglect or set aside natural language in mathematics teaching » (op. cit., 64).

2. ARTICULATION PROBLEMS BETWEEN CARTESIAN AND PARAMETRIC VIEWPOINTS IN LINEAR ALGEBRA[2]

The work presented here was the work of Marlene Alves Dias' doctoral dissertation, directed by Michèle Artigue, entitled, *'Les problèmes d'articulation entre points de vue 'cartésien' et 'paramétrique' dans l'enseignement de l'algèbre linéaire'* (*Articulation problems between Cartesian and parametric points of view in the teaching of linear algebra*), presented in May 1998 at Denis Diderot University in Paris.

Alves Dias's research, in some sense, extends Pavlopoulou's work. She starts from the point that some resistant difficulties in linear algebra, identified by research, can be formulated in terms of lack of cognitive flexibility between settings, semiotic registers, thinking modes, and points of view. She firstly develops an extensive survey of the literature dealing with advanced mathematical thinking and analyzes the place devoted in it to cognitive flexibility. Her analysis shows not only the long predominance in research of hierarchical visions of cognitive development poorly sensitive to flexibility, but also the increasing attention paid to flexibility and the diversity of constructs elaborated in order to approach it. She then focuses on one specific form of cognitive flexibility which plays an important role in linear algebra: flexibility between the parametric and Cartesian points of view (Rogalski, 1991).

Fundamental objects of linear algebra, such as vectors and subspaces, can be approached from different points of view. The distinction between the parametric and the Cartesian perspectives can be seen as a "linear conversion" of the general distinction between extensive and comprehensive characterizations one can find in set theory. For instance, a vector space is approached through the parametric viewpoint if emphasis is put on a system of generators and on the way a generic element of the subspace can be obtained and represented through the use of such generators. Similarly, a vector space is approached through a Cartesian viewpoint if emphasis is put on a linear equation or system which characterizes it. Alves Dias, in her research work, tries to understand the mathematical needs of the flexibility between these viewpoints, the real nature of the difficulties students meet at developing an efficient articulation, and the way teaching practices take charge of these needs and difficulties.

Articulation between Cartesian and parametric points of view involves articulations between settings and semiotic registers. Alves Dias is thus led to refine the distinctions introduced by Pavlopoulou and to distinguish, in the symbolic register, between:

- *Intrinsic-explicit parametric representations*, such as:

  A= lin{a,b} = {v/ v=αa+βb, with α, $\beta \in \mathbb{R}$ }

- *Table-explicit parametric representations*, such as:
  A= lin$\{(1,0,0),(0,1,0)\}$= $\{\alpha(1,0,0)+\beta(0,1,0),$ with $\alpha, \beta \in \mathbb{R}\}$
- *Equation-implicit parametric representations*, such as:
  A= lin$\{(1,-2,1,0),(1,-3,3,1)\}$
  = $\{(x,y,z,t) \in \mathbb{R}^4 / x=\alpha+\beta, y=-2\alpha-3\beta, z=\alpha+3\beta, t=\beta,$ with $\alpha, \beta\in \mathbb{R}\}$
- *Table-implicit parametric representations*, such as:
  A=$\{(\alpha,\beta,0),$ with $\alpha, \beta\in \mathbb{R}\}$
- *Intrinsic Cartesian representations*, such as:
  A=$\{v/ T(v) =0\}$, T being a linear operator
- *Explicit Cartesian representations*, such as:
  A=$\{(x,y,z) \in \mathbb{R}^3 / z=0$ and $2x+y=0\}$

However, her main hypothesis is that cognitive flexibility is not a purely semiotic process; that it is of a more complex nature.

In order to test this hypothesis, she firstly develops an epistemological and mathematical study, investigating the historical sources of the articulation and the mathematical tasks and techniques which, today, support and require its mastering in a first course of linear algebra.

In finite dimensional spaces, the conversion between parametric and Cartesian representations can be easily dealt with through algorithms linked to the solving of linear systems. Such a fact tends to hide the cognitive complexity of articulation processes.

The epistemological analysis carried out, based on the work developed by Dorier on the genesis of the notion of rank (Dorier,1993), restores, in some sense, the real complexity. It shows the long time required for the development of a flexible articulation between the two viewpoints and the dissymmetry of this historical genesis: the transition from the Cartesian to the parametric being achieved about two hundred years before the reverse transition. It also shows the dependence of this achievement on the emergence of the fundamental concepts of rank and duality, well known as difficult concepts for students.

The analysis of the conceptual and technical mathematical needs of flexibility in elementary linear algebra, is organized around the identification of elementary tasks (13 tasks). These tasks are associated to notions and each task is described by a set of variables (7 variables). Moreover, for each task, Alves Dias identifies the potential and necessary needs in terms of flexibility between Cartesian and parametric viewpoints and the associated knowledge.

Let us give a very simple example of a task about subspace operations: to find a basis and the dimension of the following subspaces : V, U, W, U∩V, V+W and U+V+W for:

$U = \{(x,y,z) \in \mathbb{R}^3 / x=0\}$

$V = \{(x,y,z) \in \mathbb{R}^3 / y-2z=0\}$

$W = $ lin$\{(1,1,0), (0,0,2)\}$

For this type of task, the mathematical needs of articulation between Cartesian and parametric viewpoints depend on the definition of the concerned subspaces. If one subspace is given by a parametric representation, finding its dimension and a basis can be obtained within the same parametric viewpoint, by determining the rank of the given system of generators and finding a free generating system. If it is given by a Cartesian representation, articulation is necessary but can function at different levels.

The variables of the task in this example are the following :

- *type of space*: $\mathbb{R}^3$
- *type of given representations*: explicit Cartesian representation, table-explicit parametric representation, intrinsic symbolic definition ($U \cap V$, $V+W$ and $U+V+W$)
- *type of required representation*: table-explicit parametric representation
- *space and subspace dimensions*: 3, 1 and 2
- *potential/necessary flexibility*: If we assume that the questions are answered in the proposed order, flexibility is necessary but can rely on parametric reading of the given equations. For instance, the definition of V can be reinterpreted as V being the set of vectors (x,2z,z) for any x and z, and then as x(1,0,0)+z(0,2,1). It can also rely on geometrical interpretations. For instance, $U \cap V$ is 1-dimensional, since $U \cap V$ is the intersection of two different vector planes; from there, finding a generator is not difficult. W and V are two distinct planes therefore their sum is the space $\mathbb{R}^3$, and this gives the answer to the last question.
- *Flexibility knowledge*: here its appears tightly linked to knowledge of the relationships between the rank of a linear system and the dimension of the subspace of its solutions.

Results obtained show that very few tasks do require flexibility between Cartesian and parametric viewpoints and that a careful choice of variables is necessary in order to foster more than a technical treatment of the articulation. The mathematical space available to develop flexibility competences in a first linear algebra course is thus reduced. However, at the end of such a course, students are expected to efficiently move between the two points of view in finite dimensional spaces and, even, to be able to use such a flexibility in a more metaphoric way when dealing with infinite dimensional spaces.

These preliminary analyses are then used in order to investigate the way teaching practices tackle flexibility. For this purpose, Alves Dias analyzes, in a very detailed way, 12 textbooks from various countries and periods: France (1984-1996 / 7 textbooks), Brazil (1980-1995 / 3 textbooks), USA (1976-1993 / 2 textbooks). Analysis shows a great diversity in the approaches to linear algebra and in the structuration of courses, even within similar approaches. But, if one exempts some very particular cases of texts linked to didactic research, whatever be the approach, textbooks are not very sensitive to flexibility issues between the two points of view. These are not explicitly taken into account in the course. One cannot find specific comments in the part of the course devoted to the solving of linear systems. When duality is introduced, no link is made with this problem. Tasks offered to students are very limited in terms of flexibility. In fact, they rarely go beyond its technical aspects ; in particular, the issue of control via the rank or duality is practically never raised or necessary.

On the basis of these analyses, Alves Dias builds a series of exercises involving an articulation between the Cartesian and parametric viewpoints. These exercises are close to the usual tasks encountered in textbooks, but they differ from them in that they require the student to show cognitive flexibility and to use explicit control via the concepts of rank and duality.

For instance:

Given the following vectors in $\mathbb{R}^3$: a = (2,3,-1), b = (1,-1,-2), c = (5,0,-7) *and* d = (0,0,1), *find a Cartesian representation for the intersection of the subspaces generated by* {a,b} *and* {c,d}.

She experiments these exercises on mathematics students, both in France and in Brazil, at different university levels: first and fourth year in France, masters level in Brazil. A specific design had been elaborated in the Brazilian context. Firstly, students had to work by groups of two or three without any material and produce a collective answer in a limited amount of time. Then, they worked at home on the same test and produced an individual answer. Finally, interviews were carried out with selected students on the basis of their two results.

Analysis of the students' responses demonstrate a variety of difficulties. For instance, students often identify one type of representation exclusively through semiotic characteristics (a representation with x's and y's is obviously Cartesian) without questioning the meaning of the representation (this was one of the reasons which led Alves Dias to refine the Pavlopoulou's categories). Concerning their means of control over the validity of statements and their anticipation of answers, she finds that a theorem like, dim E = dim Kerf + dim Imf, although known and correctly used by many students, is very seldom used for those purposes; even in cases in which it would immediately bring up a contradiction with the result obtained, or in cases in which it offers valuable information in order to anticipate the correct answer.

For the great majority of students, even the most advanced, the resolution of tasks is only managed at a technical level, through the treatment of numerical tables associated with systems of vectors, matrices or linear systems. These tables are, a priori, efficient semiotic instruments, but students tend to operate with them automatically, without taking into account the mathematical meaning of the transformations they make meaning which is not independent from the kind of object the table represents. This tendency generates a lot of formal skids, leading to absurd results or contradictions students generally do not notice. More generally, interpreting results in terms of variables and parameters, associating vectors and equations, proves to be extremely problematic for most students. Their geometrical knowledge does not seem very helpful and the kind of reasoning we have developed above, about the intersection task, although very elementary, was quite marginal both in French and Brazilian answers.

Alves-Dias's research shows, on a specific part of linear algebra, yet emblematic of many others, that the question of cognitive flexibility is complex. This cannot be reduced to questions of registers or settings, even if these points are important. In usual institutional practices, its development is left to the private responsibility of students: algorithms linked to the articulation are officially taught through the techniques for solving linear systems, but educational systems seem to consider that it is enough for ensuring flexibility competences. Research results prove that it is far from being the case and that even advanced students have not developed basic competences in that area. It is thus necessary to help students to develop such a cognitive flexibility and, in order to do that, teachers have to be well aware of the necessity of carefully choosing the tasks and their variables, especially if they want their students to develop the anticipation and control abilities necessary to be in charge of the technical management of articulation and to avoid formal skids. Indeed, a small change in the formulation of a problem may sometimes have important implications in terms of the cognitive processes involved in its solution.

3. STUDIES WITH FIRST-YEAR UNIVERSITY STUDENTS IN VANNES[3]

These studies were carried out over the two academic years of 1991/92 and 1992/93 at the *Centre Universitaire Scientifique de Vannes* (CUSV), with first-year science students. They were carried out by the IREM[4] research group in Rennes (Bardy et al. 1993).

In an introductory chapter, the authors analyze what is included in the university teaching of linear algebra, based on the official national curriculum set out for high school seniors. Their conclusion underlines just how new this field is for students coming out of secondary school since they have received no formal preparation. Among the difficulties identified in linear algebra are: the number of new words to learn, comparable to a foreign language, the totally new methods of exposition and demonstration, and the nature of the 'universal' model of linear algebra.[5]

To better understand the nature of these difficulties, the authors carried out several investigations. First-year science students at the CUSV sit a mid-term exam which is used to orient students with insufficiently high grades toward a branch of complementary training, which prepares them either for a two-year degree over three years or for another orientation altogether. The teaching of linear algebra (approximately 12 hours of lecture classes and 20 hours of practical classes) is completed over the months of November and December, entirely before this first examination.

3.1. The 1991/92 study

Twenty-three students in the above-mentioned *complementary training* (CT) were tested along with 23 students majoring in *physics/math* (PM). The questionnaire was given in January, one month after the linear algebra teaching was completed. Then, the CT students received a new linear algebra course (totaling 48 hours: half in lectures, half in practical courses). One activity of model recognition was experimented during this phase. They then received a questionnaire on the subject of matrices at the end of the year, whose results were analyzed by the research team.

> The objective of these questionnaires is always the same. We try to see to what extent such and such an elementary notion in linear algebra has been learned, understood, or seems to have been assimilated. What we ask, in general, is to check the boxes which correspond to proposed answers, and to give short explanations or, for the first questionnaire, to give definitions. In this way, comprehension and meaning are favored rather than the aptitude to complete a repetitive calculation without mistakes, where the meaning of what is calculated is often uncertain. (op. cit., 19)

The two notions targeted in the first questionnaire are kernel and generator system. This investigation demonstrates that although the notion of kernel has been moderately well understood, the notion of generator system has not been understood at all. Moreover, the analyses bring out the misuse of terminology: students do not hesitate, for example, to speak of the kernel of a family of vectors or of a subspace. The last part of the questionnaire asked students to interpret geometrical figures in the language of vector geometry, and then to give the equivalent statements in polynomial terms. Students demonstrated very little success in this activity. This shows, in particular, that affine and vector aspects are not clearly distinguished. Moreover, weaker students (CT) were not able to pass to polynomials and even among the PM students, few of them succeeded at this task. Comparison of the two populations (CT and PM) shows that, although for the latter the definitions and certain automatic reflexes seem better acquired, which explains their better success rate, the activities described by the authors as 'comprehension' produce, globally, a low success rate for both populations.

Student errors are explained as follows. The model recognition activity given to the CT students, consisted in doing 3 linear algebra exercises within three different settings: $\mathbb{R}^3$, $\mathbb{R}_2[X]$ and a functional subspace of dimension three. These three exercises were given as homework. They correspond to a single model from the formal theory point of view, and lead therefore to identical calculations (show that an application is linear, give bases of the kernel and the image, calculate pop). Students did not know that these three exercises were similar and, generally speaking, carried them out well. Since the correction was done in class, the teacher also asked how these three exercises were similar, and then asked students to do a new exercise in the same vein beginning with « let E be any finite dimensional vector space... ». The results of this activity showed that students were not able to recognize the underlying model and experienced great difficulty in expressing and formalizing the mathematical activities in which they had succeeded previously. Referring to Chapter IV of the second part of this book, we recognize here an attempt to propose a meta-type activity. From this point of view, we see, however, that the meta part of the activity has remained very implicit, requiring further work to make this activity operational. This work will not bear fruit unless meta is seen as part of the whole approach at several levels.

Finally, the matrix questionnaire given at the end of the year shows that although students may have truly acquired the knowledge and know-how, these are still shaky. « Difficult questions are inadequately treated by students and answers are often ridiculous: what is written is not always under perfect control. The questions that are not asked in an academic fashion and require a little thinking receive inadequate treatment as well. Only 2 or 3 students seem to have gone beyond the level of technique and acquired full comprehension of the concept taught. » (op. cit., 35).

3.2. The 1992/93 study

In the 1992/93 academic year the authors tested students just before the December mid-term exams, at the end of the coursework on linear algebra, but before the orientation of certain students toward other options. A true/false test requiring some short explanations was given and graded. This grade was combined with the December exam grade. The answers of 76 students were analyzed. The analyses were carried out with various statistical methods, including CHIC software designed by S. Ag Almouloud to give « the hierarchial classification of Lerman similitudes » or « the hierarchical classifications of Gras and Larher implications. » In addition to the test variables, variables reflecting student origin, grades on a first assignment in November (which did not involve linear algebra), and on the mid-term exam (which took place after the test analyzed here) were also taken into account. It would be too long to detail these analyses here and we will simply present the conclusions.

Several points can be summarized:

a)There was a significant difference in behavior between those who had had a year of linear algebra and those who were discovering the discipline for the first time: the latter were clearly less successful and seemed disoriented.

b) There was a difference in behavior and average scores between those specializing in math and physics and those specializing in social science and economics (which we have already observed many times). The latter seemed more disoriented than the former faced with this new subject (nearly zero correlation between the November test and the present test or the mid-term test).

c) Although, on the whole, the notion of dimension of a vector space or a vector subspace was fairly well understood, the notion of linear independence was not. Many students probably knew the definition and its translation in terms of solving a linear system, but concretely, if they were given 3 linearly dependent vectors to which was added a fourth one, independent of the three others, the majority of students (not repeating the course) found that the 4 vectors were linearly independent.

d) Students had no concept of what it means to analyze a proof . This question was asked at the end of the test[6]; however, the number of non-responses and the small number of observations containing even a glimmer of truth show that students understood virtually nothing.[7]

e) A last point (more pedagogical) concerns the nature of the mistakes. Clearly, certain mistakes were isolated events within a globally positive environment; others, on the contrary, were part of a conglomeration of errors which contributed to global incomprehension. This is all the more enlightening in that it is often the way students respond wrongly to a single question that discriminates between these different behaviors. (op. cit., 58)

Following this study, the authors suggested improvements for the teaching of linear algebra. These proposals can be outlined in the eight points below.

1. Set objectives for the comprehension of a few basic notions (in other words, limit the content and postpone the formalization of certain notions that can only be initially studied with the help of examples).
2. Limit the vocabulary and symbols to those strictly necessary and adequate (to avoid overloading students).
3. Use examples already seen by students in secondary school   (geometry and equations). Objects such as functions and polynomials are too complex for beginners.
4. Design exercises and tests that test true comprehension and not merely mechanical manipulations.
5. Use this time to teach students how to properly formulate a proof.
6. Give students representations.
7. Spread this early teaching of linear algebra over time.
8. Integrate results of research in didactics into classroom teaching.

The above points have been discussed in earlier chapters. Limiting content (point i) is a general preoccupation. Point ii) is broached differently depending on the case. The research of Dorier et al. has pointed out the difficulties linked to using logic and the language of set theory, but instead of recommending a reduction of formalism, it recommends better management. Moreover, it appears that linear algebra lends itself well to teaching students certain notions of logic and the language of set theory. Should one avoid the difficulty or overcome it with the necessary means? Point iii) has been discussed in the work of Dorier et al. and in Harel. Point iv) directly echoes the use of the meta lever for unifying and generalizing concepts, such as those discussed in chapters 1 to 4 in Part 2 of this book. Point v) is more closely related to Harel's work, point vi) to the research of Hillel and Sierpinska, Pavlopoulou and Dias, and point vii), with long-term time management, is presented in chapter 2. Finally, what has just been said shows that point viii) finds even more support now than at the moment this research was carried out.

The present research is not strictly speaking a fundamental didactic analysis: the resulting suggestions remain partially explored hypotheses. Nevertheless, it presents a precise observation of student skills in linear algebra. In particular, the last study investigates a rich test and the statistical analysis is well-documented, using complex statistical tools adapted to didactic investigation, thereby reinforcing the relevance of the results obtained. This is invaluable information for more in-depth studies in didactics, a source of inspiration and a database for a preliminary verification of certain hypotheses. Lastly, its format is important in the dissemination of certain ideas to teachers which could lead them to read more specific works. It is, however, unfortunate that the choice of test questions is not clearly explained. For example, there are few elements of a preliminary analysis of the exercises given to students, limiting the generalizability of the results and possibly compromising their relevance (cf. the preceding note).

4. STRUCTURING OF KNOWLEDGE[8]

In his doctoral dissertation, Ahmed Behaj (1999) is interested in what he calls the 'structuration du savoir' ('structuring of knowledge') for teachers as well as for students (see also Behaj and Arsac, 1998). He bases his approach on the hypothesis that, to be communicated, knowledge has to be structured. The structuring is therefore a transitory stage, which reveals an encapsulating process from the teacher and an expanding process from the student.

Concerning the teachers, Behaj used the framework of the 'didactical transposition' as introduced by Chevallard (1985). According to this theoretical approach, when a teacher starts preparing a class, he or she only has a limited range of possible actions. The process of didactical transposition - fixing the structure and content of what has to be taught - is mostly done prior to what teachers do in the classroom: when preparing their classes they can only make local decisions. Behaj's hypothesis is that, since teachers at university level, where there are fewer institutional constraints, may make more of their own choices, then these choices may be very different and have a great (*or greater*) influence on their students' way of learning. From the students' point view, this work aimed at pointing out the relation between the way students may structure the knowledge in their memory and the way they use it when solving problems. Behaj also aimed at evaluating the way a piece of knowledge evolves for one student during all his university studies, and especially, what remains of the first course a student ever had on a certain element of knowledge, after a few years.

The research developed according to two axes : theoretical and experimental. We will not refer much to the first axis here, but mainly to the second part which is related to linear algebra. The experimental part dealt with the following set of concepts: vector (sub)space, linear (in)dependence, generators, basis, rank and dimension. They were chosen because they offer a great variability: each can be defined in different ways and there are several possible choices in their order of presentation. For instance, a basis is an ordered set of independent generators, but it is also a maximal ordered set of independent vectors, or a minimal ordered set of generators, or an ordered set of vectors, of which each vector of the space can be written as a linear combination in a unique way . These four viewpoints, characterizing the same notion, all have their own importance in various situations. They reflect different ways of organizing the connections between the notions of generators, independence and basis; in other words, different structuring of the knowledge related to the set of notions. This structuring is not fixed once and for all after the first course; it is constantly changing when students learn new notions or solve new problems. Moreover, variation in the structuring may be due to more local choices. For instance, let us consider the four statements :

1. The vectors V_1, V_2, ...V_n are independent iff

$$\forall \lambda_1, \lambda_2, ... \lambda_n; \lambda_1 V_1 + \lambda_2 V_2 + ... + \lambda_n V_n = 0 \Rightarrow \lambda_1 = \lambda = ... = \lambda_n = 0.$$

2. The vectors V_1, V_2, ...V_n are dependent iff

$$\exists (\lambda_1, \lambda_2, ..., \lambda_n) \neq (0, 0, ..., 0) \text{ such that: } \lambda_1 V_1 + \lambda_2 V_2 + ... + \lambda_n V_n = 0.$$

3. The vectors V_1, V_2, ...V_n are independent iff none of them can be written as a linear combination of the others.

4. The vectors V_1, V_2, ...V_n are dependent iff one is a linear combination of the others.

1. or 3. can be used as the definition of linear independence, the other being a property. Similarly, for the dependence, 2. or 4. can be chosen as the definition, the other being a property. Moreover, dependence is the contrary of independence. One of the two has to be defined first. That leaves quite a few possibilities in structuring the text of knowledge about these two notions. For instance, the "bourbakist" viewpoint, which has influenced most textbooks since the 1950s in France and in other countries, like Morocco, is to start with definition 1., then definition 2., 4. is given as a property, and 3. is barely worth being noticed. This puts the emphasis on the formal aspect versus the more intuitive aspect. But it also gives straight away the most practical tools for solving problems.

Another choice may be to start with 4. as a definition. This can be connected with the question of reducing sets of generators to a minimum, which is a way to give more meaning to the new concept. Then 3. is as the negation in the register of natural language of 4.. This start gives an intuitive approach to the pair of opposite notions. However, the definitions are not quite operational when solving problems. Indeed, to check whether vectors are independent, one has to check one after the other that each vector cannot be written as a linear combination of the others. It is then important to have a characterization that gives an equal role to each vector, that is essentially what is given with property 2.Property 1. is then the negation within the register of formal language of 2. This is more technical, but the question is raised when students already have put a meaning behind the concepts. It is clear that the two different ways of structuring the definitions and properties, related to the concepts of independence and dependence, reflect different epistemological choices that are likely to influence the learning process. This is the type of question that Behaj is interested in.

In the experimentation, teachers were individually interviewed and the interviews audio-taped. They were asked to describe how they usually present these notions to their students, and they had to give a schema of their course. Then they were given two different proposals of presentation, and were asked to react. They were also asked to explain the role they assign to examples, exercises and proofs in the different parts of their course.

The results of the experiment showed a great variability. Moreover, it showed that teachers have different priorities at different times of their teaching. At the beginning of a course they may be very concerned with rigor and logical order in the presentation of the concepts. But, closer to examinations, they may be more concerned with efficiency. On the other hand, it was found that teachers' beliefs about how students learn, have a great influence on their way of structuring their courses, not only in relation to the order of presentation of concepts, but also concerning the emphases assigned to examples, exercises or proofs. Teachers have very different attitudes at various moments of their teaching, therefore it is impossible to separate a course like linear algebra into quasi-independent items. On the long term, local choices made by a teacher have to be evaluated as a whole, as a dynamical system with complementarity, but also contradictions and side effects.

The student side of the experiment concerned students from second to fifth years of university mathematics and science programs in France and Morocco. They were interviewed in pairs and were asked to write, in common, a plan of a lesson presenting the concepts for first year students. They were also asked to explain the role they would assign to examples, exercises and proofs in the different parts of their course. Finally, they were asked more personal direct questions, about their understanding of linear algebra and its evolution during their studies.

This experiment is interesting in that it gives an idea of what remains one, two, three or four years after the course. Moreover, students who are put in the position of a teacher preparing a course have to reflect on their knowledge in a different way. Through the recordings of the interviews, it is possible to trace a part of the maturity they have acquired in their knowledge and experience in linear algebra. Most often they try to remember the type of teaching they had in first year to use it as a model, but they also criticize it according to more recent experiences while using linear algebra, in more advanced mathematical courses or, less often, in other subjects. It is quite frequent to see students suddenly making connections, for the first time, between two moments of their mathematical past, sometimes very distant. Having to build a course for other students makes them suddenly realize that they had finally understood things that were obscure at the beginning. It appears, as well, that several students explain that they had to wait until the very end of the first year, while preparing the exam, or even until they were in second year, to overcome their difficulties with formalism and finally understand something in linear algebra.

These results are important as regards the question of the evaluation of the long term teaching. Maturity is a necessity; it may take more time for some students than for others. Moreover, at university, official assessment is not continuous, and students tend to have not only selective, but also intense periods of individual work; these moments are often essential for their understanding. Stating that any evaluation, apart from final or mid-term exams, cannot, therefore, be consistent would be extreme. However, the question of the evaluation of long term teaching design, taking into account students' individual work and institutional constraints, is a real theoretical challenge for research in mathematics education.

5. USING GEOMETRY TO TEACH AND LEARN LINEAR ALGEBRA[9]

The work described here is our doctoral dissertation, still in progress, directed by Dorier. It aims at studying how geometry can be used in the teaching and learning of linear algebra. A first difficulty lies in the variable meaning of the word "geometry", which can refer to elementary Euclidean geometry as well as to theoretical affine geometry, grounded itself on linear algebra. We encounter the same problem with "geometrical intuition": many teachers mention it as helpful to learn and work in linear algebra, but they confer various meanings to that expression.

We found a useful tool, in order to avoid ambiguities, in Fischbein's work on intuition in mathematics. According to Fischbein (1987), intuition offers behaviorally meaningful representations, allowing the reasoning activity to rely upon apparently certain conceptions. An important factor of intuition is the use of models ; if a notion is not representable intuitively, one tends to produce (deliberately or unconsciously) a model which can replace the notion in the reasoning process.

Fischbein distinguishes analogical and paradigmatic models. In the case of analogical models, the model and the original belong to two distinct conceptual systems. In the case of a paradigmatic model, the model is a subclass of the original.

A geometrical model, being directly related to physical space, offers to the reasoning activity certain conceptions (and can therefore be considered as intuitive). It can inspire solving strategies, confirm meaningfulness of solutions; but it can also smuggle uncontrolled components into the reasoning process, especially because it is associated to a pictorial representation.

Different geometrical models can be used in different parts of linear algebra ; these models can be either analogical (if they stem from elementary Euclidean geometry, independent of linear algebra) or paradigmatic (if they stem from a part of geometry relying on linear algebra).

The use of models in linear algebra has epistemological as well as psychological aspects.

We can therefore use Fischbein's theory to analyze:
- the role of geometry in the birth and the evolution of linear algebra
- how geometry is used in linear algebra textbooks
- how geometry is involved in students' and teachers' practice in linear algebra.

Many mathematicians have taken part in the rise and evolution of linear algebra ; in our work, we examined works of some of them in which the role of geometry seemed important (the first part of this book was our main source). Leibniz, Hamilton, Grassmann, Schmidt, Riesz : all of them resorted to geometrical models (according to Fischbein's theory) ; but the contents and the uses of these models are very different. Moreover, one mathematician can use geometry in different ways at different stages of his work. In particular, it can be very different in the research process and in the final written report. We will relate here two brief examples to point out different ways of using geometrical models.

In his research about Quaternions, Hamilton's first aim was to develop an intrinsic geometric analysis, with an algebraic point of view. However, looking for the definition and properties of the multiplication of triplets led him to interpret the triplets as directed line segments in space. Eventually, the model of rotations in space led him to consider quadruplets instead of triplets, and thus to discover quaternions. The model was used in order to choose the structure, the properties of the operations, but also in order to legitimize the existence of quaternions (the same thing happened with the geometrical representation of complex numbers).

In the Riesz's work, the main factor of analogy seems to be the existence of a dual (analytic/synthetic) aspect, in analysis as well as in geometry: representation by series of functions is the analytic aspect. Riesz tried to find a common, symbolical model for both domains.

The historical analysis shows that the relations between linear algebra and geometry are very complex ; this allows us to suppose that using geometry to teach linear algebra will raise important didactical problems.

In France, introducing linear algebra in the secondary school syllabus (1969) was linked with redesigning the teaching of geometry ; linear algebra took a predominant position, and geometry became a mere application of it. In the eighties, linear algebra was expelled from secondary school ; nowadays, students encounter it for the first time during the first year of university (see the last chapter of the first part of this book).

Many university teachers recommend that some geometry lectures should precede linear algebra. But we observed various opinions concerning the content of such lectures, and about the use of figures in a linear algebra course. Teachers also seem to have different expectations about the use of geometry in students' works, especially when figures are involved. Some teachers accept a mere figure as a correct answer to a linear algebra exercise, while others ask for formal proofs for the same task. During three years spent at the university, a single student meets several teachers ; we can thus expect to observe various uses of different geometrical models when students work in linear algebra.

We submitted a questionnaire to post-graduate students, with exercises and questions about linear algebra concepts that present strong links with geometry : basis, symmetry, projection, etc. We also asked them to draw or interpret some figures, especially in critical cases, leading to questions such as : is it possible to illustrate with figures more than three-dimensional linear algebra situations ? Is it possible to represent polynomials as vectors ?

The analysis of their answers shows that some of the students use a geometrical model stemming from secondary school geometry: they refer to central symmetries, coordinate axis (both notions are not studied at the university), and produce affine figures, with points instead of vectors, when asked to give geometrical representations.

On the other hand, some students seem to have a very formal attitude; they use properties such as : "a symmetry is a transformation characterized by sos=Id", or " E = Ker(s-Id) $\oplus$ Ker(s+Id)" ; "a projector is a transformation characterized by pop=p". They do not seem to associate any geometrical representation with these concepts.

Between these two "extreme" attitudes, we observed students using figures with vectors only in some particular cases: when they try to recall a property, or when a particular notion is involved, like orthonormal basis, orthogonal projection, or more generally, concepts belonging to Euclidean vector spaces.

The use of a secondary school geometrical model is clearly correlated with misconceptions in linear algebra. Students who seem to be acquainted with linear algebra concepts are able to associate them with coherent vectorial representations, even when they do not use these representations in reasoning processes. Finding common conventions about the use of geometry in a linear algebra course could be a first factor of improvement concerning students' practices ; but choosing these conventions is a delicate matter.

Is it possible to find, in the geometry taught in secondary school, attainments useful in linear algebra, avoiding misconceptions stemming from affine geometry ? Is it possible to determine a "geometrical part" of linear algebra, associated to pictorial representations, that can be used as a model to help students in their reasoning processes in linear algebra? We are going on further investigations, in order to provide elements of answers for such questions.

6. NOTES

[1] A work by Kallia Pavlopoulou (IREM, 10 Rue du Général Zimmer, 68000 Strasbourg, France) reported by Jean-Luc Dorier.

[2] A work by Marlene Aves Dias (IREM de Paris 7, Université de Paris 7, case 7018, 75 251 Paris Cedex 05, France) reported by Michèle Artigue.

[3] A work by P. Bardy, D. Le Bellac, R. Le Roux, J. Memin, and D. Saby (IREM, Université de Rennes 1, Campus de Beaulieu, 35042 Rennes Cedex, France) reported by Jean-Luc Dorier.

[4] Institut de Recherche sur l'Enseignement des Mathématiques. The IREM were created in the 1970s during the modern math reforms. There is roughly one per region. They serve as a place of exchange and training between secondary school and university teachers.

[5] This last idea is comparable to the unifying and generalizing concepts developed in the first two chapters of this book. The authors cite Dorier's dissertation, which seems to be the only research document from the Dorier-Robert-Robinet-Rogalski group at their disposal.

[6] « We propose the following problem with a partial proof:
Problem: Let F be a vector subspace of a vector space E. There is, then, a single vector subspace G of E such that: F + G = E.
Proof of unicity: Suppose that there are G and G', two different vector subspaces of E such that F + G = E and F + G' = E. We therefore deduce that G = E - F and G' = E - F therefore G = G', contradicting the hypothesis.
The problem is true/false
Theproof given is correct/ incorrect. »

[7] We could perhaps moderate the argument given by the authors by asking whether the comprehension of the notion of the sum of two subspaces is stronger here than the analysis aspect of a demonstration. After all, with the erroneous conception of the sum, the demonstration could be legitimately found irreproachable.

[8] A work by Ahmed Behaj (Groupe de Recherche en Didactique des Mathématiques, Université Sidi Mohamed Ben Abdellah, Faculté des Sciences, Dahr Meraz, B.P. 1796, Atlas, Fès, Morocco) reported by Jean-Luc Dorier.

[9] A work by Ghislaine Chartier (Laboratoire de Didactique des Mathématiques, Université de Rennes 1, Campus de Beaulieu, F-35042 Rennes Cedex, France) reported by herself.

REFERENCES - PART II

The references specific to linear algebra are indicated with (*).

(*)Alves Dias, M. (1993). *Contribution à l'Analyse d'un Enseignement Expérimental d'Algèbre Linéaire en DEUG A Première Année.* Mémoire de DEA, Université de Paris 7, 1993.

(*)Alves Dias, M & Artigue, M. (1995). Articulation problems between different systems of symbolic representations in linear algebra, in *The proceedings of the 19th annual meeting of the international group for the Psychology of Mathematics Education,* Universidade Federal de Pernambuco, Recife, Brésil, 3 vols, 2:34-41.

(*)Alves Dias, M. (1988). *Problèmes d'articulation entre points de vue "cartésien" et "paramétrique" dans l'enseignement de l'algèbre linéaire,* Doctorial dissertation, Université Paris 7.

Bachelard, G. (1983). La formation de l'esprit scientifique, Paris: Presses Universitaires de France (First published in 1938).

Balacheff, N. (1990). Towards a Problematique for research on mathematics teaching, *Journal for Research in Mathematics Education* **21**, 258-272.

(*)Bardy, P., Le Bellac, D., Le Roux, R., Mermin, J. & Saby, D. (1993). *Les débuts de l'algèbre linéaire en DEUG A,* IREM de Rennes.

Bazanska, T., Karwacka, I. & Nykowska, M. (1978). *Zadania z Matematyki. Podrecznik dla Studiow Ekonomicznych,* Warszawa: Panstwowe Wydawnictwo Naukowe.

(*)Behaj, A. (1999). *Eléments de structurations à propos de l'enseignement et l'apprentissage à long terme de l'algèbre linéaire,* doctorial dissertation, Dhar El Mehraz Université, Fès Morocco.

(*)Behaj, A. & Arsac, G. (1998). La conception d'un cours d'algèbre linéaire, *Recherches en Didactique des Mathématiques* **18(3)**, 333-370.

Boero, P., Pedemonte, B., & Robotti, E. (1997). Approaching theoretical knowledge through voices and echoes: a Vygotskian perspective, in *Proceedings of the 21st Conference of the International Group for the Psychology of Mathematics Education,* Lahti, Finland, 1997, vol. 2, 81-88.

Bosch, M. & Chevallard, Y. (1999). La sensibilité de l'activité mathématique aux ostensifs. Objet d'étude et problématique, *Recherches en Didactique des Mathématiques* **19(1)**, 77-123.

Brousseau, G. (1983). Les obstacles épistémologiques et les problèmes en mathématiques, *Recherches en Didactique des Mathématiques* **4(2)**, 164-198.

Brousseau, G. (1984). The crucial role of the didactical contract in the analysis and construction of situations in teaching and learning mathematics, in Steiner et al (eds.), *Theory of Mathematics Education,* Occasional Paper 54, Bielefeld, IDM, pp. 110-119.

Brousseau, G. (1997). *Theory of didactical situations in mathematics,* edited and translated by Balacheff, N., Cooper, M., Sutherland, R. and Warfield, V., Dordrecht: Kluwer.

Boschet, C. & Robert, A. (1985). *Acquisition des premiers concepts d'analyse sur R dans une section ordinaire de première année de DEUG*, Cahier de Didactique des Mathématiques n° **7**, IREM de Paris VII.

(*)Carlson, D. (1993a). Teaching linear algebra: must the fog always roll in*?*, *College Mathematics Journal* **24** *(special issue about linear algebra)*, 29-40.

(*)Carlson, D. Johnson, C., Lay, D. & Porter, A. (1993b). The Linear Algebra Curriculum Study Group recommendations for the first course in linear algebra, *College Mathematics Journal* **24** *(special issue about linear algebra)*, 41-46.

(*)Carlson, D. (1994). Recent developments in the teaching of linear algebra in the United States, Aportationes Matemáticas. XXVI *Congreso Nacional de la Sociedad Matematica Mexicana*. Serie Comunicaciones **14**, 371-382.

Chevallard, Y. (1985) *La transposition didactique*, Grenoble: La Pensée Sauvage ed. Reed. 1991.

Daintith, J. & Nelson, R.D. (eds) (1989). *Dictionary of Mathematics*, London: The Penguin Books.

(*)Dieudonné, J. (1964). *Algèbre linéaire et géométrie élémentaire*, Paris: Hermann.

(*)Dorier, J.-L. (1990a). *Contribution à l'Étude de l'Enseignement à l'Université des Premiers Concepts d'Algèbre Linéaire. Approches Historique et Didactique*, Doctorial dissertation, Université J. Fourier - Grenoble 1, France.

(*)Dorier, J.-L. (1990b). *Analyse dans le Suivi de Productions d'Étudiants de DEUG A en Algèbre Linéaire*, Cahier DIDIREM n°6, IREM de Paris VII.

(*)Dorier, J.-L. (1990c). *Analyse Historique de l'Émergence des Concepts Élémentaires d'Algèbre Linéaire*, Cahier DIDIREM n°7, IREM de Paris VII.

(*)Dorier, J.-L. (1991). Sur l'enseignement des concepts élémentaires d'algèbre linéaire à l'université, *Recherches en Didactique des Mathématiques* **11(2/3)**, 325-364.

(*)Dorier, J.-L. (1992). *Illustrer l'Aspect Unificateur et Généralisateur de l'Algèbre Linéaire*, Cahier DIDIREM n°14, IREM de Paris VII.

(*)Dorier, J.-L. (1993). Premières approches pour l'étude de l'enseignement de l'algèbre linéaire à l'université, *Annales de Didactiques et de Sciences Cognitives* **5**, Strasbourg: IREM, 95-123.

(*)Dorier, J.-L., Robert A., Robinet J. & Rogalski M. (1994a). L'enseignement de l'algèbre linéaire en DEUG première année, essai d'évaluation d'une ingénierie longue et questions, In Artigue M. et al. (eds), *Vingt ans de Didactique des Mathématiques en France*, Grenoble : La Pensée Sauvage, pp. 328-342.

(*)Dorier, J.-L, Robert, A., Robinet, J. & Rogalski, M. (1994b). The teaching of linear algebra in first year of French science university, in the *Proceedings of the 18th conference of the international group for the Psychology of Mathematics Education*, Lisbonne, 4 vols., 4: 137- 144.

(*)Dorier, J.-L. (1995a). A general outline of the genesis of vector space theory, *Historia Mathematica* **22(3)**, 227-261.

(*)Dorier, J-L. (1995b). Meta level in the teaching of unifying and generalizing concepts in mathematics, *Educational Studies in Mathematics* **29(2)**, 175-197.

(*)Dorier J.-L. (1998). The role of formalism in the teaching of the theory of vector spaces, *Linear Algebra and its Applications* (275) **1(4)**, 1998, 141-160.

(*)Dorier J.-L. (1998b). État de l'art de la recherche en didactique des mathématiques à propos de l'enseignement de l'algèbre linéaire, *Recherches en didactique des mathématiques* **18(2)**, 191-230.

(*)Dorier J.-L., Robert A., Robinet J. & Rogalski M. (2000). On a research program about the teaching and learning of linear algebra in first year of French science university, *International Journal of Mathematical Education in Sciences and Technology* **31(1)**, 27-35.

Douady, R. (1986). Jeu de cadres et dialectique outil-objet, *Recherches en Didactique des Mathématiques* **7(2)**, 5-31.

Dreyfus,T. ,& Hillel, J. (1998). Reconstruction of meanings for function approximation, *International Journal of Computers for Mathematical Learning,* **3**, 93-112.

Dubinsky, E. (1991). Reflective abstraction in advanced mathematical thinking, in *Advanced Mathematical Thinking*, D. Tall (ed.), Dordrecht: Kluwer, pp. 95-123.

Duval, R. (1993). Registre de représentation sémiotique et fonctionnement cognitif de la pensée, *Annales de Didactiques et de Sciences Cognitives* **5**, Strasbourg: IREM, 37-65.

Duval, R. (1995). *Semiosis et pensée humaine. Registres sémiotiques et apprentissages intellectuels*, Bern: Peter Lang.

Eisenberg, T. & Dreyfus, T. (1968). On visual versus analytical thinking in mathematics, in *The Proceeding of the tenth International Conference of the Psychology of Mathematics Education*, University of London, 153-158.

Fischbein, E. (1987). *Intuition in science and Mathematics, An Educational Approach*, Dordrecht/Boston/Lancaster/Tokyo: D.Reidel Publishing Company.

(*)Fletcher, T. J. (1972). *Linear Algebra through Applications*, London: Van Nostrand Reinhold.

Friedberg, S., Insel, A. & Spence, L. (1979). Linear Algebra, New Jersey: Printice-Hall.

Garuti, R., Boero, P. & Chiappini, G. (1999). Bringing the Voice of Plato in the Classroom to Detect and Overcome Conceptual Mistakes, in *Proceedings of the 23rd Conference of the International Group for the Psychology of Mathematics Education*, Haifa, Israel, July 25-30, III: 9-16.

Goldin, G. (1987). Cognitive Representational Systems for Mathematical Problem Solving, in C. Janvier (ed.), *Problems of representation in the teaching and learning of mathematics*, Hillsdale, N.J.: Lawrence Erlbaum, 125-144.

Granger, G; (1979). *Langages et épistémologie*, Paris: Klinksieck.

Greeno, G. J. (1983). Conceptual entities, in D. Gentner & A. L. Stevens (eds.), *Mental Models*, Hilsdale, N-J: Lawrence Erlbaum, pp. 227-252.

(*)Haddad, M. (1999). *Difficulties in the Teaching and Learning of Linear Algebra - A Personal Experience*. Master's Thesis. Montréal: Concordia University.

Hamilton, W.R. (1837). Theory of conjugate functions, or algebraic couples; with a preliminary and elementary essay on algebra as the science of pure time, *Transactions of the Royal Irish Academy* **17**, 293-422. In H. Halberstam and R.E. Ingram (Eds*), The mathematical papers of Sir William Rowan Hamilton*, Vol. III, Algebra. Cambridge: Cambridge University Press.

(*)Harel, G. (1985). *Teaching Linear Algebra in High School*, Unpublished doctoral dissertation, Ben-Gurion University of the Negev, Beer-Sheva, Israel.

(*)Harel, G. (1989a). Learning and teaching linear algebra: Difficulties and an alternative approach to visualizing concepts and processes, *Focus on Learning Problems in Mathematics* **11**, 139-148.

(*)Harel, G. (1989b). Applying the principle of multiple embodiments in teaching linear algebra: Aspects of familiarity and mode of representation, *School Science and Mathematics* **89**, 49-57.

(*)Harel, G. (1990). Using geometric models and vector arithmetic to teach high school students basic notions in linear algebra, *International Journal for Mathematics Education in Science and Technology* **21**, 387-392.

Harel, G. & Kaput, J. (1991). The role of conceptual entities in building advanced mathematical concepts and their symbols, in Tall, D. (ed), *Advanced Mathematical Thinking*, Dordrecht : Ridel, pp. 82-94.

(*)Harel, G. (1997). The linear algebra curriculum study group recommendations: moving beyond concept definition, in Carlson, D. et al. (eds), *Ressousrces for teaching linear algebra*, pp. 107-126, MAA Notes, Vol. 42.

Harel, G. & Sowder, L. (1998). Students' proof schemes, in Dubinsky, E., Schoenfeld, A. & Kaput, J. (Eds.), *Research on Collegiate Mathematics Education*, Vol. III. AMS, pp. 234-283.

Harel, G. (1999). Students' understanding of proofs: a historical analysis and implications for the teaching of geometry and linear algebra, Linear algebra and its applications **302-303**, 601-613.

Johnson-Laird, P.N. & Wason, P.C. (eds.) (1977). *Thinking. Readings in Cognitive Science*, Cambridge: Cambridge University Press.

Kielbasinski, A. & Schwetlick, H. (1992). *Numeryczna Algebra Liniowa. Wprowadzenie do Obliczen Zautomatyzowanych*, Warszawa: Wydawnictwa Naukowo-Techniczne.

(*)Lay, D.C. (1994). *Linear Algebra and its Applications*. Massachusetts: Addison-Wesley.

Legrand, M. (1990). Circuit, ou les règles du débat mathématique, in *Enseigner autrement les mathématiques en DEUG Première Année*, Commission inter-IREM université, pp. 129-161.

Leron (1993). *Undergraduate students' conceptions and misconceptions of group isomorphism*, paper presented at the joint AMS/MAA/CMS meeting, July 1993, Vancouver, Canada.

(*)Lipschutz, S. (1991). *Linear Algebra*. (2nd edition) Schaum's Outline Series, New York: McGraw-Hill.

Marody, M. (1987). *Technologie Intelektu. Jezykowe Determinanty Wiedzy Potocznej i Ludzkiego Dzialania*. Warszawa: Panstwowe Wydawnictwo Naukowe.

Otte, M. (1990). Arithmetic and geometry: some remarks on the concept of complementarity, *Studies in Philosophy and Education* **10**, 37-62.

(*)Ousman, R. (1996). *Contribution à l'enseignement de l'algèbre linéaire en première année d'université*, Doctorial dissertation, Université de Rennes I, France.

Pallascio, R. (1992). Le développement de la connaissance rationnelle et l'outil mathématique au primaire, in M. Schleifer (ed.), *La formation du jugement. Montréal: Éditions Logiques*, pp. 159-171.

(*)Pavlopoulou, K. (1993). Un problème décisif pour l'apprentissage de l'algébre linéaire: la coordination des registres de représentation. *Annales de Didactiques et de Sciences Cognitives* **5**, Strasbourg: IREM, 67-93.

(*)Pavlopoulou, K. (1994). *Propédeutique de l'algèbre linéaire : la coordination des registres de représentation sémiotique*, Doctorial dissertation, Université Louis Pasteur (Strasbourg 1), prépublication de l'Institut de Recherche Mathématique Avancée.

Pépin, Y. (1994). Savoirs pratiques et savoirs scolaires: une réprésentation constructiviste de l'education, *Revue des Sciences de l'Education* **20(1)**, 63-86.

Piaget, J., Inhelder & Szeminska (1960). *The Child's Conception of Geometry*, New-York: Basic Books.

Piaget, J. & Beth, E. W. (1961). *Epistémologie mathématique et psychologie (Essai sur les relations entre la logique formelle et la pensée réelle)*, Bibliothèque Scientifique Internationale, Etude d'épistémologie génétique n° XIV, Presses Universitaires de France.

Piaget, J. & Garcia, R. (1989). *Psychogenesis and the history of science*, New York: Columbia University Press.

Pirie, S. & Kieren, T. (1994). Beyond metaphor: formalizing in mathematical understanding within constructivist environments, *For the Learning of Mathematics* **14(1)**, 39-43.

Praslon, F. (1994). *Analyse de l'aspect méta dans un enseignement de DEUG A concernant le concept de dérivée. Étude des effets sur l'apprentissage*, mémoire de DEA de l'université de Paris VII.

Robert, A. (1987). *De quelques spécificités de l'enseignement des mathématiques dans l'enseignement post-obligatoire*, Cahier de didactique des mathématiques n° 47, IREM de Paris VII.

(*)Robert, A., Robinet, J. & Tenaud, I. (1987). *De la géométrie à l'algèbre linéaire*, Brochure 72, IREM de Paris VII.

Robert, A. (1988). *Rapport enseignement/apprentissage (début de l'analyse sur R). Fascicule 1, Analyse d'une section de DEUG A première année (les connaissances antérieures et l'apprentissage)*, Cahier de didactique des mathématiques n°181, IREM de Paris VII.

Robert, A. & Tenaud, I (1988). Une expérience d'enseignement de la géométrie en Terminale C, *Recherches en Didactiques des Mathématiques* **9(1)**, 31-70.

(*)Robert, A. & Robinet, J. (1989). *Quelques résultats sur l'apprentissage de l'algèbre linéaire en première année de DEUG*, Cahier de Didactique des Mathématiques n°53, IREM de Paris VII.

Robert, A. (1990). *Un projet long d'enseignement (algèbre et géométrie - licence en formation continuée)*, Cahier DIDIREM n°9, IREM de Paris VII.

Robert, A. (1992). Projets longs et ingénierie pour l'enseignement universitaire : questions de problématique et de méthodologie. Un exemple : un enseignement annuel de licence en formation continue, *Recherches en Didactique des Mathématiques* **12(2/3)**, 181-220.

Robert, A. (1993). Analyse du discours non strictement mathématiques accompagnant les cours de mathématiques dans l'enseignement post-obligatoire, *Educational Mathematics Studies* **28(1)**, 73-86.

Robert, A. & Robinet, J. (1993). *Prise en compte du "méta" en didactique des mathématiques*, Cahier de DIDIREM n°21, IREM de Paris VII.

(*)Robert, A, & Robinet, J. (1996). Prise en compte du méta en didactique des mathématiques, *Recherches en didactique des mathématiques* **16(2)**, 145-176.

(*)Robert, A. (1998). Outil d'analyse des contenus mathématiques enseignés au lycée et à l'université) État de l'art de la recherche en didactique des mathématiques à propos de l'enseignement de l'algèbre linéaire, *Recherches en didactique des mathématiques* **18(2)**, 191-230.

(*)Robinet, J. (1986). *Esquisse d'une genèse des concepts d'algèbre linéaire*, Cahier de Didactique des Mathématiques n°29, IREM de Paris VII.

(*)Rogalski, M. (1990a). Pourquoi un tel échec de l'enseignement de l'algèbre linéaire?, in *Enseigner autrement les mathématiques en DEUG Première Année*, Commission inter-IREM université, pp. 279-291.

(*)Rogalski, M. (1991). *Un enseignement de l'algèbre linéaire en DEUG A première année*, Cahier de Didactique des Mathématiques n°53, IREM de Paris VII.

(*)Rogalski, M. (1994). L'enseignement de l'algèbre linéaire en première année de DEUG A, *La Gazette des Mathématiciens* **60**, 39-62.

(*)Rogalski, M. (1995). Que faire quand on veut enseigner un type de connaissances tel, que la dialectique outil/objet ne semble pas marcher, et qu'il n'ait pas apparemment de situation fondamentale ? L'exemple de l'algèbre linéaire. *Séminaire DidaTech n°169*, Grenoble, 127-162.

(*)Rogalski, M. (1996). Teaching linear algebra : role and nature of knowledge in logic and set theory which deal with some linear problems, in L. Puig et A. Guitierrez (eds), *Proceedings of the XX° International Conference for the Psychology of Mathematics Education*, 4 vol., Valencia : Universidad, 4: 211-218.

Selden, A. & Selden, J. (1987). Errors and misconceptions in college level theorem proving, in *Proceedings of the Second International Seminar on Misconceptions and Educational Strategies in Science and Mathematics*, Cornell University, 3 vol., 3: 456-471.

Sfard, A. (1987). Two Conceptions of Mathematical Notions: Operational and Structural. In J.C. Bergeron, N. Herscovics, and C. Kieran (Eds.), *Proceedings of the 11th Conference of the International Group for the Psychology of Mathematics Education*, Montréal, Canada, July 19-25, 1999, 1:162-169.

Sfard, A. (1991). On the dual nature of mathematical conceptions: reflections on process and objects as different sides of the same coin, *Educational Studies in Mathematics* **22(1)**, 1-36.

Sfard, A. (1994). The gains and the pitfalls of reification - the case of algebra, *Educational Studies in Mathematics* **26**, 191-228.

Sierpinska, A. (1992). On Understanding the Notion of Function. In G. Harel and E. Dubinsky (Eds.), *The Concept of Function. Aspects of Epistemology and Pedagogy*. Mathematical Association of America. MAA Notes, Volume 25, pp. 25-58.

Sierpinska, A. (1995). Mathematics: 'in context', 'pure', or 'with applications'?, *For the Learning of Mathematics* **15(1)**, 2-15.

Sierpinska, A. (1995). The diachronic dimension in research on understanding in mathematics - usefulness and limitations of the concept of epistemological obstacle, in M. Otte, N. Knoche, & H.N. Jahnke (eds), *Interaction between History of Mathematics and Mathematics Learning*, Göttingen: Vandenhoeck & Ruprecht.

Sierpinska, A. (1996). Interactionnisme et Théorie des Situations. Format d'Interaction et Contrat Didactique, *Séminaire DidaTech n°172*, Grenoble, 5-37.

Sierpinska, A. (1997). Formats of Interaction and Model Readers. *For the Learning of Mathematics* **17.2**, 3-12.

(*)Sierpinska, A., Dreyfus, T. & Hillel, J. (1999). Evaluation of a Teaching Design in Linear Algebra: The Case of Linear Transformations, *Recherches en Didactique des Mathématiques* **19(1)**, 7-41.

(*)Sierpinska, A., Trgalová, J., Hillel & J., Dreyfus, T. (1999). Teaching and Learning Linear Algebra with Cabri. In O. Zaslavsky (Ed.), *Proceedings of the 23rd Conference of the International Group for the Psychology of Mathematics Education*, Haifa, Israel, July 25-30, 1999, 1: 119-134.

Skemp, R.K. (1978). Relational and Instrumental Understanding, *Arithmetic Teacher* **26.3**, 9-15.

Sowder, L., & Harel, G. (1998). Types of students' justifications, *Mathematics Teacher*, **91**, 670-675.

Spalt, D.D. (1992). Le continu de l'Analyse Classique dans la perspective du Résultatisme et du Genésiologisme, in J.-M. Salanskis and H. Sinaceur (eds), *Le Labirynthe du Continu. Colloque de Cerisy*, Paris: Springer, pp. 85-95.

(*)Strang , G. (1988). *Linear Algebra and its Applications*, 3rd ed., Harcourt Brace Jovanich.

(*)Strang, G. (1993). Graphs, Matrices, and Subspaces, *The College Mathematics Journal*, Special Issue on Linear Algebra, **24/1**, 20-28.

Tait, W.W. (1992). Reflections on the concept of a priori truth and its corruption by Kant, in M. Detlefsen (ed.), *Proof and Knowledge in Mathematics*, London/New York: Routledge.

Tenaud, I.(1991). *Une expérience d'enseignement de la géométrie en Terminale C : enseignement de méthodes de travail en petits groupes,* Thèse de Doctorat de l'université de Paris VII.

Turowicz, A. (1973). *Teoria Macierzy*, Skrypt Uczelniany Nr 284, Krakow: Akademia Gorniczo-Hutnicza.

Vergnaud, G. (1990). La théorie des champs conceptuels, *Recherches en Didactiques des Mathmatiques* **10(2/3)**, 133-170.

Vermersch, P. (1991). l'entretien d'explicitation, *Les Cahiers de Beaumont* **52bis/53**, 63-70.

Vinner, S.(1983). Concept definition, concept image, and the notion of function, *International Journal of Mathematics Education, Science and Technology* **14**, 293-305.

Vygotsky, L.S. (1987). *The Collected Works of L.S. Vygotsky*, Volume 1, Problems of General Psychology, Including the Volume Thinking and Speech. New York and London: Plenum Press.

Walker, R.J. (1978). *Algebraic Curves*, New York/Heidelberg/Berlin: Springer.

CONCLUSION

As mentioned in the introduction, this book does not aim to provide a miraculous way of teaching linear algebra. Moreover, the products of the research presented are not, in general, statements of the type, 'Such and such a way of teaching linear algebra will result in a better understanding, by all students, of the basic concepts of the theory'. On the other hand, this does not mean that the results of research cannot be useful for teaching linear algebra. They give a diagnosis of students' difficulties along with possible explanations within theoretical frameworks involving didactical, epistemological or cognitive elements. In fact, the most reliable results of the research are, in a sense, 'negative' in nature. But this book also presents some innovative teaching experiments which succeeded in attaining some of their goals in order to improve locally the teaching and learning of linear algebra. They lead to some well founded 'recommendations' or 'good advice' concerning the practice of teaching, but the advice is not to be confused with an infallible recipe; in fact, the recommendations are but conjectures that are still open to questioning.

We hope that the material provided reflects the richness of the various research works on the teaching of linear algebra. In this conclusion, we will now try to put forward the connections and the unity between these different works, through the main issues they address.

1. DIAGNOSES - EPISTEMOLOGICAL ISSUES

There is a wide consensus to say that both learning and teaching linear algebra are difficult. Dorier, Robert, Robinet and Rogalski have pointed out the resistance of the obstacle of formalism, which Hillel and Sierpinska have also experienced in their studies. Bardy et al. insist on the feeling of many students of being overwhelmed with new definitions or just words to learn, as if they had to learn a new language. These difficulties are certainly common to all higher algebra domains. However, in most countries, linear algebra is the first contact of students with such a field, without any preparation for formalism. The epistemological analysis of the historical development of linear algebra, presented by Dorier, enlightens us on the reason for these difficulties, and gives us some keys to deal with this problem. It suggests that formalism is inherent to linear algebra since the theory of vector space is a unifying and generalizing theory which emerged more from organizational necessities than just new problems to be solved. On the educational side, three slightly different, yet complementary, answers are given.

- Dorier et al. show the necessity for students to get involved, along with their mathematical work, in a reflexive analysis on the objects they manipulate, in order to understand the unifying and generalyzing aspect of linear algebra concepts. This is what leads these researchers to the concept of 'meta' activities.
- Harel suggests a progressive approach according to three pedagogical principles. Concepts to be modeled in terms of linear algebra should first acquire a status of conceptual entities in the eyes of students (Concreteness Principle). Then, students should see a need for the building of linear algebra concepts (Necessity

273

Principle), while instruction should allow and encourage the generalizibility of concepts (Generalizibility Principle).

- Based on a series of experimentations of teaching projects, Sierpinska suggests that students' tendency to rely on practical rather than theoretical thinking in linear algebra is a reason behind many of their difficulties, especially with the structural aspects of the theory.

The three approaches offer didactical tools in order to make students understand the challenge of 'linear algebra thinking'. Nevertheless, success is only partial in the three works and suggests that teaching the structural approach specific to higher algebra may be restricted to stringently mathematically oriented curricula, in which students may have a sufficient motivation for formalism. In this restricted context, some teaching designs have proven their ability in making students able to understand the unifying and generalizing aspect of linear algebra.

In these latter approaches, the issue of logic and set theory is discussed. Skills in these fields are required in order to complete tasks within formal linear algebra. However, Dorier et al. proved that rather than a preliminary course in logic and set theory disconnected from any specific mathematical content , it is more effective to put an emphasis on these questions within linear algebra itself, in connection with more intuitive and meaningful prior elements of knowledge (especially in geometry or systems of linear equations).

On the other hand, Harel suggests that an entry through geometry should be privileged. Nevertheless, he remarks that there is no smooth progression from $\mathbb{R}^2$ and $\mathbb{R}^3$ to $\mathbb{R}^n$ and then more abstract linear structures. Rogalski's experiment offers a complex design in which geometrical preliminary activities along with activities in various other mathematical fields (linear equations, magic squares, polynomials, recurrent sequences, etc.) can be capitalized through a meta level reflection initiating the work at the structural level of linear algebra. Finally, Chartier's work shows that the use of geometry as a privileged entry into linear algebra must be carefully planned.

2. COGNITIVE FLEXIBILITY

Linear algebra is an 'explosive compound' of languages, settings and systems of representation. There is the geometric language of lines and planes, the algebraic language of linear equations, n-tuples and matrices, the 'abstract' language of vector spaces and linear transformations. There are the settings of geometry, of algebra, but also of graphical representations which allow a metaphoric use of geometry in higher dimensional spaces. There are the 'graphical', the 'tabular' and the 'symbolic' registers of the languages of linear algebra. There are also the 'Cartesian' and the 'parametric' representations of subspaces. Teachers and texts constantly move between these languages, registers and modes of representation without allowing for the time necessary to make the conversions and discuss their validity. They appear to assume that these conversions are 'natural' and obvious and do not need any conceptual work at all. These issues are especially discussed in Hillel's, Pavlopoulou's and Alves Dias's works, but they are also important in Sierpinska's and Rogalski's approaches.

Pavlopoulou and Hillel show the importance, in the understanding of linear algebra, of the students' ability to translate from one register or language into another. However, Alves Dias proves that this flexibility is not sufficient: control at a more conceptual level is required. Her conclusions are confirmed by Sierpinska's and Dorier et al's approaches.

On the most general level, a good understanding of linear algebra requires a fair amount of 'cognitive flexibility' in moving between the various languages (e.g. the language of matrix theory and the language of the vector space theory), viewpoints (Cartesian and parametric) and semiotic registers. Understanding of linear algebra requires also that the students encapsulate into identifiable conceptual structures a substantial range of what was, previously, individual objects and actions on these objects (e.g. functions have to be viewed as objects in themselves, elements of a vector space rather than procedures of assigning numbers to other numbers). Moreover, understanding of linear algebra requires the ability to resort to 'theoretical thinking' as a means of control over the 'intuitions', mental imagery and dependence on the context, characteristic of 'practical thinking'. 'Practical thinking' is, however, necessary to avoid a situation where linear algebra is no more than a foreign, cryptic and formal language which can be written but not used to think with.

3. LONG-TERM AND EVALUATION

The teaching of linear algebra in the first year of university usually covers a semester and a fair amount of hours. Moreover, the novelty of the approach, compared to secondary school mathematics, makes the issue of maturity especially important. Therefore, the 'long-term' dimension appears as a central didactical variable, that several works presented in this book have taken into account. Behaj's contribution shows how students' understanding of linear algebra evolves in their minds one, two or three years after the first course on the subject. This research points out the change in perspective due to various applications outside linear algebra but also the recollection of the first approach. Moreover, it also affords information on teachers' practice regarding the 'long term' variable: several types of structuring of the knowledge are presented by one teacher, depending on the time of the teaching. A first sequence on the subject may put the emphasis on the rigor and the logic of mathematical language, while, with time, the structuring is more oriented towards solving exercises. The necessity for assessment may also influence the structuring especially towards the end of the term.

In Rogalski's experiment, the 'long-term' variable is central. The 'meta' dimension must take into account a necessary maturing in the reflexive analysis made by students on their mathematical activity. Moreover, the novelty of this type of activity makes the difference between students' time of response even sharper. On the theoretical plane, this makes the issue of evaluation of the effects of the teaching design quite complex. This is a real challenge for research in education (certainly not only in mathematics), to be able to take into account long-term teaching projects and not only local experiments. Indeed, in a long-term teaching design, global choices have to be made, but it is difficult to describe in detail how they have to be implemented as local decisions in different phases of the experiment. Moreover, students' individual homework between classes is difficult to take into account in the theoretical framework.

4. PERSPECTIVES

Many questions have been left open for new research works along the preceding chapters of this book. Some refer to local issues specific to linear algebra. We will not try to recall them here because they would lose their accuracy if taken out of the context from which they emerged. Rather, we will point out two more global issues which, in our eyes, have not yet been investigated substantially. Not that they do not deserve our attention, but because they are complex and also because they raise methodological difficulties that could not be taken into account before a sufficient number of research works in mathematics education at university level had been made.

• Linear algebra is connected to the students' previous knowledge in various mathematical fields. What can the researcher assume about the state of this knowledge? Is it possible to refer only to the official curriculum? Moreover there is a change of 'culture' between secondary and university teaching. How can this be taken into account in the research issues? Robert has introduced the notion of 'levels of conceptualization', in order to approach this problem. Chartier applied it in her work on the relation between geometry and linear algebra: What is geometry? What do students know about it? Does it mean the same for primary school, secondary school or university teachers? What are the connections?

• In research works in mathematics education at the university level, usually the teacher experimenting the teaching project is one of the researchers in charge of the project. Therefore, there is a kind of intrinsic control of many local parameters. This leaves open the question of the transmission of the teaching project to other teachers, especially concerning a long-term teaching project, like for linear algebra.

As can be seen from these conclusions, many questions remain open, and several avenues for new research can be proposed. We hope that reading this book will have challenged the curiosity of the reader and encouraged her or him to explore new ideas, whether in teaching linear algebra or engaging in a research project.

Jean-Luc Dorier
Aline Robert
Anna Sierpinska

NOTES ON CONTRIBUTORS

Jean Luc Dorier is a Professor in a University Teacher Training Institute (I.U.F.M.) in Lyon and head of a Research Team in the field of Didactics of Mathematics, in Grenoble. His work focuses on the teaching and learning of Linear Algebra at University level. His interest in this didactical question led him to an extensive study of the history of Linear Algebra and he developed a research program which includes theoretical issues on the connection between research in History and in Didactics of Mathematics. His other fields of interest include : the teaching of vectors (in Maths and in Physics, as well as their use in Technological fields), the modelling issue in Mathematics especially in Economics oriented curricula, the teaching of arithmetics at College level, etc.

Address:

Jean-Luc Dorier
Equipe DDM
Laboratoire LEIBNIZ
46, ave F. Viallet
38 031 Grenoble Cedex
France

e-mail: Jean-Luc.Dorier@imag.fr

Aline Robert is a Professor in a University Teacher Training Institute (I.U.F.M.) in Versailles and head of the Research Team DIDIREM in Paris 7 University. She started her research work in Didatics of Mathematics with the teaching of Analysis in University, and then in Linear Algebra. She is also leading a research program about teachers' practises in secondary education. This program involves theoretical and experimental issues, as well as developments towards teacher training.

Address:

Aline Robert
IUFM
45 ave des Etats-Unis
B.P. 171
78 001 Versailles
France

e-mail: robert@math.uvsq.fr

Jacqueline Robinet is Senior Lecturer in Paris 7 University and part of the Research Team DIDIREM. She worked on the concept of *didactical engineering* and then specialized in research on tertiary level math education. In France, she was, with Aline Robert, the first to investigate the teaching of Linear Algebra.

Address:

Jacqueline Robinet
IREM de Paris 7
Université de Paris 7
case 7018
2 place Jussieu
75 251 Paris Cedex 05
France

e-mail: robinet@gauss.math.jussieu.fr

Marc Rogalski is a Professor in Lille 1 University and part of the research team in Analysis in Paris 6 University as well as of DIDIREM. His interests in Didactics of Mathematics cover all the subjects taught in the two first years of University. He is also involved in teacher training activities.

Address:

Marc Rogalski
Equipe d'Analyse
Université de Paris 6
2, place Jussieu
75 252 Paris Cedex 05
France

e-mail: mro@ccr.jussieu.fr

Guershon Harel is a Professor at the University of California, San Diego. His area of research is cognition and epistemology of mathematics and their application in mathematics curricula and teacher education, focusing on the concept of mathematical proof, the learning and teaching of linear algebra, and the development of proportional reasoning.

Address:

Guershon Harel
Department of Mathematics
University of California, San Diego
San Diego, CA 92093-0112
USA

e-mail: gharel@math.ucsd.edu

Joel Hillel is a Professor of Mathematics and the Chair of the Department of Mathematics and Statistics at Concordia University. Since 1992 he has been involved in an on-going research project examining the teaching and learning of Linear Algebra at the university level. This research, in collaboration with Anna Sierpinska, has been funded both by Canada's SSHRC grants and Quebec's FCAR grants. Professor Hillel is particularly interested in students' constructions, and sense of linear algebra notions, when they arise in computer environments such as MAPLE or CABRI-GEOMETRY.

Address:

Joel Hillel
Department of Mathematics and Statistics
Concordia University
7141 Sherbrooke St W.
Montreal
Quebec H4B 1R6
Canada

e-mail: jhillel@alcor.concordia.ca

Anna Sierpinska is a Professor in the Department of Mathematics and Statistics at Concordia University in Montreal, Canada. Her interest in mathematics education started with secondary school students' problems with understanding the notions of limits and infinity. Since she moved from her native Poland to Canada in 1990, she has been engaged in the research on undergraduate students' difficulties in linear algebra. She has also collaborated in projects related to more general questions, such as the notion of understanding in mathematics, language and communication in the mathematics classroom, and meta-questions on the nature of mathematics education as a research discipline.

Address:

Anna Sierpinska
Department of Mathematics and Statistics
Concordia University
7141 Sherbrooke St W.
Montreal
Quebec H4B 1R6
Canada

e-mail: SIERP@VAX2.CONCORDIA.CA

Michèle Artigue is Professor in Paris 7 University and head of the Institute for Research in Teaching Mathematics (IREM) in this university. She developped a theoretical framework centered on the concept of *didactical engineering*. She also led a research program about the teaching and learning of Differential Equations at university. Among her other fields of interest are the teaching of Analysis, including in computer environments, and Linear Algebra.

Address:

Michèle Artigue
IREM de Paris 7
Université de Paris 7
case 7018
2 place Jussieu
75 251 Paris Cedex 05
France

e-mail: artigue@math.jussieu.fr

Ghislaine Chartier is teaching in Rennes 1 University. She has been teaching Linear Algebra for several years, in first and second year. She is now finished her doctorate about the relations between Linear Algebra and Geometry.

Address:

Ghislaine Chartier
Laboratoire de Didactique des Mathématiques
Université de Rennes 1
Campus de Beaulieu
35 042 Rennes Cedex
France

e-mail: chartier@maths.univ-rennes1.fr

INDEX

281

Mathematics Education Library

Managing Editor: A.J. Bishop, Melbourne, Australia

18. N. Bednarz, C. Kieran and L. Lee (eds.): *Approaches to Algebra*. Perspectives for Research and Teaching. 1996 ISBN 0-7923-4145-7; Pb 0-7923-4168-6

19. G. Brousseau: *Theory of Didactical Situations in Mathematics*. Didactique des Mathématiques 19701990. Edited and translated by N. Balacheff, M. Cooper, R. Sutherland and V. Warfield. 1997 ISBN 0-7923-4526-6

20. T. Brown: *Mathematics Education and Language*. Interpreting Hermeneutics and Post-Structuralism. 1997 ISBN 0-7923-4554-1

21. D. Coben, J. O'Donoghue and G.E. FitzSimons (eds.): *Perspectives on Adults Learning Mathematics*. Research and Practice. 2000 ISBN 0-7923-6415-5

22. R. Sutherland, T. Rojano, A. Bell and R. Lins (eds.): *Perspectives on School Algebra*. 2000 ISBN 0-7923-6462-7

23. J.-L. Dorier (ed.): *The Teaching of Linear Algebra in Question*. 2000 ISBN 0-7923-6539-9

KLUWER ACADEMIC PUBLISHERS – DORDRECHT / BOSTON / LONDON